CAMBRIDGE International Examinations

IGCSE
Mathematics

Karen Morrison

CAMBRIDGE
UNIVERSITY PRESS

CAMBRIDGE UNIVERSITY PRESS
Cambridge, New York, Melbourne, Madrid, Cape Town, Singapore, São Paulo

Cambridge University Press
The Water Club, Beach Road, Granger Bay, Cape Town 8005, South Africa

www.cambridge.org
Information on this title: www.cambridge.org/9780521011136

First published 2002
Reprinted 2003, 2004, 2005

Printed in the United Kingdom at the University Press, Cambridge

ISBN-13   978-0-521-01113-6 paperback
ISBN-10   0-521-01113-2 paperback

. . . .

Acknowledgements
We would like to acknowledge the contributions that the following people
and institutions have made to the manuscript: the Cambridge Open
Learning Project (COLP), Lisa Greenstein and Carsten Stark.

Contents

Module 6 – Statistics

Module 7 – Probability

Module 8 – Transformations

Introduction

Mathematics: IGCSE has been designed and written to help prepare students for the IGCSE examinations. Each topic in the examination syllabus is covered in detail and concepts are explained, practised and then revised to help students achieve mastery. The grid on pages v to x shows where each section of the syllabus is covered in this book.

The IGCSE syllabus is divided into core and extended (core plus supplementary) topics. Both are covered in this book and students should aim to cover all material. However, for those wishing to study only the core material, all extended topics are clearly indicated by a coloured line in the margin.

The structure of this book

Mathematics: IGCSE is divided into eight modules, each focusing on a key mathematical area. The modules are sequential and build on skills covered in earlier modules.

Each module contains:
- **explanatory text** which is easy for students to read on their own
- **worked examples** which demonstrate the processes required to solve the problem and show students how to lay out their work in the necessary steps to achieve top results
- **graded exercises** that allow students to practise skills and apply them in a range of contexts that are modern, meaningful and culturally inclusive
- **a wide range of question styles** based on the ways in which problems can be posed in examinations to help students avoid the trap of rote-learning
- **hint boxes** that suggest tried-and-tested methods and short-cuts for solving problems
- **remember boxes** which refer students back to concepts previously covered, alerting them to the need for revision and highlighting the cumulative nature of mathematical skills and concepts
- **exam tips** which highlight activities and give suggestions to help students succeed in examinations
- **'Check your progress'** activities at the end of each module, which provide a mini-examination with questions on all topics covered in the module.

In addition, answers to all exercises and Check your progress problems are given at the end of the book. These are provided to make it easier for students to get into the habit of assessing their own work and trying to find and correct their own mistakes. In order to avoid copying, no model solutions are given.

Should you feel that students need more practice or that they are experiencing difficulties in a particular area, you will find that most exercises can be easily adapted by changing the given numbers.

Syllabus coverage in *Mathematics: IGCSE*

Theme or topic	Core topics	Supplementary topics
1. Number, set notation and language pp 1–8 pp 9–14 Throughout *Nov 24*	Use natural numbers, integers, prime numbers, common factors and multiples, rational and irrational numbers, real numbers; continue a given number sequence; recognise patterns in sequences and relationships between different sequences; generalise to simple algebraic statements relating to such sequences	Use language, notation and Venn diagrams to describe sets and represent relationships between sets
2. Squares, square roots and cubes p 14 Throughout	Calculate squares, square roots and cubes of numbers	
3. Directed numbers pp 15–17 Throughout	Use directed numbers in practical situations	
4. Vulgar and decimal fractions and percentages pp 18–23 pp 28–30 Throughout	Use the language and notation of simple vulgar and decimal fractions and percentages in appropriate contexts; recognise equivalence and convert between these forms	
5. Ordering pp 7–9 Throughout	Order numbers by magnitude and demonstrate familiarity with the symbols $=$, $\neq$, $<$, $>$, 7, 8	
6. Standard form pp 64–65	Use the standard form $A \times 10^n$ where n is a positive or negative integer, and $1 \leq A < 10$	

Theme or topic	Core topics	Supplementary topics
7. The four rules p 4 Throughout	Use the four rules for calculations with whole numbers, decimal fractions, and vulgar (and mixed) fractions, including correct order of operations and use of brackets	
8. Estimation pp 23–25 pp 38–41 Throughout	Make estimates of numbers, quantities and lengths; give approximations to specified numbers of significant figures and decimal places and round off answers to reasonable accuracy in the context of given problems	
9. Limits of accuracy pp 39–41	Give appropriate upper and lower bounds for data given to a specified accuracy	Obtain upper and lower bounds to solutions of simple problems given data to a specified accuracy
10. Ratio, proportion, rate pp 26–28 pp 62–63	Demonstrate an understanding to the elementary ideas and notation of ratio, direct and inverse proportion and common measures of rate; divide a quantity in a given ratio; use scales in practical situations, calculate average speed	Express direct and inverse variation in algebraic terms and use this form of expression to find unknown quantities; increase and decrease a quantity by a given ratio
11. Percentages pp 28–32	Calculate a given percentage of a quantity; express one quantity as a percentage of another; calculate percentage increase or decrease	Carry out calculations involving reverse percentages
12. Use of an electronic calculator p 3 Throughout	Use an electronic calculator efficiently; apply appropriate checks of accuracy	
13. Measures pp 32–35	Use current units of mass, length, area, volume and capacity in practical situations and express quantities in terms of larger or smaller units	
14. Time p 36	Calculate times in terms of the 24-hour and 12-hour clock; read clocks, dials and timetables	

Theme or topic	Core topics	Supplementary topics
15. Money p 35	Calculate using money and convert from one currency to another	
16. Personal and household finance pp 31–32 pp 44–47	Use given data to solve problems on personal and household finance involving earnings, simple interest, discount, profit and loss; extract data from tables and charts	
17. Graphs in practical situations pp 88–98 pp 211–217	Demonstrate familiarity with Cartesian coordinates in two dimensions; interpret and use graphs in practical situations including travel graphs and conversion graphs; draw graphs from given data	Apply the idea of rate of change to easy kinematics involving distance–time and speed–time graphs, acceleration, and deceleration; calculate distance travelled as area under a linear speed–time graph
18. Graphs of functions pp 87–88 pp 98–99 pp 103–115	Construct tables of values for functions of the form $ax + b$, $\pm x^2 + ax + b$, $\frac{a}{x}$ ($x \neq 0$) where a and b are integral constants; draw and interpret such graphs; find the gradient of a straight line graph; solve linear and quadratic equations approximately by graphical methods	Construct tables of values and draw graphs of the form ax^n where a is a rational constant and $n = -2$, $-1, 0, 1, 2, 3$ and simple sums of not more than three of these and for functions of the form a^x where a is a positive integer; estimate gradients of curves by drawing tangents; solve associated equations approximately by graphical methods
19. Straight line graphs pp 99–103		Calculate the gradient of a straight line from the coordinates of two points on it; calculate the length of a straight line segment from the coordinates of its end points; interpret and obtain the equation of a straight line graph in the form $y = mx + c$
20. Algebraic representation and formulae pp 49–51 pp 56–58 pp 76–78	Use letters to express generalised numbers and express basic arithmetic processes algebraically; substitute numbers for words and letters in formulae; transform simple formulae	Construct equations from given situations; transform more complicated formulae

Theme or topic	Core topics	Supplementary topics
21. Algebraic manipulation pp 70–75 Throughout	Manipulate directed numbers; use brackets and extract common factors	Expand products of algebraic expressions; factorise expressions; manipulate algebraic fractions; factorise and simplify expressions
22. Functions pp 116–120		Use function notation to describe simple functions and their inverses; form composite functions as defined by $gf(x) = g(f(x))$
23. Indices pp 64–69	Use and interpret positive, negative and zero indices	Use and interpret fractional indices
24. Solutions of equations and inequalities pp 54–55 pp 78–80 pp 82–84	Solve simple linear inequalities in one unknown; solve simultaneous linear equations in two unknowns	Solve quadratic equations by factorisation and either by use of formulae or by completing the square; solve simple linear inequalities
25. Linear programming pp 120–127		Represent inequalities graphically and use this representation in the solution of simple linear programming problems
26. Geometrical terms and relationships pp 133–138 pp 154–156 pp 161–162	Use and interpret geometrical terms; use and interpret vocabulary of triangles, quadrilaterals, circles, polygons and simple solid figures including nets	Use the relationship between areas of similar triangles, with corresponding results for similar figures and extensions to volumes and surface areas of similar solids
27. Geometrical constructions pp 145–150 pp 169–170	Measure lines and angles; construct a triangle given three sides using rule and compasses only; construct other simple geometrical figures from given data using protractors and set squares as necessary; construct angle bisectors and perpendicular bisectors using straight edges and compasses only; read and make scale drawings	

Theme or topic	Core topics	Supplementary topics
28. Symmetry pp 163–166	Recognise rotational and line symmetry in two dimensions and properties of triangles, quadrilaterals and circles directly related to their symmetries	Recognise symmetry properties of the prism and pyramid; use the following symmetry properties of circles • equal chords are equidistant from centre • the perpendicular bisector of a chord passes through the centre • tangents from an external point are equal in length
29. Angle properties pp 133–138 pp 166–168	Calculate unknown angles using the following geometrical properties: • angles at a point • angles formed within parallel lines • angle properties of triangles and quadrilaterals • angle properties of regular polygons • angles in a semi-circle • angle between the tangent and radius of a circle	Use the following geometrical properties: • angle properties of irregular polygons • angle at the centre of a circle is twice the angel at circumference • angles in the same segment are equal • angles in opposite segments are supplementary
30. Locus pp 172–173	Use loci and the method of intersecting loci for sets of points in two dimensions	
31. Mensuration pp 150–154 pp 157–160	Carry out calculations involving the perimeter and area of a rectangle and triangle, the circumference and area of a circle, the area of a parallelogram and a trapezium, the volume of a cuboid, prism and cylinder and the surface area of a cuboid and a cylinder	Solve problems involving the arc length and sector area as fractions of the circumference and area of a circle, the surface area and volume of a sphere, pyramid and cone
32. Trigonometry pp 169–171 pp 176–209	Interpret and use three-figure bearings measured clockwise from north; apply Pythagoras' theorem and the sine, cosine and tangent ratios for acute angles to the calculation of a side or an angle of a right-angled triangle	Solve trigonometrical problems in two dimensions involving angles of elevation and depression; extend sine and cosine functions to angles between 90° and 360°; solve problems using the sine and cosine rules for any triangle an formula area of triangle $= \frac{1}{2} ab \sin C$; solve simple trigonometrical problems in three dimensions including angle between a line and a plane

Theme or topic	Core topics	Supplementary topics
33. Statistics pp 210–233	Collect, classify and tabulate statistical data; read, interpret and draw simple inferences from tables and statistical diagrams; construct and use bar charts, pie charts, pictograms, simple frequency distributions and histograms with equal intervals; calculate the mean, median and mode for individual and discrete data and distinguish between the purposes for which they are used	Construct and read histograms with equal and unequal intervals; construct and use cumulative frequency diagrams; estimate the median, percentiles, quartiles and inter-quartile range; calculate an estimate of the mean for grouped and continuous data; identify the modal class from a grouped frequency distribution
34. Probability pp 236–244	Calculate the probability of a single event as either a fraction or a decimal	Calculate the probability of simple combined events, using possibility diagrams and tree diagrams where appropriate
35. Vectors in two dimensions pp 252–258	Describe a translation by using a vector represented by $\binom{x}{y}$, **AB** or **a**; add vectors and multiply a vector by a scalar	Calculate the magnitude of a vector $\binom{x}{y}$ as $\sqrt{x^2 + y^2}$ represent vectors by directed line segments; use the sum and difference of two vectors to express given vectors in terms of two coplanar vectors; use position vectors
36. Matrices pp 265–276		Display information in the form of a matrix of any order; calculate the sum and product of two matrices; calculate the product of a matrix and a scalar quantity; use the algebra of 2×2 matrices including the zero and identity 2×2 matrices; calculate the determinant and inverse $\mathbf{A}^{-1}$ of non-singular matrix $\mathbf{A}$
37. Transformations pp 247–254 pp 262–265 pp 265–276	Reflect simple plane figures in horizontal or vertical lines; rotate simple plane figures about the origin, vertices or mid-points of edges of the figures, through multiples of 90; construct given translations and enlargements of simple plane figures; recognise and describe reflections, rotations, translations and enlargements	Use the following translations of the plane: reflection (M); rotation (R); translation (T); enlargement (E); shear (H); stretching (S) and their combinations; identify and give precise descriptions of transformations connecting given figures; describe transformations using coordinates and matrices

Working with numbers

We all work confidently with numbers every day, often without really thinking about them. Professor Brian Butterworth, in his book *The Mathematical Brain*, estimates that he processes about 1 000 numbers per hour – that's about 6 million numbers a year – without doing anything special! Butterworth is not special, nor is he a mathematician. The numbers he processes are all found in daily life – 51 numbers on the first page of a newspaper (prices, dates, amounts); numbers on the radio (stations, frequencies, news, hit parades); sporting results; time; labels on food; instructions for cooking; money; addresses; car registration plates; bar codes; page numbers in a book, and many, many more examples. What is more, these numbers are different: some are whole numbers, some are fractions, some are decimals, some are in order, some are random. His point is simple – we depend on numbers every day, and we need to understand number systems and how to use them.

Remember

When a list of numbers is followed by … it means that the list does not end with the last number written. It can continue infinitely.

Number systems

Different sets of numbers can be defined in the following ways:

Natural numbers ($\mathbb{N}$)	1, 2, 3, 4, 5, …
Whole numbers	0, 1, 2, 3, 4, 5, …
Integers (also called directed numbers) • Positive integers • Negative integers	…, –5, –4, –3, –2, –1, 0, 1, 2, 3, 4, 5, … 1, 2, 3, 4, 5, … (positive numbers are written without a + sign) –1, –2, –3, –4, –5, …
Even numbers	2, 4, 6, 8, 10, … (can be divided by 2)
Odd numbers	1, 3, 5, 7, 9, … (not divisible by 2)
Prime numbers – natural numbers that are divisible by 1 and the number itself – these numbers have only two factors	2, 3, 5, 7, 11, 13, … 1 is not a prime number as it only has one factor 2 is the only even prime number The largest known prime has thousands of digits
Square numbers – when natural numbers are multiplied by themselves you get a square number	1, 4, 9, 16, 25, 36, …
Fractions – these are parts of whole numbers, also called rational numbers	$\frac{1}{2}$, $\frac{1}{4}$, $2\frac{1}{2}$ (these are vulgar or common fractions; the top line is called numerator, bottom line is called denominator) 0.5, 0.4, 0.335 (these are decimal fractions) Percentages (%) are fractions of 100

These numbers all fit into the larger set of real numbers ($\mathbb{R}$). You can see how they fit on the diagram on the next page.

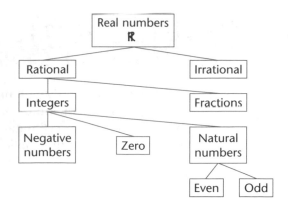

Mathematical symbols

In addition to numbers, symbols are used to write mathematical information. Make sure you recognise the following symbols:

+ Plus or positive	= Equal to	< Less than	∴ Therefore
− Subtract or minus	≠ Not equal to	> Greater than	∞ Infinity
× Multiply	≈ Approximately equal to	≤ Less than or equal to	‖ Parallel to
÷ Divide	: Is to / such that	≥ Greater than or equal to	√ Square root

Exercise

1. Which of these numbers are natural numbers?
   3, −2, 0, 1, 15, 4, 5
2. Which of the following are integers?
   −7, 10, 32, −32, 0
3. Which of these are prime numbers?
   21, 23, 25, 27, 29, 31
4. Write down the next two prime numbers after:
   a) 30   b) 80.
5. Are all prime numbers odd?
6. Are all odd numbers prime?
7. If you add two odd numbers, what kind of number do you get?
8. What kind of number will you get when you add an odd and an even number?
9. Write down:
   a) four square numbers bigger than 25
   b) four rational numbers smaller than $\frac{1}{2}$
   c) the set of negative integers between −7 and 0
   d) the decimal equivalents of $\frac{1}{2}$, 75% and $1\frac{1}{2}$.

Using a calculator

The table shows you some of the functions that are found on most calculators. However, all calculators are different and you must read the instruction manual and make sure you know how to use the model you have.

Key	Function	Key	Function
INV or **2ndF**	Used to access some functions	**EXP** or **EE**	This is the standard form button
C	Cancels the last number entered	**$a^{b}/_{c}$**	This is the fraction key; to enter $\frac{2}{3}$, press 2 $a^{b}/_{c}$ 3
AC	Cancels all data entered	**Min** or **STO**	Stores the displayed number in memory
x^2	Calculates the square of a number	**MR** or **RCL**	Recalls whatever is in the memory
x^3	Calculates the cube of a number	**M+**	Adds the display value to memory
$\sqrt{}$	Calculates the square root of a number	**M−**	Subtracts the display value from memory
$\sqrt[3]{}$	Calculates the cube root of a number	**Mode**	Gives the mode for calculation (refer to your manual)
+/−	Reverses the sign (changes positive to negative or negative to positive)	**DRG**	Changes units to degrees, radians or grads; your calculator should be normally set in degrees
x^y	Calculates any power; to calculate 2^4, you key in 2 x^y 4	**sin** **cos** **tan**	Calculates sine, cosine or tangent values (trigonometry)

Exercise

Use your calculator to work out:
1. 5^2
2. $\sqrt{2.6}$
3. $\frac{3}{4} - \frac{1}{5}$
4. $(3 \times 10^7) - (2 \times 10^6)$
5. 3^7
6. $-7 + 3 \div 2$.

Calculation rules

Mathematics is governed by a set of rules that has been developed to avoid confusion when working with complicated operations. These rules tell you about the order of operations – in other words, what you should do first in a sum like this one: $3 \times 4 + 14 \div 2$.

One way to remember the order of operations is to use a system called BODMAS. This is a mnemonic that stands for:

B – Brackets. Work out anything in brackets first. When there is more than one set of brackets, work from the outside to the inner sets.

O – Of. Change 'of' to '×' and work it out.

D – Divide.

M – Multiply. When there are only × and ÷ signs in a sum, you can work in any order.

A – Add.

S – Subtract. When there are only + and – signs in a sum, you can work in any order.

Most modern calculators are programmed to follow BODMAS rules.

These examples can help you to see how BODMAS works:

■ Examples

B
O
D
M
A
S

1. $3 \times 4 + 14 \div 2$ $(\div)$
   $= 3 \times 4 + 7$ $(\times)$
   $= 12 + 7$ $(+)$
   $= 19$

2. $18 - 14 \div (3 + 4) + 2 \times 3$ (Brackets)
   $= 18 - 14 \div 7 + 2 \times 3$ ($\div$ and $\times$)
   $= 18 - 2 + 6$ ($-$ and $+$)
   $= 22$

Exercise

Evaluate:

1. $(16 - 10) \div 2$
2. $16 - 10 \div 2$
3. $(4 + 3) \times 2$
4. $4 + 3 \times 2$
5. $(14 - 5) \div (20 - 2)$
6. $30 + 132 \div 11$
7. $5 \times 5 + 6 \div 2$
8. $2 + 5 \div 3 \times 6$.

Fill in signs to make these expressions true:

9. $5 \,\square\, 3$
10. $-5 \,\square\, -3$
11. $5.7 \,\square\, \frac{5}{7}$
12. $2\frac{1}{2} \,\square\, 2.5$
13. $4 = \boxed{16}$
14. $\frac{1}{3} \,\square\, 0.333$
15. $3.14 \,\square\, \frac{22}{7}$
16. $-7 \,\square\, \frac{1}{2}$
17. $3.333 \,\square\, \frac{1}{3}$
18. $-2^2 \,\square\, -4$

Factors

A *factor* of a number is a number that will divide exactly into it.

Consider the number 12.
The factors of 12 are: 1, 2, 3, 4, 6 and 12. That is, 1, 2, 3, 4, 6 and 12 will divide exactly into 12.
   Of these factors, 2 and 3 are called the *prime factors*, because they are also prime numbers.

Prime factors

Prime factors of a number are factors of the number that are prime numbers.
   You can write every number as the product of two or more numbers. For example, $12 = 4 \times 3$. But notice that 4 is not a prime number. You can break 4 down further. So now you have $12 = 2 \times 2 \times 3$. Since 2 and 3 are both prime numbers, you can say that you've written 12 as the *product of prime factors*.

Writing numbers as the product of prime factors

To write a number as a product of its prime factors, first try to divide the given number by the first prime number, 2. Continue until 2 will no longer divide into it exactly. Then try the next prime number, which is 3, then 5 and so on, until the final answer is 1.

Hint

To find the product of two or more numbers means you must *multiply* those numbers together.

■ Examples

1. Write 60 as a product of prime factors.

Divide 60 by 2 as shown	2	60
Again, divide by 2	2	30
Now, you can divide 15 by 3	3	15
Finally, divide by 5	5	5
		1

   So $60 = 2 \times 2 \times 3 \times 5$

2. Write 1 617 as a product of prime factors.
   1 617 cannot be divided exactly by 2, but it will divide exactly by the next prime number, which is 3.

539 cannot be divided exactly by 3	3	1 617
or 5 but it will divide exactly by 7	7	539
Divide by 7 again	7	77
Finally, divide by 11	11	11
		1

   So $1\,617 = 3 \times 7 \times 7 \times 11$

Multiples

A *multiple* of a number is the product of that number and an integer.
Multiples of 3 are 3, 6, 9, 12, ... How do you find them?
$3 \times 1 = 3, 3 \times 2 = 6, 3 \times 3 = 9 ...$

So 6 is a *multiple* of 3 because $6 = 3 \times 2$. Also, 3 is a *factor* of 6 because $6 \div 3 = 2$.

Lowest common multiple (LCM)

The *lowest common multiple* (LCM) of two or more numbers is the smallest number that is a multiple of each of them. 12 is a multiple of 3 and 4. It is also the smallest number that *both* 3 and 4 will divide into. So 12 is the LCM of 3 and 4.

There are two methods of finding the LCM of numbers. The first method is to list the multiples of each number and then pick out the lowest number that appears in every one of the lists. That is, the *lowest* number *common* to all the lists.

The second method is to express each of the numbers as a product of prime factors and then work out the smallest number which includes each of these products.

Some examples will make these methods clearer to you.

■ Example

Find the LCM of 12 and 15.

Method A: The multiples of 12 are 12, 24, 36, 48, ⑥⓪, 72, ...
The multiples of 15 are 15, 30, 45, ⑥⓪, 75, 90, ...
The lowest number that appears in both these lists is 60.

So the LCM of 12 and 15 is 60.

Method B: Expressing each number as a product of prime numbers:
$12 = 2 \times 2 \times 3$
$15 = 3 \times 5$

Any number that is a multiple of 12 must have at least two 2s and one 3 in its prime factor form.
Any number that is a multiple of 15 must have at least one 3 and one 5 in its prime factor form.
So any common multiple of 12 and 15 must have at least two 2s, one 3 and one 5 in its prime factor form.
The lowest common multiple of 12 and 15 is $2 \times 2 \times 3 \times 5$.

So the LCM of 12 and 15 is 60.

Exercise

1. Write the following numbers as a product of prime factors.
   a) 18   b) 16   c) 64   d) 81   e) 100
   f) 36   g) 21   h) 11   i) 45   j) 108

In each of the following questions, find the LCM of the given numbers.

2. a) 9 and 12   b) 12 and 18   c) 15 and 24
   d) 24 and 36   e) 3 and 5

3. a) 4, 14 and 21   b) 4, 9 and 18   c) 12, 16 and 24
   d) 6, 10 and 15

Patterns and sequences

Look at the following patterns of numbers.
1, 2, 3, 4, 5, ...
5, 10, 15, 20, 25, ...
What are the next three numbers in each pattern?

6, 7, 8 are the next three numbers in the first pattern and 30, 35, 40 are the next three numbers in the second pattern.

There is a special name for such patterns of numbers. They are called *sequences*. Each number in a sequence is called a *term* of the sequence. Each sequence is a group of numbers that has two important properties. The numbers are listed in a particular order and there is a rule that enables you to continue the sequence. Either the rule is given, or the first few terms of the sequence are given and you have to work out the rule.

■ Examples

Write down the next two terms in each of the following sequences:
1. 2, 6, 10, 14, 18, ...
2. 2, 6, 18, 54, 162, ...
3. 27, 22, 17, 12, 7, ...
4. 2, 3, 5, 8, 12, 17, ...

1. In the sequence, each term is 4 more than the previous term. The next two terms are 22 (that is 18 + 4) and 26.
2. In the sequence, each term is 3 times the previous term. The next two terms are 486 (that is 162×3) and 1 458.
3. In the sequence, each term is 5 less than the previous term. The next two terms are 2 (that is 7 − 5) and −3.
4. The increases from one term to the next are 1, 2, 3, 4, 5. The next two increases will be 6 and 7, so the next two terms are 23 (that is 17 + 6) and 30 (that is 23 + 7).

Exercise

1. Write down the next two terms in each of the following sequences:
   a) 2, 4, 6, 8, ...
   b) 3, 6, 9, 12, ...
   c) 2, 3, 5, 7, 11, 13, ...
   d) 9, 6, 3, 0, −3, ...
   e) 3, 4, 6, 9, 13, ...
   f) 1, 3, 4, 7, 11, 18, 29, ...

2. Find two square numbers, each less than 100, which are also cube numbers.

3. Study the sums of consecutive odd numbers shown below.

$$1 = 1$$
$$1 + 3 = 4$$
$$1 + 3 + 5 = 9$$
$$1 + 3 + 5 + 7 = 16$$
$$1 + 3 + 5 + 7 + 9 = 25$$

   a) Write down the next two lines of this pattern.
   b) Work out the sum of the first 10 odd numbers.
   c) Work out the sum of the first 210 odd numbers.
   d) Try to write a general rule for finding the sum of n consecutive odd numbers.

Hint

Consecutive means that the numbers are next to one another in the sequence.

4. Six dots can be arranged in a triangle $\quad \cdot \quad \cdot$ so 6 is a *triangle number*.

   The sequence of triangle numbers starts with 1, 3, 6, 10, ...

   a) Write down the next two triangle numbers.
   b) Study the sums of pairs of consecutive triangle numbers.
   $$1 + 3 = 4$$
   $$3 + 6 = 9$$
   $$6 + 10 = 16$$
   (i) Write down the next two lines of this pattern.
   (ii) What special type of numbers are these sums of consecutive triangle numbers?
   (iii) Work out the sum of the 10th triangle number and the 11th triangle number.

5. In each of the following, one number must be removed or replaced to make the sequence work. Work out which number this is and rewrite the sequences.
   a) 1, 7, 13, 16, 25, 31, ...
   b) 1, 4, 7, 9, 16, 25, ...

Sets

A set is a well-defined collection of objects that usually have some connection with each other. Sets can be described in words. For example: set A is a set of the oceans of the world; set B contains natural numbers less than or equal to 10. Sets can also be listed between curly brackets { } or braces. For example:

A = {Indian, Atlantic, Pacific, Arctic, Antarctic}

B = {1, 2, 3, 4, 5, 6, 7, 8, 9}

Objects that belong to a set are called elements and are indicated by the symbol ∈. ∈ means 'is an element of'. In the examples above, we can say Atlantic ∈ A or 2 ∈ B.

∉ means 'is not an element of'. Again using our examples, we can say Mount Everest ∉ A and 11 ∉ B.

The sets described and listed above are finite sets – they have a fixed number of elements.

Sets that do not have a fixed number of elements are infinite sets. The set of natural numbers greater than 10 is an example of an infinite set. This can be listed as {10, 11, 12, 13, ...}.

A set may also have no elements. Such a set is called an empty set. The symbols { } or ∅ indicate an empty set. An example of an empty set would be women over 6 m tall, or square circles. The number of elements in an empty set is 0 but {0} is not an empty set – it is a set containing one element, 0.

Set builder notation

This is a method of defining a set when it is difficult to list all the elements or when the elements are unknown.

■ Examples

1. The infinite set of even natural numbers can be listed like this: {2; 4; 6; 8; 10; ...}. This set can be described in set builder notation like this: $\{x: x \in \mathbb{N}, x$ is an even number$\}$. You read this as 'the set of all elements x such that x is an element of the set of natural numbers and is even'.

2. The finite set of natural numbers between 9 and 90 can be listed as {10; 11; 12; 13; ...; 89} or written in set builder notation as $\{x: x \in \mathbb{N}, 9 < x < 90\}$. You read this as 'the set of elements x such that x is an element of the set of natural numbers and x is bigger than 9 and smaller than 90'.

Exercise

1. List these sets.
   a) The set of persons living in your home.
   b) The set consisting of the first five odd numbers.
   c) The set whose objects are the last four letters of the English alphabet.
   d) The set of even numbers between 1 and 7.

2. Describe these sets.
   a) $A = \{2, 3, 5, 7\}$ 　　　　　　　　　b) $P = \{s, t, u, v, w, x, y, z\}$
   c) $Q = \{5, 10, 15, 20, 25\}$
3. $A = \{1, 2, 3, 4, 5, 6\}$, $B = \{2, 4, 6, 8, 10\}$, $C = \{3, 5, 7, 9, 11\}$.
   List the sets given by:
   a) $\{x: x \in A, x > 3\}$ 　　　　　　　　b) $\{x: x \in B, x \le 6\}$
   c) $\{x: x \in C, 5 < x < 12\}$ 　　　　　d) $\{x: x = 2y + 1, y \in B\}$
   e) $\{(x, y): x \in B, y \in C, x = 2y\}$.
4. Express in set builder notation the set of $\mathbb{N}$:
   a) greater than 5 　　　　　　　　　　　b) less than 10
   c) between 3 and 11 　　　　　　　　　　d) which are prime.

Relationships between sets

Equal sets

Sets that contain exactly the same elements are said to be equal.
Consider these sets:
   A = The set of letters in the word END = $\{E, N, D\}$
   B = The set of letters in the word DEN = $\{D, E, N\}$
We can say $A = B$.
The order of the elements in the set does not matter.

Subsets

If every element of set A is also an element of set B, then A is a subset of B.
This is written as $A \subset B$. $\subset$ means 'is a proper subset of'. $\not\subset$ means 'not a proper subset of'. Proper subsets always contain fewer elements than the set itself.

The proper subsets of $\{D, E, N\}$ are:
$\{D\}$　$\{E\}$　$\{N\}$　$\{D, E\}$　$\{D, N\}$　$\{E, N\}$
Trivial subsets of $\{D, E, N\}$ are $\{\}$ (the empty set) and $\{D, E, N\}$.

If a set has n elements, it will have 2^n subsets. For example, a set with 3 elements will have 2^3 subsets. That is $2 \times 2 \times 2 = 8$ subsets.
　The set of elements from which to select to form subsets is called the universal set. The symbol $\mathscr{E}$ is used to denote the universal set. You should remember that the universal set could change from problem to problem.

Exercise

1. Find a universal set for each of the following sets.
   a) the set of people in your class that have long hair
   b) the set of vowels
   c) $\{2, 4, 6, 8\}$
   d) {goats, sheep, cattle}
2. If $\mathscr{E}$ is the set of students at your school, define five subsets of $\mathscr{E}$.

Intersection and union of sets

The intersection of two sets refers to elements that are found in both sets. For example:

A = {a, b, c, d}

B = {c, d, e, f}

The intersection of these two sets is {c, d}. You write this as A ∩ B. Remember that A ∩ B = B ∩ A.

When two sets have no elements in common, they are called disjoint sets. The intersection of disjoint sets is the empty set or ∅.

The elements of two or more sets can be combined to make a new set. This is called the union of the sets. For example:

A = {1, 2, 3, 4}

B = {4, 5, 6}

C = {1, 2, 3, 4, 5, 6}

Set C is the union of set A and set B. We write this as A ∪ B = C.

Exercise

1. For each of the following, list the set which is the intersection of the two sets.
   a) {1, 2, 3, 4, 5, 6} and {4, 5, 8, 9, 10}
   b) {a, b, c, d} and {w, x, y, z}
   c) {1, 2, 3, 4, 5, ...} and {1, 3, 5, 7, ...}
   d) {m, n, o, p, q} and {d, o, g}
2. $\mathscr{E}$ = {1, 2, 3, 4, 5, 6}, A = {1, 2, 3, 4}, B = {3, 4, 5}, C = {5}.
   List the sets A ∩ B, B ∩ C, A ∩ C and $\mathscr{E}$ ∩ B.
3. Write down the union of the following sets.
   a) A = {a, b, c} and B = {d, e, f}
   b) A = {x} and B = {y}
   c) P = {2, 4, 6, 8, 10} and Q = {5, 6, 3, 1}
   d) {0, 1, 2, 3} and {4, 5, 6, ...}
4. State whether the following statements are true or false.
   a) P ∪ Q = Q ∪ P   b) M ∪ M = 2M
   c) B ∪ ∅ = ∅   d) B ∪ B = B^2

The complement of a set

The complement of a set A is the set of those elements that are in the universal set but are not in A. The complement of set A is written as A′.

For example, if $\mathscr{E}$ = {1, 2, 3, 4, 5} and A = {2, 4, 5}, then all the members of $\mathscr{E}$ that are not in A make the subset {1, 3}. This subset is the complement of A, so A′ = {1, 3}.

A set and its complement are disjoint. A ∩ A′ = ∅.

The union of a set and its complement is the universal set. A ∪ A′ = $\mathscr{E}$.

Exercise

1. $\mathcal{E} = \{p, q, r, s, t, v\}$ and $Y = \{p, r, v\}$. List the members of Y'.
2. $\mathcal{E} = \{1, 2, 3, \ldots, 20\}$. P is the set of prime numbers in $\mathcal{E}$.
   a) List the elements of the set P.
   b) List the elements of the set P'.
3. $W = \{$whole numbers$\}$ and subset $T = \{$even numbers$\}$.
   Describe in words the complement of T with respect to W.
4. The universal set is the set of all integers.
   What is the complement of the set of negative numbers?

Venn diagrams

Sketches used to illustrate sets and the relationships between them are called Venn diagrams. You need to understand the basics of Venn diagrams before you can use them to help you solve problems involving sets.

The rectangle represents $\mathcal{E}$

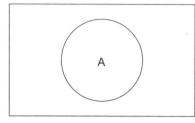

The circle represents set A

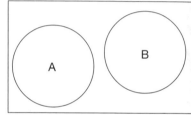

Set A and set B are disjoint

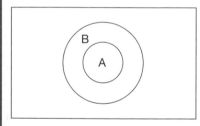

$A \subset B$

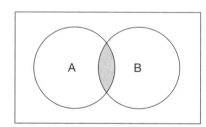

$A \cap B$ is the shaded portion

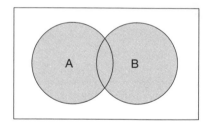

$A \cup B$ is the shaded portion

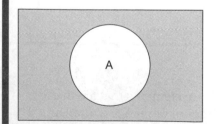

A' is the shaded portion

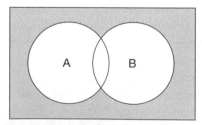

$(A \cup B)'$ is the shaded portion

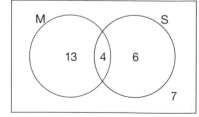

Venn diagrams can also be used to show the number of elements $n(A)$ in a set.
In this case:
M = {students doing maths}
S = {students doing science}

Exercise

1. Use the given Venn diagram to answer the following questions.
   a) List the elements of A and B.
   b) List the elements in A ∩ B.
   c) List the elements in A ∪ B.

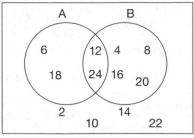

2. Use the given Venn diagram to answer the following questions.
   a) List the elements that belong to:
      (i) P
      (ii) Q.
   b) List the elements that belong to both P and Q.
   c) List the elements that belong to:
      (i) neither P nor Q
      (ii) P but not Q.

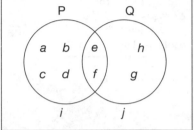

3. Draw a Venn diagram to show the following sets and write each element in its correct space.
   a) The universal set is {a, b, c, d, e, f, g, h}.
      A = {b, c, f, g}   B = {a, b, c, d, f}
   b) $\mathscr{E}$ = {whole numbers from 20 to 36 inclusive}
      A = {multiples of 4}   B = {numbers greater than 29}

4. The universal set is {students in a class}.
   V = {students who like volleyball} and S = {students who play soccer}.
   There are 30 students in the class. The Venn diagram shows numbers of students.
   a) Find the value of x.
   b) How many students like volleyball?
   c) How many students in the class do not play soccer?

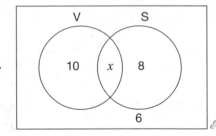

5. Shade the region in the Venn diagram which represents the subset A ∩ B′.

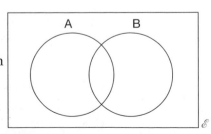

Powers and roots

Make sure you can find
the correct buttons on your
calculator to raise numbers
to powers and to find the
roots of powers.

Remember

- 3^2 is another way of writing 3×3.
- You write the square root of 64 as $\sqrt{64}$.
- 6^3 means $6 \times 6 \times 6$.
- $\sqrt[3]{216}$ means the cube root of 216.

When a number is multiplied by itself one or more times, the answer is a power of the first number. The number you started with is called a root of the power. For example: $5 \times 5 = 5^2 = 25$.

In this example, 5 is raised to the power of 2 or five is squared. The root of the power is 5. Because the root was squared, we say 5 is the *square root*.

Square numbers and roots

If a natural number is multiplied by itself, you get a *square* number. It may also be called a *perfect square*.

For example, $3 \times 3 = 3^2 = 9$ so 9 is a square number.

3 is said to be the *square root* of 9.

The first five square numbers are 1, 4, 9, 16 and 25.

■ Examples

1. $8 \times 8 = 64$ so 64 is a square number and its square root is 8. A shorter way of writing this is $8^2 = 64$ and $\sqrt{64} = 8$.
2. $12 \times 12 = 144$ so 144 is a square number and its square root is 12. Can you write this in a shorter way?

Cube numbers and roots

If the square of the natural number is multiplied again by the same natural number, the result is called a *cube number*.

For example, $6 \times 6 = 36$ and $36 \times 6 = 216$ so 216 is a cube number.

This is usually written as $6 \times 6 \times 6 = 216$ or $6^3 = 216$.

6 is said to be the *cube root* of 216.

The first five cube numbers are 1, 8, 27, 64 and 125.

■ Examples

1. $9 \times 9 \times 9 = 729$ so 729 is a cube number and its cube root is 9.
2. $20^3 = 8\,000$ so $\sqrt[3]{8\,000} = 20$

Exercise

1. Add four more terms to each of these two sequences.
   a) 1, 4, 9, 16, 25, ...
   b) 1, 8, 27, 64, 125, 216, ...
2. Use your calculator to work out:
   a) $\sqrt{49}$; $\sqrt{121}$; $\sqrt{256}$; $\sqrt{8}$
   b) $\sqrt[3]{125\,000}$; $\sqrt[3]{1\,728}$; $\sqrt[3]{729}$
3. Calculate:
   a) 13^2
   b) 9^3
   c) $(-4)^2$
   d) -3^3
   e) 100^2
   f) 100^3.

Directed numbers

Look at the temperature of each object shown in the picture. Notice that some of the colder objects have temperatures that are below 0 °C. These are negative temperatures and are marked with a – sign. Notice also that temperatures above 0 °C are written without a + sign. In mathematics, we accept that numbers are positive unless they are marked as negative. Positive and negative numbers are also called directed numbers.

Temperature °C

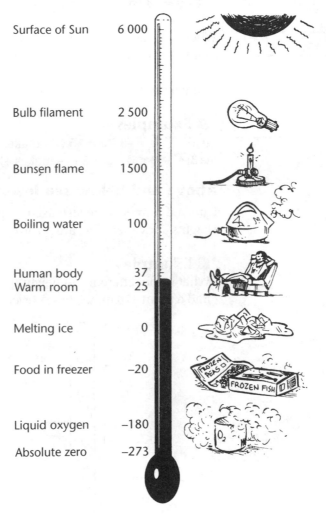

Surface of Sun	6 000
Bulb filament	2 500
Bunsen flame	1 500
Boiling water	100
Human body	37
Warm room	25
Melting ice	0
Food in freezer	–20
Liquid oxygen	–180
Absolute zero	–273

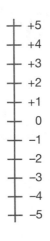

When you are working with directed numbers, it is helpful to represent them on a number line. Your number line can be either horizontal or vertical.

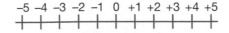

$$-5 \quad -4 \quad -3 \quad -2 \quad -1 \quad 0 \quad +1 \quad +2 \quad +3 \quad +4 \quad +5$$

The number line can be used to add or subtract directed numbers.
Find your starting position and then move left or right or up or down.
The sign of the number indicates how you should move on the number
line. +4 means four places to the right (or up). –4 means four places to
the left (or down).

■ Examples

1 + 2 = +3 (start at 1 and move 2 places to the right; end at 3)
2 – 5 = –3 (start at 2 and move 5 places to the left; end at –3)
–2 + 4 = 2 (start at –2 and move 4 places to the right; end at +2)
–1 – 3 = –4 (start at –1 and move 3 places to the left; end at –4)

When two signs appear together, you can replace them by one sign, using
mathematical rules. You will learn more about these rules in Module 2.

Change the signs using the rules before you work with the number line.

■ Examples

(–3) + (–2) = –3 – 2 = –5 (+ – makes –)
(+3) – (–2) = 3 + 2 = 5 (– – makes +)

Above and below sea level

Directed numbers are also used to give heights above and below sea level.
Heights above sea level are positive. Heights below sea level are negative.

■ Example

What is the difference in height between a point 230 m above sea level
and a point 170 m below sea level?

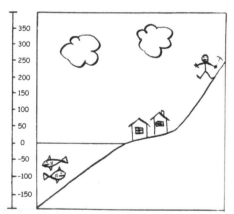

A rough sketch can be helpful when you are trying to solve problems like this one.

Remember

The rules are:
+ + makes +
+ – makes –
– + makes –
– – makes +

230 m above sea level means +230 m.
170 m below sea level means –170 m.

Difference in height = 230 – (–170)
= 230 + 170
= 400 m (remember to include the units of measurement in your answer!)

Exercise

1. Fill in the correct signs (< or >) in place of the * in each example to show which of these heights above and below sea level is the greater.
   a) 150 m * 75 m
   b) 50 m * –75 m
   c) –65 m * 30 m
   d) 0 m * –20 m
   e) –10 m * –20 m
   f) –25 m * 25 m

2. Find the value of:
   a) (+5) + (+3)
   b) (–4) + (+7)
   c) (+6) + (–2)
   d) (–1) + (–2)
   e) (+3) – (+8)
   f) (+2) – (–3)
   g) (–5) – (+1)
   h) (–5) – (–6)
   i) (–4) + (–3) – (+7).

3. On a certain day at a certain time in Siberia, the temperature was –33 °C. On the same day at the same time, the temperature in Brazil was 33 °C. What is the difference between these two temperatures?

4. The photograph on the left shows you a minimum and maximum thermometer marked in °C. The liquid in the tube pushes an indicator, which remains in place to show the minimum and maximum temperature for a given period. The knob at the side is used to reset the indicators.
   a) What was the temperature at the time this photograph was taken?
   b) What was the minimum temperature experienced on that day?
   c) What was the maximum temperature experienced on that day?
   d) What is the difference between the minimum and maximum temperatures?

5. A submarine is 50 m below sea level.
   a) It goes down a further 280 m to point P. Write down the depth of point P below sea level.
   b) From point P it rises 110 m to point Q. What is the depth of Q below sea level?

6. The temperature on a freezer thermometer shows that food is being stored at –20 °C.
   a) What would the temperature be if it was raised by 5 °C?
   b) What would the temperature be if it was lowered by 0.5 °C every hour for 12 hours?

Fractions

You learnt earlier that fractions are rational numbers that represent parts of a whole. Common or vulgar fractions are written with a numerator and denominator. The line dividing the two parts of the fraction means divide. The same fraction can be expressed in many different ways. For example:

$\frac{1}{2} = \frac{2}{4} = \frac{3}{6} = \frac{4}{8} = \frac{5}{10}$ and so on.

These are called equivalent fractions because they are all equal in value. To change a fraction to an equivalent fraction, you multiply the numerator and denominator by the same number.

$\frac{1}{2} = \frac{1}{2} \times \frac{2}{2} = \frac{2}{4}$ (multiply the numerator and denominator by 2)

$\frac{1}{2} = \frac{1}{2} \times \frac{4}{4} = \frac{4}{8}$ (multiply the numerator and denominator by 4)

You can get an equivalent fraction back to its simplest form by dividing the numerator and denominator by the same number.

In mathematics, fractions are normally written in their simplest form. When the numerator and denominator of a fraction have no common factors, the fraction is said to be in simplest form.

The fractions $\frac{1}{6}$, $\frac{2}{3}$ and $\frac{5}{7}$ are all in simplest form.

Remember

Any number divided by itself is 1: $\frac{4}{4} = 1$; $\frac{100}{100} = 1$. So, in fact, to make equivalent fractions, you are simply multiplying by 1.

Note

Changing a fraction to simplest form by dividing is sometimes called cancelling. For example: $\frac{\cancel{8}}{\cancel{15}} = \frac{1}{3}$.

■ Examples

1. Express in simplest form: $\frac{90}{120}$

$\frac{90}{120} = \frac{90 \div 3}{120 \div 3} = \frac{30}{40} = \frac{3}{4}$

or $\frac{9\cancel{0}}{12\cancel{0}} = \frac{9 \div 3}{12 \div 3} = \frac{3}{4}$

2. Fill in the missing numbers:

$\frac{1}{2} = \frac{\square}{16}$ $\qquad\qquad$ $\frac{4}{\square} = \frac{12}{15}$

Note: $2 \times 8 = 16$ $\qquad$ Note: $12 \div 3 = 4$

$\therefore \frac{1}{2} \times \frac{8}{8} = \frac{8}{16}$ $\qquad\qquad$ $\therefore \frac{4}{5} = \frac{12 \div 3}{15 \div 3}$

Exercise

1. Write three different equivalent fractions for each of the following given fractions.

   a) $\frac{1}{3}$ $\qquad$ b) $\frac{2}{5}$ $\qquad$ c) $\frac{5}{7}$

2. Express each of the following fractions in its simplest form.

   a) $\frac{5}{20}$ $\qquad$ b) $\frac{16}{24}$ $\qquad$ c) $\frac{56}{280}$

3. Find the missing numbers.

   a) $\frac{3}{5} = \frac{\square}{15}$ $\qquad$ b) $\frac{3}{\square} = \frac{24}{56}$ $\qquad$ c) $\frac{1}{\square} = \frac{25}{100}$

Operations on fractions

Adding or subtracting fractions

When adding or subtracting fractions, the denominators have to be the same. To make denominators the same, you find the LCM of the denominators. Then you add or subtract the numerators and write the answer over the common denominator.

■ Examples

1. $\frac{3}{7} + \frac{2}{7} = \frac{5}{7}$

2. $\frac{2}{9} + \frac{4}{9} + \frac{1}{9} = \frac{7}{9}$

3. $\frac{1}{2} + \frac{2}{5} = \frac{5}{10} + \frac{4}{10} = \frac{9}{10}$

4. $\frac{7}{12} - \frac{5}{12} = \frac{2}{12} = \frac{1}{6}$

5. $\frac{5}{8} - \frac{1}{20} = \frac{25}{40} - \frac{2}{40} = \frac{23}{40}$

6. $2\frac{1}{4} - 1\frac{7}{10} = \frac{9}{4} - \frac{17}{10} = \frac{45-34}{20} = \frac{11}{20}$

Multiplying fractions

When multiplying fractions, first multiply the numerators. Then multiply the denominators. Write the product of the numerators above the product of the denominators.

■ Examples

1. $\frac{1}{4} \times 3$

$= \frac{1}{4} \times \frac{3}{1}$

$= \frac{3}{4}$

2. $\frac{2}{3} \times \frac{9}{16}$

$= \frac{{}^1\cancel{2} \times \cancel{9}^3}{{}^1\cancel{3} \times \cancel{16}^8}$ (cancel)

$= \frac{3}{8}$

3. $\frac{2}{3}$ of $1\frac{4}{5}$

$= \frac{2}{{}_1\cancel{3}} \times \frac{\cancel{9}^3}{5}$

$= \frac{6}{5}$

$= 1\frac{1}{5}$

Exercise

Evaluate the following.

1. $\frac{4}{7} + \frac{2}{7}$

2. $\frac{7}{9} + \frac{5}{9}$

3. $\frac{3}{8} + \frac{1}{4}$

4. $2\frac{1}{7} + 1\frac{3}{6}$

5. $1\frac{4}{9} + 3\frac{5}{12}$

6. $3\frac{2}{9} + 2\frac{1}{3} + 2\frac{7}{12}$

7. $\frac{6}{15} - \frac{4}{15}$

8. $\frac{5}{8} - \frac{1}{4}$

9. $\frac{1}{3} - \frac{1}{5}$

10. $2\frac{5}{8} - 1\frac{1}{4}$

11. $3\frac{1}{4} - 1\frac{3}{8}$

12. $5 - \frac{3}{7}$

13. $5 \times \frac{6}{7}$

14. $\frac{1}{5} \times \frac{2}{3}$

15. $\frac{2}{3} \times \frac{6}{7}$

16. $4\frac{2}{3} \times 1\frac{1}{2}$

17. $\frac{2}{3}$ of 81

18. $\frac{1}{4}$ of $\frac{1}{2}$

19. $2\frac{1}{2}$ of 80

20. $3\frac{1}{7} \times 4\frac{3}{8}$

21. $\frac{2}{3} + \frac{4}{8} \times \frac{1}{2}$

22. $3\frac{1}{2} - 4\frac{1}{3}$

23. $\frac{12}{100} \times \frac{1}{8}$

24. $12 + 3\frac{1}{4}$

Dividing fractions

Hint

The quick method of finding a reciprocal is to turn the fraction upside down. So the reciprocal of $\frac{3}{4}$ is $\frac{4}{3}$ and the reciprocal of $\frac{1}{4}$ is $\frac{4}{1}$. To find the reciprocal of a whole number, write it as a fraction first. $5 = \frac{5}{1}$ and so the reciprocal of 5 is $\frac{1}{5}$.

In order to divide fractions, you need to understand the concept of the reciprocal. The reciprocal or multiplicative inverse of 3 is $\frac{1}{3}$ because $3 \times \frac{1}{3} = \frac{3}{1} \times \frac{1}{3} = 1$. The product of reciprocals is always 1.

To divide a fraction by another fraction, you multiply the first fraction by the reciprocal of the second fraction.

■ Examples

$$\frac{1}{2} \div \frac{1}{4}$$
$$= \frac{1}{2} \times \frac{4}{1}$$
$$= \frac{4}{2}$$
$$= 2$$

$$\frac{5}{6} \div 6$$
$$= \frac{5}{6} \times \frac{1}{6}$$
$$= \frac{5}{36}$$

$$2\frac{1}{2} \div 3$$
$$= \frac{5}{2} \div \frac{3}{1}$$
$$= \frac{5}{2} \times \frac{1}{3}$$
$$= \frac{5}{6}$$

$$4\frac{1}{2} \div 6\frac{1}{4}$$
$$= \frac{9}{2} \div \frac{25}{4}$$
$$= \frac{9}{\cancel{2}_{1}} \times \frac{\cancel{4}^{2}}{25}$$
$$= \frac{18}{25}$$

Exercise

Evaluate the following.

1. $\frac{2}{3} \div 3$
2. $\frac{3}{4} \div 3$
3. $2\frac{4}{5} \div 7$
4. $7 \div 1\frac{3}{4}$
5. $\frac{7}{12} \div \frac{14}{15}$
6. $4\frac{1}{8} \div 2\frac{3}{4}$
7. $\frac{18}{28} \div \frac{3}{4}$
8. $2\frac{1}{7} \div \frac{2}{14}$
9. $9\frac{1}{3} \div \frac{4}{7}$

Understanding fractions

A newspaper reports that $\frac{1}{3}$ of 45 million people are illiterate in a country. What does this mean? If you remember that 'of' means 'multiply', this is easy!

$$\frac{1}{3} \times \frac{45\,000\,000}{1} = 15\,000\,000.$$

The newspaper also reports that 40 out of every 50 people interviewed supported the national football team. What fraction is this?

$$\frac{\cancel{40}}{\cancel{50}} = \frac{4}{5}$$

The same newspaper reports that 200 people, $\frac{1}{3}$ of all the visitors to a local zoo, complained about the small cages. How many visitors were there altogether?

$\frac{1}{3}$ is 200

This means that $\frac{3}{3}$ (the total) of the number is 3×200.

So the total number is 600.

Finally, another report showed that, on average, consumers spent $60 or $\frac{2}{25}$ of their income on transport every month. What is the average monthly income according to this?

$\frac{2}{25}$ of the income is $60

So $\frac{1}{25}$ of the income is $30.

This means that $\frac{25}{25}$ of the income is 30×25.

So, the average monthly income in this example is $750.

Exercise

1. If $\frac{1}{5}$ of a number is 150, find the number.

2. 300 is $\frac{1}{4}$ of what number?

3. A man spends $300 on food. This is $\frac{3}{5}$ of his salary. Calculate his salary.

4. A water tank was $\frac{5}{8}$ full when it contained 275 ℓ of water.
   How many litres would it contain when full?

5. Evaluate $2\frac{1}{2} + 1\frac{3}{8}$.

6. Calculate $\frac{1}{5} + \frac{4}{15}$.

7. Calculate $27 \div 4\frac{1}{2}$.

8. $\dfrac{1}{f} = \dfrac{1}{u} + \dfrac{1}{v}$. Find the value of f when $u = \frac{1}{4}$ and $v = \frac{2}{3}$, giving your answer as a fraction.

9. Three friends, Alfred, Bianca and Carlos, decide to buy a car. Alfred pays $\frac{1}{4}$ of the cost. Bianca pays $\frac{1}{3}$ of the cost and Carlos pays the rest.
   a) What fraction of the cost does Carlos pay?
   b) Bianca pays $500 more than Alfred. Calculate the cost of the car.

10. Geoff spent $\frac{2}{5}$ of his money on food and $\frac{1}{3}$ on CDs.
    a) What fraction did he spend altogether?
    b) What fraction did he have left over?
    c) If he had $500, how much was left?

11. How many skirts can you make from 22 metres of fabric if each skirt requires $2\frac{1}{4}$ metres?

12. A bucket holds $14\frac{1}{2}$ litres of water. How many tins with a capacity of $\frac{3}{4}$ litres can you fill from the bucket?

13. A group of athletes travel to the Olympic Games. $\frac{1}{20}$ travel by aeroplane, $\frac{1}{12}$ by train, and $\frac{2}{5}$ by coach. The rest travel by car.
    What fraction travelled by car?

Decimals

Decimal numbers are based on powers of ten. The number 5 268 (five thousand two hundred and sixty eight) means:

5 000 + 200 + 60 + 8

The position of the numbers is important. In the example above, we have thousands, hundreds, tens and units.

What about this number: 5.268?

In this case, the decimal point has been used to show where the units end and fractions begin. The fractions are also expressed in powers of ten:

$\frac{2}{10}, \frac{6}{100}, \frac{8}{1\,000}$.

Changing fractions to decimals

All common fractions can be changed to decimals by dividing the numerator by the denominator. For example: $\frac{2}{5} = 2 \div 5 = 0.4$

Adding and subtracting decimals

When you add or subtract decimals, you can:

• change them to common fractions and add them
• place them above each other with the decimal points in line and add them normally
• use a calculator.

■ Examples

$0.5 + 0.3$

$= \frac{5}{10} + \frac{3}{10}$

$= \frac{8}{10} = 0.8$

$0.34 + 0.02$

$\begin{array}{r} 0.34 \\ + \ 0.02 \\ \hline 0.36 \end{array}$

$8.64 - 5.6034$

8.6400 (fill places)

$\begin{array}{r} - \ 5.6034 \\ \hline 3.0366 \end{array}$

Multiplying and dividing decimals

When you multiply or divide decimals, you can:

• change to common fractions and apply the rules you know
• count the number of decimal places; add them for multiplication, subtract them for division; work as usual and insert the decimal point when you are done
• use a calculator.

■ Examples

0.4×0.3

$= \frac{4}{10} \times \frac{3}{10}$

$= \frac{12}{100}$

$= 0.12$

5.408×3.2

$5\,408 \times 32 = 173\,056$

(4 decimal places)

$\therefore 5.408 \times 3.2$

$= 17.3056$

$1.144 \div 0.02$

$= \frac{0.144}{0.02}$

$= \frac{0.144 \times 100}{0.02 \times 100}$

$= \frac{14.4}{2}$

$= 7.2$

Exercise

1. Work out the following.
   a) $5.87 + 1.03 + 0.1$ b) $9.99 + 0.03$
   c) $7.92 - 0.97$ d) $10 - 0.918$
   e) 3.87×4 f) $3.88 \div 4$
   g) $0.208 \div 5$ h) 0.8×0.09
   i) $2.391 \div 0.03$
2. Write as decimals:
   a) $\frac{3}{4}$ b) $\frac{2}{5}$
   c) $\frac{27}{100}$ d) $\frac{5}{8}$.

Recurring decimals

You will notice that when you divide by certain numbers, you often get an unending decimal value. For example, when you divide 8 by 15 you get a result 0.53333 … without end. Try writing $\frac{8}{15}$ as a decimal yourself. The figure 3 recurs, and we call this a *recurring decimal*. We represent this symbolically by putting a dot over the repeating figure.

So $\frac{8}{15} = 0.5\dot{3}$

Similarly, $\frac{2}{3} = 0.6666666$ … which is written as $0.\dot{6}$.

If you change $\frac{1}{7}$ to a decimal, you'll get 0.142857142857142857 …

Here '142857' recurs and we put a dot over the first and last digits of the part that recurs.

So $\frac{1}{7} = 0.\dot{1}4285\dot{7}$

Similarly, $\frac{5}{11} = 0.45454545$ … which is written as $0,\dot{4}\dot{5}$.

If the division is unending, you can write the answer as a recurring decimal or you can *round* it after a suitable number of decimal places.

Rounding numbers

Look at these statements:
   'The population of Durban is about 750 000.'
   'It takes light around 8.65 years to travel from the star Sirius to the Earth.'
   'The thickness of one of my hairs is roughly 0.0075 centimeters.'

Each of the statements contains a number which is not exact. The population of Durban is not exactly 750 000, although this figure gives you a good idea of its size. The second and third statements are about measurements of time and length, which are approximate. In each of the statements, the number is correct to a particular degree of accuracy. We say that the number has been *rounded* or *corrected* to that degree of accuracy.

Note

Irrational numbers
Some decimals do not end nor recur. Such numbers are irrational. $\sqrt{2}$, $\sqrt{3}$, $\sqrt{5}$, $\sqrt{7}$, $\sqrt{8}$ are all irrational. In fact, the square root of any number that is not a perfect square is irrational. Try to find $\sqrt{2}$ on your calculator.

Rounding to the nearest ten

Consider the number 273. This number lies between 270 and 280. It is nearer to 270 than it is to 280. We write 273 = 270 to the nearest ten.

The number 518 is between 510 and 520. It is nearer to 520 than it is to 510. We write 518 = 520 to the nearest ten.

How would you round 845 to the nearest ten? 845 is exactly half way between 840 and 850. In such cases, we always round *up* for the nearest ten. So 845 = 850 to the nearest ten.

In a similar way, you can round numbers to the nearest hundred, or nearest thousand, or nearest million and so on.

Rounding to the nearest unit

What about rounding a number like 7.63 to the nearest unit? 7.63 is between the numbers 7 and 8. Do you round up or down? 7.63 is closer to 8, so 7.63 = 8 rounded to the nearest unit.

Rounding to decimal places

In the same way that you could be asked to give an answer rounded to the nearest unit or nearest ten, you could be asked to give an answer rounded to a certain number of decimal places. The answer is said to be rounded (or corrected) to the number of decimal places.

Here is a method you can use:
Work out the answer to one more place than you need. If the extra number is 5 or more, add 1 to the number before it. If the extra number is less than 5, leave the number before it as it is.

■ Examples

1. Write 43.2976 correct to 1 decimal place.
    You need to correct to 1 decimal place. So look at the figure in the second decimal place. This is 9, which is more than 5. So you must add 1 to the 2 in the first decimal place.
    So 43.2976 = 43.3 correct to 1 decimal place.
2. Take the same number, 43.2976. This time write it correct to 2 decimal places.
    At the second decimal place we have 9. The figure after this is 7, which is more than 5. So you must add 1 to 9.
    So 43.2976 = 43.30 correct to 2 decimal places.
3. Write 9.9999 correct to 3 decimal places. Note that adding 1 in this case changes all the preceding digits to 10.
    The answer is thus 10.000 correct to 3 decimal places.

Rounding to significant figures

Remember

'Significant figures' is often shortened to s.f.

If you read a number from left to right, ignoring the decimal point, the first significant figure is the first number that is not zero. All figures after that are also significant.

5.143 has 4 significant figures.

0.0003056 also has 4 significant figures.

You may be asked to round numbers to a certain number of significant figures. Look at these examples carefully to see how this differs from rounding off to decimal places.

■ Examples

Write the following numbers correct to 3 significant figures.

1. 4 768 000
2. 7 471
3. 367.82
4. 6.781
5. 0.002178

1. To correct to 3 significant figures, look at the fourth significant figure. The fourth significant figure is 8. This is more than 5. So add 1 to the third figure, 6. Thus, 4 768 000 = 4 770 000 correct to 3 significant figures.

    Don't forget to write the zeros.

    Don't write 4 768 000 = 477!

2. The fourth significant figure is 1, which is less than 5. So leave the third figure as it is.

    Hence 7471 = 7470 correct to 3 significant figures.

3. 367.82 = 368 correct to 3 significant figures.

4. 6.781 = 6.78 correct to 3 significant figures.

5. The first significant figure is 2. The fourth significant figure is 8. So 0.002178 = 0.00218 correct to 3 significant figures.

Exercise

1. Write these numbers correct to 3 decimal places.
    a) 29.712
    b) 1.62815
    c) 202.9157
    d) 4.6798
    e) 0.003527
    f) 1000.5645
    g) 0.6254

2. Do question 1 again but this time write the numbers correct to 3 significant figures.

3. Write these fractions as decimals, correct to 3 decimal places.
    a) $\frac{2}{3}$
    b) $\frac{5}{7}$
    c) $\frac{1}{6}$
    d) $\frac{8}{11}$
    e) $\frac{4}{9}$
    f) $\frac{1}{8}$

Ratio and proportion

Weston News

The richest man in Weston died yesterday, leaving $9 million to be shared amongst his three children, Willem, Jane and Zoe in the ratio 2 : 3 : 4.

Ratio

The news article on the left talks about ratio. A ratio is a comparison between two or more amounts. A ratio has no units. If you wish to compare quantities that are in different units, you must convert them to the same units before making them into a ratio.

A ratio of $3 to $12 is written as 3 : 12. The : sign is read as 'to'.

Ratios can also be written as fractions. $3 : 12 = \frac{3}{12} = \frac{1}{4}$. This means that the ratio 3 : 12 is the same as 1 : 4.

■ Examples

1. Share $24 between Tony and Joan in the ratio 3 : 5.

   This means that for every 8 units, Tony gets $\frac{3}{8}$ and Joan gets $\frac{5}{8}$.

   $\frac{3}{8} \times \frac{24}{1} = \$9$

   $\frac{5}{8} \times \frac{24}{1} = \$15$

   Work out what each of the children in the news article would get using this method.

2. Simplify the ratio $50 : $75.

   $50 : 75 = \frac{50}{75} = \frac{2}{3} = 2 : 3$ (remember to remove the units)

3. Express 2 m : 75 cm in its simplest form.
   Change the measurements so that both units are in cm.

   $200 \text{ cm} : 75 \text{ cm} = 200 : 75 = \frac{200}{75} = \frac{8}{3} = 8 : 3$

Exercise

1. Express the following ratios as simply as possible.
   a) $60 : $20   b) 2 m : 40 cm
2. The areas of two fields are in the ratio 2 : 3. If the area of the larger field is 78 hectares, what is the area of the smaller one?
3. Three men invest $2 000, $3 500 and $4 500 respectively in a business and agree to share the profits in the ratio of their investments. The profits in the first year were $8 000. How much did they each receive?
4. At the battle of Trafalgar, the British fleet of 27 ships met a French fleet of 18 ships and a Spanish fleet of 15 ships. Find the ratio of the sizes of the three fleets, in its lowest terms.
5. René and Pierre share $225 in the ratio 5 : 4. How much does each receive?
6. At the battle of Waterloo 72 500 French soldiers were opposed by 25 000 British, 17 500 Dutch and 27 500 German soldiers. Find the ratio of the sizes of the four armies. Give your answer in its lowest terms.

Map scale

The scale of a map is usually given as a ratio in the form of $1 : n$. The first number in the ratio is 1, for example, $1 : 25\,000$. This means that 1 unit of measurement on the map must be multiplied by 25 000 to get the distance (in the same units) in real life.

■ Examples

1. Express these map scales in the form $1 : n$.

   a) 5 cm to 2 km
      2 km = 200 000 cm
      5 : 200 000
      $= 1 : 40\,000$

   b) 4 mm : 5 m
      5 m = 5 000 mm
      4 : 5 000
      $= 1 : 1\,250$

2. If two places are 8.4 cm apart on a map with a scale of $1 : 50\,000$, what is the real distance between them?

   1 cm on the map represents 50 000 cm in reality
   ∴ 8.4 cm on the map represents $8.4 \times 50\,000$ cm in reality
   $(8.4 \times 50\,000 \div 100\,000)$ km
   $= 4.2$ km

   The distance between the two places is 4.2 km.

Proportion

Proportion is a way of comparing the ratios of quantities.

Direct proportion

If 4 loaves of bread cost $12, what is the cost of 8 loaves? The answer is $24. In this example the prices of 4 and 8 loaves are compared. When the number of loaves is doubled, the price also doubles. If you halved the number of loaves, the price would also halve.

This type of proportion is a direct proportion because an increase or decrease in one quantity (bread) will lead to an increase or decrease of the other quantity (price) in the same proportion.

Exercise

1. Five bottles of perfume cost $200. How much would 11 bottles cost?
2. Four soft drinks cost $9. How much would you pay for three?
3. A car travels 30 km in 40 minutes. How long would it take to travel 45 km at the same speed?
4. To make 12 buns you need:

240 g flour	48 g sultanas
60 g margarine	75 mℓ milk
24 g sugar	12 g salt.

   a) How much of each ingredient would you need to make 16 buns?
   b) Express the amount of flour to margarine in this recipe as a ratio.

A palm tree grows 20 m in 25 years. How much will it grow in the next 50 years?

Answer: This problem can't be solved by direct nor inverse proportion. The rate of growth is not constant, so you cannot solve it.

Inverse proportion

It takes 6 people 8 days to do a piece of work. How long will it take 12 people to do the same piece of work, assuming they all work equally hard? In this case there are more people, so the work will take less time. The number of people doubles, so the number of days halves.

This type of proportion is called inverse proportion. In inverse proportion, one quantity (people) increases in the same proportion as the other quantity (days) decreases.

■ Examples

1. A man travelling at 30 km/h gets home from work in 24 minutes. How long would it take him if he travelled at 36 km/h?
   30 km/h takes 24 minutes.
   So, at 1 km/h it would take (30×24) minutes.
   (Remember it would take longer, so multiply)
   ∴ at 36 km/h it would take $\frac{30 \times 24}{36} = 20$ minutes

2. A woman working 6 hours per day could do a job in 4 days. How many hours must she work to do the job in 3 days?
   To do the job in 4 days takes 6 hrs/day.
   ∴ to do the job in 1 day would take $4 \times 6 = 24$ hrs/day.
   To do the job in 3 days would take $24 \div 3 = 8$ hrs/day.

Exercise

1. Four people can build a wall in 15 days. How long would it take 6 people to build the wall at the same rate?
2. A basket of chicken feed is enough for 4 chickens for 5 days. How many days would it last for 8 chickens?
3. 20 mm of rain falls in the first 7 days of April. How much rain falls in the whole month?

Rate

Sometimes we need to compare two different quantities that are measured in different units. If a quantity of one thing is considered in relation to a unit of another, we call this a *rate*. You will work in more detail with rates later in this module.

Percentages

The phrase per cent comes from the Latin 'per centuri' which means 'as per hundred'.

A percentage is a ratio in which the second quantity is always 100. In other words, a percentage is a fraction in which the denominator is always 100. $\frac{25}{100}$ is twenty-five per cent or 25%. 72% is $\frac{72}{100}$ or 72 : 100.

You can write any fraction as a percentage by changing the denominator to 100.

■ Examples

1. $\frac{1}{2} = \frac{1}{2} \times \frac{50}{50} = \frac{50}{100} = 50\%$ or $\frac{1}{2} = \frac{1}{1\cancel{2}} \times \frac{\overset{50}{\cancel{100}}}{100} = \frac{50}{100} = 50\%$

2. $\frac{2}{5} = \frac{2}{5} \times \frac{20}{20} = \frac{40}{100} = 40\%$ or $\frac{2}{5} = \frac{2}{1\cancel{5}} \times \frac{\overset{20}{\cancel{100}}}{100} = \frac{40}{100} = 40\%$

Writing a percentage as a fraction

To change a percentage into a fraction, change the % symbol into $\times \frac{1}{100}$ and simplify the fraction.

■ Examples

1. 5%

$5 \times \frac{1}{100}$

$= {}^{1}\cancel{5} \times \frac{1}{\cancel{100}_{20}}$

$= \frac{1}{20}$

2. 250%

$250 \times \frac{1}{100}$

$= \overset{5}{\cancel{250}} \times \frac{1}{\cancel{100}_{2}}$

$= \frac{5}{2}$ or $2\frac{1}{2}$

3. $7\frac{1}{2}\%$

$= \frac{{}^{3}\cancel{15}}{2} \times \frac{1}{\cancel{100}_{20}}$

$= \frac{3}{40}$

4. $33\frac{1}{3}\%$

$= \frac{\cancel{100}}{3} \times \frac{1}{\cancel{100}}$

$= \frac{1}{3}$

Exercise

Express the following fractions as percentages.

1. $\frac{1}{5}$ 2. $\frac{4}{5}$ 3. $\frac{1}{10}$ 4. $\frac{7}{10}$

5. $\frac{1}{50}$ 6. $\frac{1}{25}$ 7. $\frac{8}{25}$ 8. $\frac{1}{8}$

9. $\frac{5}{6}$ 10. $\frac{2}{3}$ 11. $\frac{5}{9}$ 12. $\frac{5}{2}$

Convert the following percentages into fractions.

13. 4%   14. 25%   15. 50%   16. 75%

17. 60%   18. 125%   19. 250%   20. $23\frac{1}{3}\%$

21. $2\frac{1}{2}\%$ 22. $66\frac{2}{3}\%$

Percentages of a given quantity

You may need to calculate a percentage of a quantity. For example, 14% VAT on $35.

To work this out, first change the percentage to a fraction. Multiply the fraction by the given number and simplify.

■ Examples

1. 5% of 600

$5 \times \frac{1}{100} = \frac{1}{20}$

$\therefore$ 5% of 600

$= \frac{1}{20} \times 600$

$= 30$

2. $2\frac{1}{2}\%$ of $200

$\frac{5}{2} \times \frac{1}{100} = \frac{1}{40}$ (change $2\frac{1}{2}$ to an improper fraction)

$\therefore \frac{1}{40} \times \frac{200}{1}$

$= \$5$

Exercise

Find the value of:

1. 10% of 150
2. 5% of 25
3. 15% of $300
4. 4% of 200 kg
5. 20% of 150 m
6. $\frac{1}{2}$% of 16 000
7. $3\frac{1}{2}$% of $400
8. $7\frac{1}{2}$% of $800
9. 26% of $50
10. 5.5% of $2 000.

Writing one number as a percentage of another

To write a given number as a percentage of another number, first write the given number as a fraction of the other number. Multiply by $\frac{100}{1}$ and simplify.

■ Examples

1. Express 15 as a percentage of 45.

    Write 15 as a fraction of 45: $\frac{15}{45}$

    $$\text{Multiply by } \tfrac{100}{1}: \frac{15}{45} \times \frac{100}{1} = 33.3\%$$

2. A factory employs 20 workers. One day 2 workers were absent from work. What percentage of workers was absent?

    Express this as a fraction: $\frac{2}{20}$

    $$\text{Multiply by } \tfrac{100}{1}: \frac{2}{20} \times \frac{100}{1} = 10\%$$

Exercise

1. Express 25 as a percentage of 50.
2. Express 12 as a percentage of 36.
3. Express $2\frac{1}{2}$ as a percentage of 50.
4. In a basket of 200 oranges, 18 are rotten. What percentage is rotten?
5. 5 workers out of a total staff of 50 were absent one day. What percentage of workers was present?
6. 4 students out of 20 failed an examination. What percentage of students passed?

The following questions are taken from IGCSE papers. Use them to test how well you have understood this section.

7. Express 35% as a fraction in its simplest form.
8. Express $3.60 as a percentage of $9.
9. A can of fruit has a mass of 530 g. The fruit has a mass of 500 g. Find the mass of the fruit as a percentage of the total mass.
10. During the first week of October, a bookshop sold 880 books. In the second week, it sold 15% fewer books. How many books did it sell in the second week?

Profit and loss

When people sell goods for a living, they try to make some money from the sale thereof. The amount of money made on the sale of an article is called the profit. If the goods are sold for less than the trader paid for them, he or she makes a loss.

The price a trader pays for an article is called the cost price. The price at which articles are sold to the public is called the selling price. If the selling price is greater than the cost price, the trader makes a profit. If the selling price is less than the cost price, the trader makes a loss.

Percentage profit and loss

Profit and loss are normally calculated as percentages of the cost price. The following formulae are used to calculate percentage profit or loss:

$$\text{Percentage profit} = \frac{\text{actual profit}}{\text{cost price}} \times 100\%$$

$$\text{Percentage loss} = \frac{\text{actual loss}}{\text{cost price}} \times 100\%$$

■ Examples

1. A shopkeeper buys an article for $500 and sells it for $600. What is the percentage profit?

   Profit = selling price – cost price $\text{Percentage profit} = \frac{\text{profit}}{\text{cost}} \times 100\%$

   $\qquad$ = $600 – $500 $\qquad\qquad\qquad\qquad\qquad = \frac{\$100}{\$500} \times 100\%$

   $\qquad$ = $100 $\qquad\qquad\qquad\qquad\qquad\qquad\quad = 20\%$

2. A person buys a car for $16 000 and sells it for $12 000. Calculate the percentage loss.

   Loss = cost price – selling price $\text{Percentage loss} = \frac{\$4\,000}{\$16\,000} \times 100\%$

   $\qquad$ = $16 000 – $12 000 = $4 000 $\qquad\qquad\qquad = 25\%$

Exercise

1. Find the actual profit and percentage profit in the following cases:
   a) cost price $20, selling price $25
   b) cost price $500, selling price $550
   c) cost price $1.50, selling price $1.80
   d) cost price 30 cents, selling price 35 cents.
2. Calculate the percentage loss in the following cases:
   a) cost price $400, selling price $300
   b) cost price 75c, selling price 65c
   c) cost price $5.00, selling price $4.75
   d) cost price $6.50, selling price $5.85.
3. A woman buys 100 oranges for $30. She sells them for 50 cents each. Calculate the percentage profit or loss made.

Discount

Sometimes traders offer a reduced price or discount for people who pay cash or who buy in bulk. Discount is calculated as a percentage of the selling price. For example, you are offered 10% discount on a car costing $20 000. How much would you pay?

10% is $\frac{10}{100} \times 20\,000 = \$2\,000$, so you would pay $20\,000 - \$2\,000 = \$18\,000$

You can also work this out as $\frac{90}{100} \times 20\,000 = \$18\,000$.

You will solve more problems with percentages later in this module (see page 42).

Measurement

Today most countries in the world use a decimal system of measurement. Decimal units of measurements are also called SI (Système International) units. The table below will help you to remember what units are used to measure length, mass, capacity, area and volume.

This table shows you only the commonly used units. However, there are some other units that we do not use often. The table on page 33 shows you the prefixes used for the units bigger than and smaller than the standard units for length, mass and capacity.

Measure	Units used	Equivalent to ...
Length – how long (or tall) something is	Millimetres (mm) Centimetres (cm) Metres (m) Kilometres (km)	10 mm = 1 cm 100 cm = 1 m 1 000 m = 1 km 1 km = 1 000 000 mm
Mass – the amount of material in an object, sometimes incorrectly called weight	Milligrams (mg) Grams (g) Kilograms (kg) Tonnes (t)	1 000 mg = 1 g 1 000 g = 1 kg 1 000 kg = 1 t 1 t = 1 000 000 g
Capacity – the inside volume of a container, how much it holds	Millilitres (ml) Centilitres (cl) Litres (l)	10 ml = 1 cl 100 cl = 1 l 1 l = 1 000 ml
Area – the amount of space taken up by a flat (two-dimensional) shape, always measured in square units	Square millimetre (mm²) Square centimetre (cm²) Square metre (m²) Square kilometre (km²) Hectare (ha)	100 mm² = 1 cm² 10 000 cm² = 1 m² 1 000 000 m² = 1 km² 1 km² = 100 ha 1 ha = 10 000 m²
Volume – the amount of space taken up by a three-dimensional object, always measured in cubic units	Cubic millimetre (mm³) Cubic centimetre (cm³) Cubic metre (m³) Millilitre (ml)	1 000 mm³ = 1 cm³ 1 000 000 cm³ = 1 m³ 1 m³ = 1 000 l 1 cm³ = 1 ml

Converting units of measurements

Sometimes you will need to change the units of a measurement. For
example, you may want to change metres to kilometres or centimetres
to millimetres. To change to a bigger unit, you have to divide. To change to
a smaller unit, you have to multiply. The amount by which you divide or
multiply will depend on how many places you are moving – the diagram
below will help you to see how this works. Each place is worth a power of
10, so if you move one place, you will multiply or divide by 10; two places,
by 100; three places, by 1 000 and so on.

Prefix	kilo	hecto	deca	standard units (m, g, ℓ)	deci	centi	milli

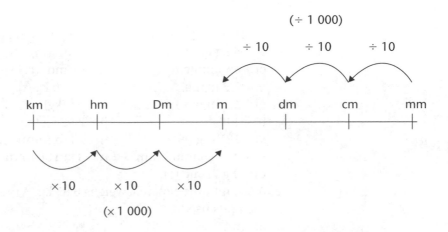

■ Examples

1. Express 5 km in metres.
$$1 \, km = 1 \, 000 \, m$$
$$So \; 5 \, km = 5 \times 1 \, 000 \, m$$
$$= 5 \, 000 \, m$$

2. Express 3.2 cm in millimetres.
$$1 \, cm = 10 \, mm$$
$$So \; 3.2 \, cm = 3.2 \times 10 \, mm$$
$$= 32 \, mm$$

3. Express 1 425 m in kilometres.
$$1 \, 000 \, m = 1 \, km$$
$$1 \, 425 \, m = (1 \, 425 \div 1 \, 000) \, km$$
$$= 1.425 \, km$$

4. Express 2 m 37 cm in centimetres.

Only the 2 m must be changed into centimetres.

$$1 \text{ m} = 100 \text{ cm}$$
$$\text{So } 2 \text{ m} = 2 \times 100 \text{ cm}$$
$$= 200 \text{ cm}$$
$$2 \text{ m } 37 \text{ cm} = 200 \text{ cm} + 37 \text{ cm}$$
$$= 237 \text{ cm}$$

5. Express 2 000 000 cm² in m².

$$10\,000 \text{ cm}^2 = 1 \text{ m}^2$$
$$2\,000\,000 \text{ cm}^2 = (2\,000\,000 \div 10\,000) \text{ m}^2$$
$$= 200 \text{ m}^2$$

Exercise

1. Express each quantity in the unit given in brackets.

   a) 4 kg (g)   b) 5 km (m)
   c) 35 mm (cm)   d) 81 mm (cm)
   e) 7.3 g (mg)   f) 5 670 kg (t)
   g) 2.1 m (cm)   h) 2 t (kg)
   i) 140 cm (m)   j) 2 024 g (kg)
   k) 121 mg (g)   l) 23 m (mm)
   m) 3 cm 5 mm (mm)   n) 8 km 36 m (m)
   o) 9 g 77 mg (g)

2. A 500 mℓ bottle of milk weighs 0.85 kg. Write in full the names of the units used.

   a) mℓ b) kg

3. Arrange in ascending order of size.

   3.22 m, $3\frac{2}{9}$ m, 32.4 cm

4. Write the following volumes in order, starting with the smallest.

   $\frac{1}{2}\ell$, 780 mℓ, 125 mℓ, 0.65 ℓ

5. How many 5 mℓ spoonfuls can be obtained from a bottle that contains 0.3 ℓ of medicine?

6. Express each quantity in the units given in brackets.

   a) 14.23 m (mm, km)   b) 19.06 g (mg, t)
   c) $2\frac{3}{4}\ell$ (mℓ, cℓ) d) 4 m² (mm², ha)
   e) 13 cm² (mm², ha)   f) 10 cm³ (mm³, m³)

Money

Working with money is the same as working with decimal fractions, because most money amounts are given as decimals. Remember, though, that when you work with money, you need to include the units ($ or cents) in your answers.

Foreign currency

The money a country uses is called its currency. Each country has its own currency and most currencies work on a decimal system (100 small units are equal to 1 main unit). This table shows you the currency units of a few different countries.

Country	Main unit	Smaller unit
USA	Dollar ($)	= 100 cents
Japan	Yen (¥)	= 100 sen
UK	Pound (£)	= 100 pence
Germany	Euro ( )	= 100 cents
France	Euro ( )	= 100 cents

Foreign exchange

When you change one currency for another, it is called foreign exchange. The rate of exchange determines how much of one currency you will get for another. Exchange rates can change daily. The daily rates are published in the press and displayed at banks. When you are asked to convert from one currency to another, you will be given a rate of exchange to work with.

■ Examples

1. Convert £50 into Botswana pula given that £1 = 9.83 pula.
   £1 = 9.83 pula
   £50 = 9.83 pula × 50
       = 491.50 pula

2. Convert 803 pesos into British pounds given that £1 = 146 pesos.
   146 pesos = £1
   So 1 peso = $\frac{£1}{146}$
   803 pesos = $\frac{£1}{146} \times \frac{803}{1}$
             = £5.50

Exercise

1. Find the cost of 8 apples at 50c each, 3 oranges at 35c each and 5 kg of bananas at $2.69 per kilogram.
2. How much would you pay for: 240 textbooks at $15.40 each, 100 pens at $1.25 each and 30 dozen erasers at 95c each?
3. If 1 dinar = £2.13, convert 4 000 dinar to pounds.
4. If US $1 = £0.9049, how many dollars can you buy with £300?
5. An American tourist visits South Africa with US $3 000. The exchange rate when she arrives is US $1 = R8.20. She changes all her dollars into rands and then spends R900 per day for seven days. She changes the rands she has left back into dollars at a rate of US $1 = R8.25. How much does she get in dollars?

Time

You have already learnt how to tell the time and you should know how to read and write time using the 12-hour and 24-hour system. The clock dial on the left shows you the times from 1 to 12 (a.m. and p.m. times). The outside dial shows what the times after 12 p.m. are in the 24-hour system.

Working with time intervals

■ Examples

1. Sara and John left home at 2.15 p.m. Sara returned at 2.50 p.m. and John returned at 3.05 p.m. How long was each person away from home?

Sara   2.50 − 2.15 = .35   Sara was away for 35 minutes.
John   3.05 − 2.15 = 2.65 − 2.15 = .50   John was away for 50 minutes.

2. A journey takes from 0535 to 1820. How long is this?

1820 − 0535 = 1780 − 0535 = 1245
The journey took 12 hours and 45 minutes.

3. How much time passes from 1935 on Monday to 0355 on Tuesday?

The easiest way to tackle this problem is to divide the time into two parts: 1935 to 2400 is one part and 0000 to 0355 the next day is the other part.

2400 − 1935 = 2360 − 1935 = 0425

0425 + 0355 = 0780 = 0820 (carry the extra 60 minutes back to the hours column)

8 hours and 20 minutes have passed.

Remember

You cannot subtract 15 from 5, so carry one whole hour over to make 65 minutes.

Remember

Always bear in mind that time is written in hours and minutes and that there are 60 minutes in an hour. This is very important when calculating time – if you put 1.5 hours into your calculator, it will assume the number is decimal and work with parts of 100.

Exercise

1. Gary started a marathon race at 9.25 a.m. He finished at 1.04 p.m. How long did he take? Give your answer in hours and minutes.

2. Yasmin's car odometer dial showed the two readings on the left before and after a journey.
   a) How far did she travel?
   b) The journey took $2\frac{1}{2}$ hours. What was her average speed in km/h?

3. Yvette records three songs on tape. The time each of them lasts is 3 minutes 26 seconds, 3 minutes 19 seconds and 2 minutes 58 seconds. She leaves a gap of 10 seconds between each of the songs. How long will it take to play the recording?

4. A journey started at 1730 hours on Friday, 7 February, and finished 57 hours later. Write down the time, day and date when the journey finished.

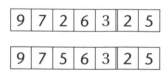

Timetables

Most travel timetables are in the form of tables with columns representing journeys. The 24-hour system is used to give the times.

■ Example

	SX	D	D	D	MO	D	SX
Aytown	0630	0745	1200	1630	1715	1800	2030
Beecity	0650	0805	1225	1650	1735	1825	2050
Ceeville	0725	0840	1315	1725	1815	1905	2125

D – daily including Sundays, SX – daily except Saturdays, MO – Mondays only

Make sure you can see that each column represents a journey. For example, the first column shows a bus leaving at 0630 every day except Saturday (6 times per week). It arrives at the next town, Beecity, at 0650 and then goes on to Ceeville, where it arrives at 0725.

Exercise

1. The timetable for evening trains between Mitchells Plain and Cape Town is shown below.

Mitchells Plain	1829	1902	1932	2002	2104
Nyanga	1840	1913	1943	2013	2115
Pinelands	1901	1931	2001	2031	2133
Cape Town	1917	1947	2017	2047	2149

   a) Shaheeda wants to catch a train at Mitchells Plain and get to Pinelands by 8.45 p.m. What is the time of the latest train she should catch?
   b) Calculate the time the 1902 train from Mitchells Plain takes to travel to Cape Town.
   c) Thabo arrives at Nyanga station at 6.50 p.m. How long will he have to wait for the next train to Cape Town?

2. The timetable for a bus service between Aville and Darby is shown below.

Aville	1030	1050	and	1850
Beeston	1105	1125	every	1925
Crossway	1119	1139	20 minutes	1939
Darby	1137	1157	until	1957

   a) How many minutes does a bus take to travel from Aville to Darby?
   b) Write down the timetable for the first bus on this service to leave Aville after the 1050 bus.
   c) Ambrose arrives at Beeston bus station at 2.15 p.m. What is the time of the next bus to Darby?

Estimation and approximation

All measurement is approximate. How accurate it is depends on the measuring instruments used. For instance, if you measure the width of a flat metal bar with an ordinary wooden ruler, you could probably measure it to the nearest millimetre. If, however, you used a micrometer (an engineer's measuring instrument), you could measure much more accurately, probably to a hundredth of a millimetre. But no measurement is completely accurate; there is always a degree of error. The degree of error can be calculated mathematically.

Estimating results of calculations

A rough idea of something is called an estimate. If you have a calculator, you can do very complicated calculations. But when you don't have a calculator with you, it is useful to be able to estimate what the answer should be, more or less. Look at these examples to see how it's done.

■ Examples

1. Estimate the value of $\frac{8.7 \times 5.2}{1.9}$.

   Because you are estimating the answer, you make estimates of the numbers involved.

   $8.7 \approx 9$
   $5.2 \approx 5$
   $1.9 \approx 2$

   So your estimate would be $= \frac{9 \times 5}{2} = 22\frac{1}{2}$.

2. Estimate the value of $\frac{18.9 \times 0.815}{2\,820}$.

   Estimate $= \frac{20 \times 0.8}{3\,000} = \frac{16}{3\,000} = \frac{5.33}{1\,000}$
   $= 0.00533$

Exercise

Estimate the values of the following fractions.

1. $\frac{23.7}{6.2}$

2. $\frac{21.9 \times 0.78}{8.91}$

3. $\frac{4.1}{0.088 \times 3.9}$

4. $\frac{12.7}{0.475 \times 3.4999}$

5. $\frac{7.23 \times 0.49}{9.26}$

Lower bound and upper bound

Suppose the length of a piece of wire is 52 cm to the nearest centimetre. As you will remember from the work on 'rounding' on page 23, this means that the wire could have any length from 51.5 cm up to, but not including, 52.5 cm. This is because the smallest number that rounds to 52, is 51.5 and the largest number that rounds to 52 is 52.4$\dot{9}$. However, as 52.4$\dot{9}$ = 52.5 rather than 52.499 or 52.49999, when the upper bound is needed in a calculation, 52.5 must be used.

52.5 is called the upper bound of the range of possible values and 51.5 is called the lower bound of the range.

If the length of the wire is denoted by x cm, then we can write:

$$51.5 \leq x < 52.5$$

less than or equal to because 51.5 would round to 52   less than, *not* equal to, because 52.5 rounds to 53

■ Examples

1. In a race, Nomatyala ran 100 m in 15.3 seconds. The distance is correct to the nearest metre and the time is correct to one decimal place. Write down the lower and upper bounds of:
   a) the actual distance Nomatyala ran
   b) the actual time taken.

   What you have to do is try to think *backwards*. What numbers will round to 100 m?
   a) The actual distance could be $\frac{1}{2}$ m below or $\frac{1}{2}$ m above 100 m because all these distances will round to 100 m.
      Lower bound of the distance = 99.5 m.
      Upper bound of the distance = 100.5 m.
   b) The smallest number that rounds to 15.3 is 15.25 and the largest number which rounds to 15.3 is just below 15.35.
      Lower bound of the time taken = 15.25 seconds.
      Upper bound of the time taken = 15.35 seconds.

2. The length of a piece of thread is 4.5 m to the nearest 10 cm. The actual length of the thread is L cm. Find the range of possible values for L, giving your answer in the form $... \leq L < ...$

   First change metres to centimetres.
   The length of the thread is measured as 450 cm to the nearest 10 cm.
   The actual length could be 5 cm above or below 450 cm.

   Lower bound = 445 cm
   Upper bound = 455 cm

   If the actual length is L cm, we can write $445 \leq L < 455$.

Exercise

1. Each of the following numbers is given to the nearest whole number. Find the lower and upper bounds of the numbers.
   a) 12   b) 8   c) 100
   d) 9   e) 72   f) 127

2. Each of the following numbers is correct to 1 decimal place. Write down the lower and upper bounds of the numbers.
   a) 2.7   b) 34.4   c) 5.0
   d) 1.1   e) −2.3   f) −7.2

3. Anne estimates that the mass of a lion is 300 kg. Her estimate is correct to the nearest 100 kg. Between what limits does the actual mass (m kg) of the lion lie? Give your answer in the form $\ldots \le m < \ldots$

4. To the nearest 10 cm, the top of a table is 130 cm long and 100 cm wide. State the range of possible values of the length (L cm) and width (W cm), giving your answers in the form $\ldots \le L < \ldots$ and $\ldots \le W < \ldots$

Sum and difference of measurements

■ Examples

1. What are the upper and lower bounds of the sum of the measurements 8 cm and 4 cm, each of which is correct to the nearest centimetre?
   8 cm lies within (8 ± 0.5) cm. That is 7.5 cm to 8.5 cm.
   4 cm lies within (4 ± 0.5) cm. That is 3.5 to 4.5 cm.
   So the upper bound of the sum = 8.5 cm + 4.5 cm = 13 cm
   and the lower bound of the sum = 7.5 cm + 3.5 cm = 11 cm.

2. Two lengths are given correct to 3 significant figures as 2.63 m and 4.75 m. Find the upper and lower bounds of the sum of these measurements.
   The first measurement lies within the range 2.625 m to 2.635 m.
   The second measurement lies within the range 4.745 m to 4.755 m.
   The upper bound of the sum = (2.635 + 4.755)m = 7.39 m.
   The lower bound of the sum = (2.625 + 4.745)m = 7.37 m.

3. What are the upper and lower bounds of the difference between the measurements 8 cm and 4 cm, each begin correct to the nearest centimetre?
   8 cm lies within the range 7.5 cm to 8.5 cm.
   4 cm lies within 3.5 cm to 4.5 cm.
   Hence the upper bound of the difference = 8.5 cm − 3.5 cm = 5 cm
   and the lower bound of the difference = 7.5 cm − 4.5 cm = 3 cm.

Product and division of measurements

■ Examples

1. The dimensions of a rectangle are 32.6 cm and 20.8 cm correct to 3 significant figures. Calculate the lower and upper bounds for the area.
   The length lies within the range 32.55 cm to 32.65 cm.
   The breadth lies within the range 20.75 cm to 20.85 cm.
   Hence the lower bound for the area is $32.55 \times 20.75 = 675.4125$ cm^2 and the upper bound of the area is $32.65 \times 20.85 = 680.7525$ cm^2.
   So the area lies between 675.4125 cm^2 and 680.7525 cm^2, or 675 cm^2 and 681 cm^2 correct to 3 significant figures.

2. Each number in the given fraction $\frac{4.2 \times 5.1}{1.6}$ is correct to 2 significant figures.
   Find, correct to 3 significant figures, the lower and upper bounds of the value of the fraction.
   The lower bound is obtained by making the numerator of the fraction as small as possible and the denominator as large as possible.
   The lower bound is $\frac{4.15 \times 5.05}{1.65} = 12.701 \ldots$
   $= 12.7$ (correct to 3 significant figures)
   The upper bound is obtained by making the numerator of the fraction as large as possible and the denominator of the fraction as small as possible.
   The upper bound is $\frac{4.25 \times 5.15}{1.55} = 14.120 \ldots$
   $= 14.1$ (correct to 3 significant figures)

Remember

The *dimensions* of a rectangle refer to the length and breadth of the rectangle.

Hint

If the denominator of the fraction is much bigger than the numerator, it will be a small number, e.g. $\frac{1}{1\,000\,000}$. If the numerator is much bigger than the denominator, it will be a big number, e.g. $\frac{1\,000\,000}{1}$.

Exercise

1. Find the lower and upper bounds of the sums of the following:
   a) 6 cm and 8 cm, each correct to the nearest centimetre
   b) 12 g and 17 g, each correct to the nearest gram
   c) 4.6 cm and 11.8 cm, each correct to 1 decimal place
   d) 1.42 kg and 0.90 kg, each correct to 2 decimal places.

2. 12 kg of sugar are removed from a container holding 50 kg. Each measurement is correct to the nearest kilogram. Find the lower and upper bounds of the mass of sugar left in the container.

3. The dimensions of a rectangle are 3.61 cm and 2.57 cm, correct to 3 significant figures.
   a) Write down the range of possible values of each dimension.
   b) Find the upper and lower bounds of the area of the rectangle.
   c) Write down the upper and lower bounds of the area correct to 3 significant figures.

4. If all the given numbers are correct to 2 significant figures, find, correct to 2 significant figures, the upper and lower bounds of:
   a) $7.8 \div 1.6$ b) $1\,300 \div 46$.

Using numbers in everyday life

Buying and selling

Shopkeepers and traders have to make decisions about prices, profit levels and whether or not to sell goods at a loss. In order to do this, they have to perform some calculations.

Calculation of selling price

You might decide that it's not worth your while to buy and sell T-shirts unless you can make at least a 50% profit. So how much would you need to sell a T-shirt for if it cost you $15?

50% of $15 $= \frac{50}{100} \times \$15 = \$7.50$. So in order to make a 50% profit, you would have to sell the T-shirt for $15 + $7.50 = $22.50.

Here are some more examples.

■ Examples

1. A shopkeeper buys an article for $20 and sells it at a profit of 30%. For how much does he sell it?

   Cost price $=$ $20   Selling price $=$ cost price + profit
        Profit $=$ 30% of the cost price $=$ $20 + $6
              $= \frac{30}{100} \times \$20$ $=$ $26
              $=$ $6

2. Find the selling price of an article that is bought for $400 and sold at a loss of 10%.

   Cost price $=$ $400   Selling price $=$ cost price – loss
          Loss $=$ 10% of the cost price $=$ $400 – $40
              $= \frac{10}{100} \times \$400$ $=$ $360
              $=$ $40

Exercise

1. Find the selling price of an article that was bought for $750 and sold at a profit of 15%.
2. Calculate the selling price of an item of merchandise bought for $3 000 and sold at a profit of $12\frac{1}{2}\%$.
3. A boy bought a bicycle for $500. After using it for two years, he sold it at a loss of 15%. Calculate the selling price.
4. It is found that an article is being sold at a loss of 12%. The cost of the article was $240. Calculate the selling price.
5. A woman makes dresses. Her total costs for 10 dresses were $377. At what price should she sell the dresses to make 15% profit?

Calculation of cost price from selling price and percentage profit

You've seen how to calculate the percentage profit if you know what the cost price and the selling price of an article are. Now you will find out how to calculate the cost price if you know the selling price and percentage profit.

It is important to remember that profit is expressed as a percentage of the cost price. It is not a percentage of the selling price.

■ Examples

1. Find the cost price of an article sold at $360 with a profit of 20%.

   To solve these problems, think of all the amounts as percentages.

$$\text{Cost percentage} = 100\%$$
$$\text{Profit percentage} = 20\% \text{ of the cost price}$$
$$\text{Selling percentage} = \text{cost percentage} + \text{profit percentage}$$
$$= 100\% + 20\%$$
$$= 120\%$$
$$\text{Selling price} = 120\% \text{ of the cost price}$$
$$120\% \text{ of the cost price} = \$360$$
$$\text{So } 100\% \text{ of the cost price} = \frac{\$360}{120} \times 100 = \$300$$
$$\text{So cost price} = \$300$$

2. If a shopkeeper sells an article for $440 and loses 12% on the sale, find his cost price.

$$\text{Cost percentage} = 100\%$$
$$\text{Loss percentage} = 12\%$$
$$\text{Selling percentage} = \text{cost percentage} - \text{loss percentage}$$
$$= 100\% - 12\%$$
$$= 88\%$$
$$\text{Selling price} = 88\% \text{ of the cost price}$$
$$88\% \text{ of the cost price} = \$440$$
$$\text{So } 100\% = \frac{\$440}{88} \times 100 = \$500$$
$$\text{So cost price} = \$500$$

Exercise

1. Find the cost price in each of the following:
   a) selling price $120, profit 20%
   b) selling price $230, profit 15%
   c) selling price $289, loss 15%
   d) selling price $600, loss $33\frac{1}{3}\%$.

2. A man's income tax is $1 500, which is 30% of his salary. Find his gross salary.

3. After a pay increase of $17\frac{1}{2}\%$, a man receives $1 363. How much did he receive before the increase?

4. An increase of 250% in the ground area of a factory took place during the last year. The area is now 15.75 hectares.
What was the area a year ago?

5. Between 1980 and 1990, a worker's weekly wage increased by 140% to $180. Calculate the weekly wage in 1980.

Simple interest

The money you pay to live in somebody else's house is called *rent*. The money you pay if you hire a car is called *rental*. The money you pay to the post office to use one of their post boxes is also called *rental*. If you borrow money from a bank, you also have to pay for the use of the money borrowed. This money you pay is called *interest*. If you deposit (or invest) money in a bank, the bank will pay you interest. The letter I stands for interest. The interest depends on the sum of money borrowed or invested. The initial amount of money borrowed or invested is called the *principal amount*. The letter P stands for the principal amount.

The interest also depends on the length of time for which the money is borrowed or invested. T is the letter used for time. Time is usually measured in years.

Finally, the interest depends on the rate of interest. This is expressed as a percentage per year and is denoted by R. For example, if a bank's interest rate is 5% per annum, it will pay R(5) per year for every R(100) invested.

So, to summarise, you invest an amount of money (P) in the bank. The bank pays you interest (I) at a fixed rate (R) per annum $\left(I = \frac{PTR}{100}\right)$. For any number of years (T) you have kept your money in the bank, you can calculate the final amount (A) of money. The final amount will be equal to the principal plus interest (A = P + I).

Hint

Per annum means 'per year' or 'for the year'.

Remember

If the amount of interest paid (or charged) is the same for each year, then it is called *simple interest*. When the interest for one year is added to the investment (or debt) and the interest for the next year is calculated on the increased investment (or debt), it is called *compound interest*.
(Note: Compound interest is not in the IGCSE syllabus).

■ Examples

1. $500 is invested at 10% per annum simple interest. How much interest is earned in 3 years?
The interest rate is 10% per annum.
10% of $500
$= \frac{10}{100} \times \$500 = \$50$
The interest every year is $50. So after 3 years, the interest is $3 \times \$50 = \$150$.

2. How long will it take for $250 invested at the rate of 8% per annum simple interest to amount to $310?
Amount = principal + interest
Interest = amount − principal
$= \$310 - \$250$
Interest = $60

So how long will it take for the interest to amount to $60?

Rate = 8% per annum

$$\frac{8}{10} \times \$250 = R20$$

The interest per year is $20.

This means that it will take 3 years to earn $60 interest.

So it will take 3 years for $250 to amount to $310 at the rate of 8% per annum simple interest.

3. A farmer gets a loan of $8 000 and clears the loan at the end of 5 years by paying $12 000. What rate percentage of simple interest did the farmer have to pay per annum?

The principal (P) is $8 000

The amount (A) is $12 000

$$
\begin{aligned}
\text{So the interest paid at the end of 5 years} &= \$12\,000 - \$8\,000 \\
&= \$4\,000 \\
\text{That is, the interest on } \$8\,000 \text{ for 5 years} &= \$4\,000 \\
\text{So the interest on } \$8\,000 \text{ for 1 year} &= \frac{\$4\,000}{5} \\
&= \$800 \\
\text{The percentage interest for 1 year} &= \frac{\$800}{\$8\,000} \times \frac{100}{100} \\
&= 10\%
\end{aligned}
$$

The rate is 10% per annum.

This work can be quite confusing if you are not completely sure of what P, A, I, R and T mean. Make sure you know these well before you attempt the following exercise. The best way to solve the problems is to make sure you understand exactly what the problem is and what you are asked to calculate. Then use your common sense and work carefully until you find the answer.

Exercise

1. Calculate the simple interest on:
   a) $250 invested for a year at the rate of 3% per annum
   b) $400 invested for 5 years at the rate of 8% per annum
   c) $700 invested for $2\frac{1}{2}$ years at the rate of 15% per annum
   d) $800 invested for 8 years at the rate of $7\frac{1}{2}$% per annum
   e) $5 000 invested for 15 months at the rate of 5.5% per annum.
2. $1 400 is invested at 4% per annum simple interest. How long will it take for the amount to reach $1 624?
3. The simple interest on $600 invested for 5 years is $210. What is the rate percentage per annum?

Hire purchase

Consumer items like foodstuffs and clothes are normally paid for in cash. But many people cannot pay cash for expensive items like television sets, furniture, cars and so on. With a *hire purchase* (HP) system or *instalment plan*, the customer usually pays a fraction (or percentage) of the price of the item and pays the remainder in a certain number of weekly or monthly instalments. The initial amount the customer pays is called the *down payment* or *deposit*. Since hire purchase involves added expense for the seller, interest is charged on outstanding balances.

 An advantage of buying on hire purchase is that the article may be used while the money is accumulated to pay for it. The goods do not become the buyer's property until all the instalments have been paid. One obvious disadvantage is that if the buyer is unable to keep paying the instalments, then the article will be *repossessed* or taken back by the seller. Some people believe it is wiser to save money in the bank and pay cash.

■ Examples

1. The cash price of a car was $20 000. The hire purchase price was $6 000 deposit and instalments of $700 per month for two years. How much more than the cash price was the hire purchase price?

$$\begin{aligned}
\text{Deposit} &= \$6\,000 \\
\text{One instalment} &= \$700 \\
\text{24 instalments (2 years)} &= \$700 \times 24 \\
&= \$16\,800 \\
\text{Total hire purchase price} &= \text{deposit} + 24 \text{ instalments} \\
&= \$6\,000 + \$16\,800 \\
&= \$22\,800
\end{aligned}$$

The hire purchase price was $2 800 more than the cash price.

2. A man buys a car for $30 000 on hire purchase. A deposit of 20% is paid and interest is paid on the outstanding balance for the period of repayment at the rate of 10% per annum. The balance is paid in 12 equal instalments. How much will each instalment be?

$$\begin{aligned}
\text{Cash price} &= \$30\,000 \\
\text{Deposit 20\%} &= \tfrac{20}{100} \times \$30\,000 \\
&= \$6\,000 \\
\text{Outstanding balance} &= \$30\,000 - \$6\,000 \\
&= \$24\,000 \\
\text{Interest of 10\% per annum} &= \tfrac{10}{100} \times \$24\,000 \\
&= \$2\,400 \\
\text{Amount to be paid by instalments} &= \text{outstanding balance} + \text{interest} \\
&= \$24\,000 + \$2\,400 \\
&= \$26\,400 \\
\text{Each instalment} &= \tfrac{\$26\,400}{12} \\
&= \$2\,200
\end{aligned}$$

Exercise

1. A shopkeeper wants 25% deposit on a bicycle costing $400 and charges 20% interest on the remaining amount. How much is:
   a) the deposit
   b) the interest
   c) the total cost of the bicycle.
2. A woman pays 30% deposit on a fridge costing $2 500 and pays the rest of the money in 1 year with interest of 20% per year. How much does she pay altogether for the fridge?

Check your progress

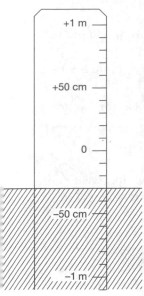

1. 8, 10, 12, 14, 15, 21, 23
   From the list of numbers above, write down:
   a) a prime number   b) a multiple of 6   c) a factor of 36.
2. What is the next number in each of the following sequences:
   a) 25, 36, 49, ...   b) 1, 8, 27, ...?
3. One day the temperature on Mount Everest changed from 28 °C to –29 °C in a few hours. By how many degrees did the temperature change?
4. The diagram alongside shows a riverside flood-warning post.
   a) What is the water level shown in the diagram?
   b) The water level now rises by 40 cm. What is the new level?
   c) During a dry month, the water level fell from +50 cm to –50 cm. By how many centimetres did it fall?
5. Work out $(17 + 28) \div 3 - 2 \times 7$.
6. $\sqrt{250}$, 25, 3.1, $2\frac{1}{2}$, 27
   Which of the numbers above is:
   a) irrational   b) square   c) prime?
7. Calculate $\frac{5}{6} (\frac{1}{4} + \frac{1}{8})$, giving your answer as a fraction in its lowest terms.
8. a) Write $\frac{1}{7}$ as a decimal, giving the first 20 places.
   b) Why is this called a recurring decimal?
9. 3 800 students took an examination.
   19% received Grade A.
   24% received Grade B.
   31% received Grade C.
   10% received Grade D.
   11% received Grade E.
   The rest received Grade U.
   a) What percentage of the students received Grade U?
   b) What fraction of the students received Grade B? Give your answer in its lowest terms.
   c) How many students received Grade A?

10. One of the signs >, = or < is missing from each of the following statements. State which sign should be placed in each box.

   a) $(-3) - (-5)$ ☐ $(-3) + (-5)$ b) $9(17 - 8)$ ☐ $9 \times 17 - 9 \times 8$

   c) $\frac{3}{4}$ ☐ $\frac{3}{5}$ d) $\frac{3}{7}$ ☐ 45%

11. Four adults and a child have a meal at a restaurant. The adults' meals all cost the same. The child's meal is half that price. The total bill is $56.25. What is the cost of one adult's meal?

12. In Canada, a bottle of mineral water costs 0.55 Canadian dollars. If 1 Euro = 1.23 Canadian dollars, how many bottles of mineral water can be bought for the equivalent of 4 Euros?

13. Maria changed $80 into Argentinian pesos when the exchange rate was $1 to 15.43 pesos. A week later, the exchange rate was $1 to 15.52 pesos. How many more pesos would Maria have received if she had waited a week before changing her dollars?

14. Ahmed bought a compact disc for $15. He sold it to Barbara, making a 20% loss.

   a) How much did Barbara pay for it?

   b) Barbara later sold the compact disc to Luvuyo. She made a 20% profit. How much did Luvuyo pay for it?

15. In 1988 there were 27 500 cases of *Salmonella* poisoning in Britain. In 1989 there was an increase of 9% in the number of cases. Calculate how many cases there were in 1989.

16. Last year, Jane's wages were $80 per week. Her wages are now $86 per week. Calculate the percentage increase.

17. Abdul's height was 160 cm on his 15th birthday. It was 172 cm on his 16th birthday. What was the percentage increase in his height?

18. What is the simple interest on $160 invested at $7\frac{1}{2}\%$ per year for 3 years?

19. Señor Vasquez invests $500 in a Government Bond, at 9% simple interest per year. How much will the Bond be worth after 3 years?

20. Klaus and Heidi plan a holiday in the USA in August.

   a) Klaus decides to change 800 Euros ( ) into dollars in January when the exchange rate is $1 =   1.68. A bank charge of 1% is then deducted. Calculate how much he receives, to the nearest dollar.

   b) (i) Heidi invests her   800 in a bank at an annual rate of 9% simple interest. Calculate the amount she has after 6 months.

   (ii) She now changes this amount into dollars. The exchange rate is $1 =   1.87, but this time there is no bank charge. Calculate how much Heidi receives, to the nearest dollar.

   c) Who made the better decision?

   d) They bring a total of $120 back with them and exchange it for   at a rate of $1 =   1.72 with no bank charge. Calculate how much they receive, to the nearest Euro.

Algebra

Algebra is a way of calculating and working out relationships between quantities using letters to represent numbers. For example, if the cost of an apple is a cents and the cost of an orange is b cents, we can write the total cost in cents of 5 apples and 4 oranges in terms of a and b as $5a + 4b$. This is an algebraic expression. The letters in algebra are called *variables* (because they can have various values).

When we know the value of a and b, we can replace them and work out the actual cost. This is called *substitution*.

The language of algebra

The statement $n + 2 = 8 - n$ tells you that if 2 is added to n, the result is the same ($=$) as taking n away from 8. There is only one possible value for n, in this case $n = 3$. Statements like this are called equations. Finding the number that makes the equation true is called solving the equation. You will solve equations later in this module.

The area of a rectangle is given as length multiplied by breadth.
This can be written in algebraic language as: $A = LB$
In this case A, B and L can stand for many numbers.

$A = LB$ is thus a formula, as it describes the relationship between the variables, but we cannot solve it like an equation because there is no one value that makes the formula true.

Algebra is like a language of mathematics, and, like any language, there are certain agreed rules. Make sure you understand that:

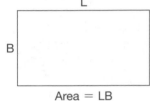

L

B

Area = LB

- $3a$ means $3 \times a$ (in algebra we do not write the $\times$ sign)
- $3ab$ means $3 \times a \times b$ (the constant is written before the variable and the variables are normally written in alphabetical order)
- a means $1 \times a$ (multiplication by 1 does not change the value, so the 1 is omitted)
- $\dfrac{a}{b}$ means a divided by b
- a^2 means $a \times a$
- a^5 means $a \times a \times a \times a \times a$
- ab^2 means $a \times b \times b$
- $(ab)^2$ means that a is multiplied by b and the result is squared, so $(a \times b) \times (a \times b)$, which is equal to $a^2 b^2$
- $5a^2$ means $5 \times a \times a$ (the power only applies to the number or variable directly before it).

■ Examples

1. When $a = 2$ and $b = 3$:
   - the value of the expression ab^2 is $2 \times 3^2 = 2 \times 9 = 18$
   - the value of the expression $(ab)^2$ is $(2 \times 3)^2 = (6)^2 = 36$.

2. Suppose you are y years old and your mother is m years old.
   If your mother is twice as old as you are, it would be written as: $2y = m$.
   If you are 25 years younger than your mother, it would be written as:
   $m - 25 = y$.

3. I think of a number, n. I double it and take away 7. My answer is 35.
   This can be written as: $2n - 7 = 35$.

Exercise

1. Find the values of the following expressions:
   a) $5 + 4a$ if $a = 3$
   b) $11 - 2a$ if $a = 4$
   c) $10x + 4$ if $x = \frac{1}{2}$
   d) $3x + 3$ if $x = 0$
   e) $16 - a^2$ if $a = 2$
   f) $x^2 + 2x$ if $x = 3$
   g) $ab - 10$ if $a = 4, b = 5$
   h) $x + \dfrac{1}{x}$ if $x = 2$.

2. If $x = 2$, $y = -3$ and $z = 4$, find the value of:
   a) $x + y + z$
   b) $3x + y$
   c) $\dfrac{xz}{z - y}$
   d) $2x + y - 3z$.

3. Given $p = 6$ and $q = 2$, find the value of:
   a) $p^2 - q^2$
   b) $(p - q)^2$.

4. Given that $x = 5$, $y = 2$ and $z = 3$, find the value of:
   a) $x - (y + z)$
   b) $x - y + z$.

5. Suppose you are y years old and your brother is b years old. Write in algebraic shorthand:
   a) your brother is older than you
   b) your brother is twice as old as you.

6. I think of a number, square it, double the answer and then take away 15. My answer is 377. Using n to represent the number I thought of, write down an equation in algebraic language.

7. Jason is x years old. His father is 3 times as old as he is. In 10 years' time, Jason wil be half his father's age. Express this as an equation. (If you like, you can try to solve it!)

8. To find the area of an ellipse, you multiply the major diameter by the minor diameter, then multiply the result by π and divide by 4. Write this formula in algebraic language. Use the diagram if you need help.

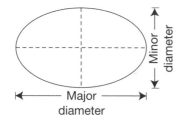
Major diameter
Minor diameter

Using formulae

Formulae are rules for finding certain quantities by combining others. You have already worked with formulae in geometry to find the area and perimeter of shapes. Formulae are also used in science, surveying and other subjects. Using a formula means replacing certain variables with numbers. This is the same as substitution.

■ Examples

1. A car is travelling at v km/h. The driver has to stop suddenly. The distance the car travels before it stops is d metres. The formula for d is:

$$d = \frac{v(v + 32)}{150}$$

v is the velocity (speed) at which the car is travelling. If the car is travelling at 48 km/h we can work out d by substituting:

$$d = \frac{48(48 + 32)}{150} = \frac{48 \times 80}{150} = \frac{3\,840}{150} = 25.6 \text{ metres}$$

2. The cost, C dollars, of printing n books is given by the formula
$C = 12n + 750$.

When n is 300, $C = 12 \times 300 + 750 = 3\,600 + 750 = \$4\,350$.

Exercise

1. Use the formula $P = 2L + 2B$ to find the value of P:
   a) when $L = 12$ and $B = 7$
   b) when $L = 4.6$ and $B = 3.5$.
2. A car is travelling at 100 km per hour. Use the formula $d = \frac{v(v + 32)}{150}$ to find its stopping distance in metres.
3. Use the formula $C = 12n + 750$ to find the value of C:
   a) when $n = 5$
   b) when $n = 500$.

4. From a point h metres above the surface of the sea, the distance to the horizon is d kilometres. The formula for d is $d = 3.55\sqrt{h}$. The top of a lighthouse is 32 m above the surface of the sea. Use the formula to calculate the distance to the horizon from the top of the lighthouse.
5. The volume of a pyramid is the area of its base times its perpendicular height divided by 3. Calculate the volume of a pyramid that has a square base of side 5 cm and a perpendicular height of 6 cm.

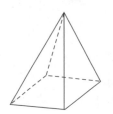

Simplifying algebraic expressions

The parts of an algebraic expression are called terms. Terms are separated from each other by + or − signs. So, $a + b$ is an expression with two terms; ab is an expression with one term; $2 + 3\frac{a}{b} - \frac{ab}{c}$ is an expression with three terms.

When an expression consists of many terms, it can be time-consuming to substitute values. For this reason, it is best to simplify expressions (at least to some extent) before you substitute.

Adding like terms

When one or more terms have exactly the same variables, they are called like terms. Like terms can be added together to simplify the expression.

■ Examples

$4a + 2a + 3a = 9a$

$4a + 6b + 3a = 7a + 6b$ (you cannot simplify further
as there are no like terms)

$5x + 2y - x = 4x + 2y$

$2p + 5q + 3q - 7p = -5p + 8q$ (this can also be written as $8q - 5p$)

Multiplying and dividing by directed numbers

Consider the multiplication table on the left. Each row of the table has a specific pattern. It is these patterns that are used to develop rules for multiplying and dividing by negative and positive numbers.

×	–3	–2	–1	0	1	2	3
3	–9	–6	–3	0	3	6	9
2	–6	–4	–2	0	2	4	6
1	–3	–2	–1	0	1	2	3
0	0	0	0	0	0	0	0
–1	3	2	1	0	–1	–2	–3
–2	6	4	2	0	–2	–4	–6
–3	9	6	3	0	–3	–6	–9

These rules are:

$(+) \times (+) = (+)$

$(+) \times (-) = (-)$

$(-) \times (+) = (-)$

$(-) \times (-) = (+)$ (there is no way to actually do this,
so the rule depends on the pattern)

Because division is the inverse of multiplication, it follows that the rules will be the same. The rules for division are:

$(+) \div (+) = (+)$

$(+) \div (-) = (-)$

$(-) \div (+) = (-)$

$(-) \div (-) = (+)$

Exercise

Apply the rules above to find the answers.

1. a) 2×-3
    b) 4×-4
    c) -3×4
    d) 5×9
    e) -2×0
    f) 8×-2
    g) 5×-3
    h) 7×-3
    i) -3×-2
    j) -7×-3
    k) 0×-4
    l) -4×-8

2. a) $-3 \div (-3)$
    b) $27 \div (-9)$
    c) $-40 \div (-4)$
    d) $-15 \div (-3)$
    e) $54 \div (-6)$
    f) $100 \div (-2.5)$
    g) $\frac{-18}{-3}$
    h) $\frac{45}{-9}$
    i) $\frac{-42}{7}$
    j) $\frac{27}{3}$
    k) $\frac{-100}{-100}$
    l) $\frac{-100}{100}$

Rules for algebra

The rules for multiplying and dividing directed numbers apply in algebra as well.

■ Examples

Given that $p = 5$, $q = -2$ and $r = -3$, work out the value of:

1. $8pq$
   $= 8 \times 5 \times (-2)$
   $= -80$

2. qr^3
   $= (-2)(-3)(-3)(-3)$
   $= 54$

Remember

An even number of negatives gives a positive answer.

3. $p^2 - 4qr$
   $= (5 \times 5) - (4 \times (-2) \times (-3))$
   $= 25 - (24)$
   $= 1$
4. $\dfrac{8qr}{-2q} = \dfrac{8(-2)(-3)}{-2(-2)} = \dfrac{8(6)}{4} = 12$

Removing brackets

Some expressions contain brackets and these usually have to be removed before the expression can be simplified. To do this, you have to multiply every term inside the bracket by the number (or terms) outside the bracket.

When a negative number is multiplied by a positive number, the product will be negative. Think of this:

$3(-2a) = -2a + -2a + -2a = -6a$ (if you apply the rule, you do not have to add like this).

The examples below are quite simple, but you will learn more about removing brackets and multiplying and dividing by directed numbers later in this module.

■ Examples

Remove the brackets and simplify the following expressions.
1. $2(3x + y) + 5(x - 2y)$
   $= 6x + 2y + 5x - 10y$ (remove brackets)
   $= 11x - 8y$ (add like terms)
2. $4(2n + 3) + 7(n + 1) = 8n + 12 + 7n + 7 = 15n + 19$
3. $3(x - 2) + 2(3 - x) = 3x - 6 + 6 - 2x = x$
4. $6(3x^2 + 2x) + 3x - 5 = 18x^2 + 12x + 3x - 5 = 18x^2 + 15x - 5$

Exercise

1. Find the value of $2n + 7n - 3n$:
   a) when $n = 8$
   b) when $n = 72$.
2. Simplify the following expressions where possible.
   a) $3p + 7q - 2p - 5q$
   b) $4n - 4 + 3n + 1$
   c) $2x^3 + 3x^2 - x^2 + 5x$
   d) $3xy - 3 + y$
   e) $2c - 5d + d - 3c$
   f) $x^2 + 4x - 3 - 7x + 9$
3. Remove the brackets in the following expressions.
   a) $4(4x - 3)$
   b) $7(2n + 5)$
   c) $8(6 - 2y)$
   d) $3(2x^2 + 3x - 5)$
   e) $6(5pq - p - 2q)$
4. Remove the brackets and collect like terms to simplify the following.
   a) $3(n + 5) + 4(2n - 3)$
   b) $7(2x - 1) + 2(3 - x)$
   c) $5y - 3 + 3(4y - 2)$
   d) $2(6c + 5d) + 6(d - 2c)$
   e) $8(x^2 + x - 3) + 5(2 - x)$

Solving linear equations

Hint

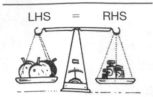

LHS = RHS

Think of an equation as a balance scale. If you change one side, you have to do the same to the other side to keep the balance.

Consider the equation $2n + 5 = 33$.

In this equation, there is only one value of n that will make the left-hand side of the equation equal to the right-hand side. When you are asked to find this value, you may be asked to 'find the value of n which satisfies the equation' or 'solve the equation'.

To solve an equation, you have to keep both sides balanced. This means that if you do something on one side of the equal sign, you have to do the same thing on the other side. The two sides of an equation will remain balanced when you:

- add the same number to both sides
- subtract the same number from both sides
- multiply both sides by the same number
- divide both sides by the same number.

What you want to do when you are solving equations like $an + b = cn + d$ (where a, b, c and d are numbers) is to use the above facts and end up with $n = \ldots$

If we look at the first equation: $2n + 5 = 33$
Subtract 5 from both sides: $2n + 5 - 5 = 33 - 5$
$$2n = 33 - 5$$
$$2n = 28$$

But we are looking for n alone.
So we divide both sides by 2: $n = 14$

■ Examples

1. Solve the equation $3x - 8 = 39$.

$$3x - 8 + 8 = 39 + 8 \quad \text{(add 8 to both sides)}$$
$$3x = 47 \quad \text{(divide both sides by 3)}$$
$$x = 15\tfrac{2}{3}$$

2. Solve the equation $n + 2 = 8 - n$.

$$n + 2 = 8 - n$$
$$n + 2 - 2 = 8 - n - 2 \quad \text{(subtract 2)}$$
$$n = 6 - n \quad \text{(add } n\text{)}$$
$$2n = 6 \quad \text{(divide by 2)}$$
$$n = 3$$

3. Find the value of y in $2(y - 4) = 18$.

$$2y - 8 = 18 \quad \text{(remove brackets)}$$
$$2y - 8 + 8 = 18 + 8 \quad \text{(add 8)}$$
$$2y = 26 \quad \text{(divide by 2)}$$
$$y = 13$$

4. Solve the equation $3(n-4) + 2(4n-5) = 5(n+2) + 16$.

$$3n - 12 + 8n - 10 = 5n + 10 + 16 \quad \text{(remove brackets)}$$
$$\therefore 11n - 22 = 5n + 26 \quad \text{(simplify)}$$
$$\therefore 11n - 5n = 26 + 22 \quad \text{(add 22; } -5n\text{)}$$
$$6n = 48 \quad (\div 6)$$
$$n = 8$$

5. A rough rule for changing temperatures in degrees Celsius (°C) to degrees Fahrenheit (°F) is $F = 2C + 30$.

   a) Find the value of F when $C = 25$

   So, $F = 2 \times 25 + 30 = 50 + 30 = 80$

   b) Find the value of C when $F = 20$

$$20 - 30 = 2C \quad (-30)$$
$$-10 = 2C \quad (\div 2)$$
$$-5 = C$$
$$\therefore C = -5$$

Exercise

1. Solve the following equations.
   a) $3x + 1 = 16$
   b) $4x + 3 = 27$
   c) $3x - 7 = 0$
   d) $2x + 5 = 20$
   e) $200y - 51 = 49$
   f) $11n + 1 = 1$

2. Solve these equations.
   a) $7x - 3 = 3x + 8$
   b) $5x - 12 = 2x - 6$
   c) $6y - 3 = 1 - y$
   d) $8m + 9 = 7m + 8$
   e) $7x - 5 = 2x$
   f) $3n - 1 = 5 - 4n$

3. Solve these equations.
   a) $2(x + 1) = x - 5$
   b) $4(x - 2) = 2(x + 1)$
   c) $5(m - 3) = 3(m - 2)$
   d) $3(y + 2) = 2(y - 1)$
   e) $6(x + 2) = 2(x - 3)$
   f) $5(p - 3) = 2(p - 7)$
   g) $4(m + 1) = 3(2m - 1)$
   h) $5(z + 3) = 4(2z + 1)$
   i) $2(7x + 4) = (x + 2)$
   j) $3(x + 1) - 4 = 2(x + 4)$
   k) $5(x - 2) - 7 = 2(x + 4)$
   l) $2(x - 10) = 4 - 3x$
   m) $7(2x + 1) = 5 - 4(2x - 3)$
   n) $2(x + 1) = 7 - 3(x - 1)$
   o) $5(2x - 1) = 9(x + 1) - 8$

4. When I treble a certain number and add 2, I get the same answer as I do when I take the number from 50. Let n represent the number. Write an equation for n and solve it.

5. A rough rule for changing temperatures in degrees Fahrenheit (F) to degrees Celsius (C) is $C = \frac{1}{2}F - 15$.

   a) Find the value of C when F = 70.

   b) Find the value of F when C = 30.

6. Fatima is twice as old as her sister. In 5 years' time, she will be 2 years older than her sister. Make an equation and solve it to work out how old they both are now.

Changing the subject of a formula

The formula $v = u + at$ is used in mechanics.

This formula is useful for calculating v if you know the values of u, a and t. We say that v is the subject of the formula.

However, what if you know the values of v, u and a and you want to work out t?

In such a case, it is best to transform the formula so that t becomes the subject. This operation is called changing the subject of a formula and it is the same as solving an equation for the required letter. This means you can follow the same steps as for equations, keeping both sides balanced:

- add the same number to both sides (if $x - b = a$ then $x = a + b$)
- subtract the same number from both sides (if $x + b = a$ then $x = a - b$)
- multiply both sides by the same number (if $\frac{x}{b} = a$ then $x = ab$)
- divide both sides by the same number (if $bx = a$ then $x = \frac{a}{b}$)
  (provided $b \neq 0$).

You can also:

- square both sides (if $\sqrt{x} = a$ then $x = a^2$)
- take the root of both sides (if $x^2 = b$ then $x = \pm\sqrt{b}$).

■ Examples

Exam tip

The square root of any number can be either negative or positive.

1. $A = \pi r^2$

   Make r the subject of the formula.

   Divide both sides by π $\qquad \frac{A}{\pi} = r^2$

   Take the square root of both sides $\qquad \pm\sqrt{\frac{A}{\pi}} = r$

   In this case, the answer must be +, so $\qquad r = \sqrt{\frac{A}{\pi}}$

2. Change the subject of the formula $S = 2n - 4$ to n.
$$S = 2n - 4$$
$$S + 4 = 2n$$
$$\frac{S + 4}{2} = n$$

3. A formula used in electricity is $C = \frac{E}{R}$.

   Change the subject of the formula to R.
$$C = \frac{E}{R}$$
$$CR = E \quad (\div C)$$
$$\therefore R = \frac{E}{C}$$

4. a) Solve the equation $\frac{x}{2} - 3 = 5$.

   b) Change the subject of the equation $\frac{x}{a} - b = c$ to x.

   a) $\frac{x}{2} - 3 = 5$ b) $\frac{x}{a} - b = c$

        $\frac{x}{2} = 8$ $\frac{x}{a} = c + b$

        $\therefore x = 16$ $\therefore x = a(c + b)$

5. A formula used in electricity is $E = V + IR$.
   a) Find the value of R when $E = 20$, $V = 15$ and $I = 2$.
   b) Express R in terms of E, V and I.

   a) $20 = 15 + 2R$ b) $E = V + IR$

        $\therefore 5 = 2R$ $\therefore E - V = IR$

        $\therefore R = 2.5$ $\therefore \frac{E-V}{I} = R$

                                       $\therefore R = \frac{E-V}{I}$

6. Change the subject of the formula $m = \frac{y-c}{x}$ to y.

           $m = \frac{y-c}{x}$ $(\times x)$

        $\therefore mx = y - c$

   $\therefore mx + c = y$ or $y = mx + c$

7. Change the subject of the formula $f = \frac{v^2}{r}$ to v.

        $f = \frac{v^2}{r}$

   $\therefore fr = v^2$

   $\pm\sqrt{fr} = v$ so $v = \pm\sqrt{fr}$

Exercise

1. a) Solve the equation $47 = 3x + 8$.
   b) Change the subject of the formula $y = 3x + 8$ to x.
   c) Find x in part b) if $y = 7$.
2. a) Solve the equation $4x + 5 = 17$.
   b) Given that $ax + b = c$, express x in terms of a, b and c.
3. Change the subject of the formula $v = u + at$ to u.
4. A formula used for a pulley system is $P = \frac{1}{2}W + 6$.
   Change the subject of the formula to W.
5. If $2x + 3y = 12$, find a formula for y in terms of x.
6. Change the subject of the formula $C = 2\pi r$ to r.
7. Change the subject of the formula $S = \sqrt{1.5H}$ to H.
8. Change the subject of the formula $E = \frac{Wv^2}{2g}$ to v.

Using algebra to solve problems

Before you can use algebra to solve problems, you need to be able to write information in algebra. This means you have to express information in terms of variables.

■ Examples

x, y and z represent three numbers. Write expressions to show:

a) the sum of the numbers

   $x + y + z$ (sum means add)

b) the difference between the first two numbers

   $x - y$ (difference means minus)

c) the product of the numbers

   xyz (product means multiply)

d) the sum of the numbers divided by 6

   $$\frac{x + y + z}{6}$$

e) double the numbers

   $2(x + y + z)$ or $2x + 2y + 2z$

Exercise

1. A mother is p years old and her daughter is s years old. Write an expression to represent:
   a) the difference in their ages
   b) their ages in 5 years' time
   c) the sum of their ages
   d) twice the daughter's age.

2. The perimeter of a rectangle is P. The length is five times longer than the breadth. Write an expression for:
   a) the length of the shortest side
   b) the length of the longest side
   c) twice the sum of the longest and shortest sides.

Steps for solving a problem

Algebra can also help you to find the solution to everyday word problems. One way of solving problems is to express them as an equation or formula. When doing this, you should follow certain steps. These are:

- make sure you understand the problem – read it carefully
- note the facts that are given and the quantities that you have to find
- state clearly the letter (usually n or x) that you will use to represent one of the quantities you have to find
- represent the other unknown quantities in terms of the letter
- write an equation using the facts that have been given
- solve the equation
- state clearly the values of the quantities you were asked to find
- check if your answer is reasonable.

■ Examples

1. In 5 years' time Nanjula will be three times as old as she was 9 years ago. How old is she now?

   Let her present age be n years old

   In 5 years she will be $(n + 5)$ years old

   9 years ago she was $(n - 9)$ years old

   In five years time her age is also $3(n - 9)$

   So: $(n + 5) = 3(n - 9)$

   $\therefore n + 5 = 3n - 27$

   $\therefore n - 3n = -27 - 5$

   $\qquad -2n = -32 \qquad$ (divide by –2)

   $\qquad\quad n = 16$

   Her present age is 16.

2. Jane gives a quarter of her sweets to Neo and then gives 5 of her sweets to Paul. She has 7 sweets left. How many did she have to start with?

   Let the number of sweets she started with be n

   She gives $\frac{n}{4}$ to Neo: $n - \frac{n}{4} = \frac{3}{4}n$

   She then gives 5 to Paul: $\frac{3}{4}n - 5$

   She has $\frac{3}{4}n - 5$ left. This amount is equal to seven.

   So: $\frac{3}{4}n - 5 = 7$

   $\therefore \frac{3}{4}n = 12$

   $\therefore 3n = 48$

   $\therefore n = 16$

   She had 16 sweets.

3. The distance around a rectangular field is 400 m. The length of the field is 26 m more than the breadth. Calculate the length and the breadth of the field.

   Let breadth $= x$ m, so length is $(x + 26)$ m

   $x + x + x + 26 + x + 26 = 400$

   $\therefore 4x + 52 = 400$

   $\therefore 4x = 348$

   $\therefore x = 87$

   So breadth $= 87$ m and length $= 87$ m $+ 26$ m $= 113$ m.

4. A rubber costs 15 cents more than a pencil. 12 pencils cost 60 cents more than 8 rubbers. Find the cost of one pencil.

   Let one pencil cost p cents, so one rubber costs $(p + 15)$ cents

   $12p = 8(p + 15) + 60$

   $\therefore 12p = 8p + 120 + 60$

   $\therefore 12p = 8p + 180$

   $\therefore 4p = 180$

   $\therefore p = 45$

   So one pencil costs 45c.

A sketch can help you.

Exercise

1. In 7 years' time, Jan will be twice as old as he was 8 years ago. How old is he now?
2. Temba is twice as old as Sipho and Silo is 5 years younger than Sipho. The total of their ages is 31 years. How old is Sipho?
3. Marcus bought a pizza and cut it into three pieces. When he weighed the pieces, he found that one piece was 7 g lighter than the largest piece and 4 g heavier than the smallest piece. The whole pizza weighed 300 g. How much did each of the three pieces weigh?
4. James earns \$5.20 per hour for his basic 40-hour week. When he works overtime he is paid his basic rate plus \$0.80. One week he works y hours and receives \$298.00. How many hours overtime did he work?
5. Last season Dickson attended all 21 home matches of his favourite football team. Sometimes he bought a ticket for a seat costing \$45. On the other occasions he bought a ticket to stand, and this cost \$25. For the whole season, Dickson paid \$765 for his tickets. How many times did he buy a ticket for a seat?

Finding a rule for a sequence

In Module 1 you worked with patterns and sequences, finding the next number in a pattern or sequence arithmetically. Using algebra, it is possible to find a general rule for a sequence. Once you have found this rule, or formula, you can use it to work out the value of any term in the sequence or predict the next terms in a given sequence.

Think about the sequence 1, 4, 9, 16, 25 …

You should recognise this as the sequence of squared numbers.

You can write this as $n \to n^2$. This means that the input number (n) is squared for any term in the sequence. So, the 11th term in this sequence would be $11 \to n^2 = 121$.

It is not always easy to recognise the pattern and write a rule (also called a function) for it. A table of values can help you to see the rule.

■ Examples

1. What is the nth term in the sequence $\frac{1}{2}, \frac{2}{3}, \frac{3}{4}, \frac{4}{5}, \frac{5}{6}, \ldots$?

Number of term (n)	1	2	3	4	5
Term	$\frac{1}{2}$	$\frac{2}{3}$	$\frac{3}{4}$	$\frac{4}{5}$	$\frac{5}{6}$
Pattern	$\frac{1}{1+1}$	$\frac{2}{2+1}$	$\frac{3}{3+1}$	$\frac{4}{4+1}$	$\frac{5}{5+1}$

The rule is thus $\dfrac{n}{n+1}$.

2. What is the nth term in the sequence 2, 4, 6, 8, 10 ...?

Number of term (n)	1	2	3	4	5
Term	2	4	6	8	10
Pattern	1×2	2×2	2×3	2×4	2×5

So the rule is $n \to 2n$.
You might not have worked like this; you might have seen the pattern as:

$1 + 1 \qquad 2 + 2 \qquad 3 + 3 \qquad 4 + 4 \qquad 5 + 5$

In this case, your rule would be $n \to n + n$. When you simplify this, you still get $n \to 2n$. But sometimes it is less easy to see the pattern.

3. Find the nth term in the sequence 5, 8, 11, 14, 17 ...?
Before you work with the table, note that there is a difference of 3 between each term in the sequence. This means that the pattern will contain the term $3n$. Put this in the pattern row as $3(n)$ where n is the number of the term. This is the first difference. Once you have done this, you can see that the value of the term is found by adding 2 to $3n$.

Number of term (n)	1	2	3	4	5
Term	5	8	11	14	17
Pattern (first)	3(1)	3(2)	3(3)	3(4)	3(5)
(second)	+ 2	+ 2	+ 2	+ 2	+ 2

So the rule is $n \to 3n + 2$.

4. Find the formula for the nth term in the sequence –8, –3, 2, 7, 12 ...

Number of term (n)	1	2	3	4	5
Term	–8	–3	2	7	12
Pattern (first)	5(1)	5(2)	5(3)	5(4)	5(5)
(second)	– 13	– 13	– 13	– 13	– 13

So the rule is $n \to 5n - 13$.

5. Find the rule for the nth term in the sequence 3, $3\frac{1}{2}$, 4, $4\frac{1}{2}$, 5 ...

The numbers increase by $\frac{1}{2}$ each time, so the rule must contain $\frac{1}{2}n$.

But $\frac{1}{2}n$ gives the sequence $\frac{1}{2}$, 1, $1\frac{1}{2}$, 2, $2\frac{1}{2}$ ... . The actual values are $2\frac{1}{2}$ more than that, so add $2\frac{1}{2}$. The rule is $n \to \frac{1}{2}n + 2\frac{1}{2}$.

Exercise

1. Find the formula for the nth term of the sequence 5, 7, 9, 11, 13, ...
2. Find the formula for the nth term of the sequence 2, 5, 8, 11, 14, ...
3. Find the formula for the nth term of the sequence 12, 10, 8, 6, 4, ...
4. Find the formula for the nth term of the sequence $1\frac{1}{2}$, 2, $2\frac{1}{2}$, 3, $3\frac{1}{2}$, ...
5. Write down the formula for the nth term of 1, 8, 27, 64, 125, ...

Direct and inverse proportion

In Module 1 you studied direct and inverse proportion arithmetically. You can solve more complicated cases of proportion using algebraic methods.

Direct proportion

If the values of two variables are always in the same ratio, the variables are said to be in direct proportion. If the variables are P and Q, we write this as $P \propto Q$. This is read as 'P is directly proportional to Q'.

$P \propto Q$ means that $\dfrac{P}{Q}$ is constant. That is, $P = kQ$ where k is a constant.

If the constant is 2, then $P = 2Q$ means that whatever P is, Q will be double that.

We can write this as $\dfrac{P}{Q} = 2$.

Inverse proportion

If the product of the value of two variables is constant, the variables are said to be inversely proportional. If the variables are P and Q, we say $PQ = k$, where k is the constant. This means that P is inversely proportional to Q.

$PQ = k$ can also be written as $P = \dfrac{k}{Q}$.

So P is inversely proportional to Q can be written as $P \propto \dfrac{1}{Q}$.

■ Examples

1. y is directly proportional to x^3 when $x = 2$, $y = 32$.
   Find the value of y when $x = 5$.
   So $y = kx^3$, but when $x = 2$, $x^3 = 32$ $\therefore$ $32 = 8k$
   $\therefore$ $k = 4$ and $y = 4x^3$
   $\Rightarrow$ when $x = 5$, $y = 4 \times 5^3 = 4 \times 125 = 500$

2. F is inversely proportional to d^2 when $d = 3$, $F = 12$.
   Find the value of F when $d = 4$.

   $F = \dfrac{k}{d^2}$, but when $d = 3$, $F = 12$, so $12 = \dfrac{k}{9}$ and $k = 108$

   $F = \dfrac{108}{d^2}$, so when $d = 4$, $F = \dfrac{108}{16} = 6.75$

3. Some corresponding values of the variables p and q are shown in the table. Are p and q directly proportional?

q	2	5	8	12
p	2.8	7	11.2	16.8

   $\dfrac{2.8}{2} = 1.4$ $\dfrac{7}{5} = 1.4$ $\dfrac{11.2}{8} = 1.4$ $\dfrac{16.8}{12} = 1.4$.

   The values are directly proportional; $p = 1.4q$.

4.

x	3	4	5	6
y	12			

Complete the table on page 62 for:

a) $y \propto x$

b) $y \propto \dfrac{1}{x}$.

a) $k = 4$ $(y = kx)$; $12 = k(3)$); $\therefore y = 4x$

x	3	4	5	6
y	12	16	20	24

b) $y \propto \dfrac{1}{x}$ means $xy = k$; since $y = 12$ when $x = 3 \rightarrow k = 36$; $\therefore y = \dfrac{36}{x}$

x	3	4	5	6
y	12	9	7.2	6

5. The speed of water in a river is determined by a water-pressure gauge. The speed (v m/s) is directly proportional to the square root of the height (h cm) reached by the liquid in the gauge. Given that $h = 36$ when $v = 8$, calculate the value of v when $h = 18$.

$v \propto \sqrt{h}$ means that $v = k\sqrt{h}$ where k is constant.

When $v = 8$, $h = 36$ and so $8 = k\sqrt{36} = 6k$.

It follows that $k = \frac{4}{3}$ and the formula connecting v and h is $v = \frac{4}{3}\sqrt{h}$.

When $h = 18$, $v = \frac{4}{3}\sqrt{18} = \frac{4}{3} \times 4.2426 = 5.66$ to 3 significant figures.

Now try some questions on direct and inverse proportion.

Exercise

1. A is directly proportional to r^2. When $r = 3$, $A = 36$.
    Find the value of A when $r = 10$.
2. I is inversely proportional to d^3. When $d = 2$, $I = 100$.
    Find the value of I when $d = 5$.
3.

q	2	5	8	12
p	75	30	20	15

Some corresponding values of the variables p and q are given in the table. Are p and q inversely proportional?
Justify your answer.

4. An electric current I flows through a resistance R. I is inversely proportional to R and, when $R = 3$, $I = 5$. Find the value of I when $R = 0.25$.

5.

s	2	6	10
t	0.4	10.8	50

Some corresponding values of the variables s and t are given in the table. Which of the following types of proportion fits these values?
$t \propto s$, $t \propto s^2$, or $t \propto s^3$

Understanding indices

In Module 1 you worked with powers of 2 and 3 – squares and cubes. You also know that you get powers to any value. For example,

$$2^5 = 2 \times 2 \times 2 \times 2 \times 2$$
$$5^2 = 5 \times 5$$
$$6^3 = 6 \times 6 \times 6$$

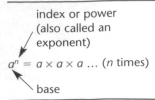
Powers of 10

Scientists, technicians and engineers often have to deal with numbers expressed as powers of 10. For example:

$$10^2 = 10 \times 10$$
$$10^3 = 10 \times 10 \times 10$$
$$10^6 = 10 \times 10 \times 10 \times 10 \times 10 \times 10$$

Both numbers above 0 and numbers below 0 (decimals) can be written as powers of 10. Look at this table and notice what happens to the index for numbers below 0.

10^4	10^3	10^2	10^1	10^0	10^{-1}	10^{-2}	10^{-3}
10 000	1 000	100	10	1	$\frac{1}{10}$	$\frac{1}{100}$	$\frac{1}{1\,000}$
					(0.1)	(0.01)	(0.001)

Exercise

1. Write these numbers out in full.
    a) 10^3 b) 10^2 c) 10^0 d) 10^{-3}
2. Write as powers of 10.
    a) 10 000   b) 1 000 000 000   c) 0.000001   d) $\frac{1}{10}$
    e) $\frac{1}{100}$ f) $\frac{1}{1\,000}$ g) 10 million   h) 3.1 million

Standard form

Sometimes we have to work with large or small numbers. Consider:
- the distance from Jupiter to the Sun is 778 000 000 km
- the area of the Pacific Ocean is 165 384 000 km²
- the wavelength of green light is 0.0000005 m
- the current in an electronic circuit is 0.00000000006 amps.

In mathematics and science, very large and very small numbers are often written as a number between 1 and 10 multiplied by a power of 10. We can write this as $a \times 10^n$ where n is an integer and $1 \leqslant a < 10$.
Writing numbers in this way is called standard form.

■ Examples

$$700\ 000 = 7 \times 10^5$$
$$2\ 500\ 000 = 2.5 \times 10^6$$
$$876\ 000\ 000 = 8.76 \times 10^8$$
$$0.432 = 4.32 \times 10^{-1}$$
$$0.0095 = 9.5 \times 10^{-3}$$
$$0.000461 = 4.61 \times 10^{-4}$$

Exercise

1. Express the following in standard form.
   a) 78 900
   b) 347.9
   c) 0.000058
   d) 0.1234
2. Write the following numbers in their ordinary form.
   a) 1.2×10^2
   b) 3.14×10^{-2}
   c) 7.605×10^{-4}
   d) 2.8×10^9
3. Work out the following, giving your answers in standard form.
   a) $(4 \times 10^4) \times (1.2 \times 10^6)$
   b) $(2.4 \times 10^8) \times (6 \times 10^{-5})$
   c) $(5.6 \times 10^6) \div (3.5 \times 10^2)$
   d) $(1.25 \times 10^4) \div (2.5 \times 10^{-7})$
   e) $(6.9 \times 10^5) + (3.8 \times 10^4)$
   f) $(9 \times 10^{-1}) + (8.3 \times 10^{-2})$
   g) $(2.7 \times 10^4) - (4.3 \times 10^3)$
   h) $(8.5 \times 10^{-2}) - (7 \times 10^{-4})$

The laws of indices

In mathematics, there are rules or laws that enable us to work with indices in calculations without calculating the actual value of the power. You will be expected to apply these laws, but it is important to understand where they come from.

The first law of indices

Consider:
$$2^3 \times 2^5 = 2 \times 2 \times 2 \times 2 \times 2 \times 2 \times 2 \times 2$$
$$= 2^8$$

So, $2^3 \times 2^5 = 2^{3+5}$

This gives the rule $a^m \times a^n = a^{m+n}$.

The second law of indices

Consider:

$$7^5 \div 7^2 = \frac{7 \times 7 \times 7 \times 7 \times 7}{7 \times 7}$$

$$= 7 \times 7 \times 7$$

$$= 7^3$$

So $7^5 \div 7^2 = 7^{5-2}$

The second rule is $a^m \div a^n = a^{m-n}$ (or $\dfrac{a^m}{a^n} = a^{m-n}$).

The third law of indices

This rule relates to calculations that involve two indices.

This is also called raising a power.

For example: $(2^3)^4$

This means $2^3 \times 2^3 \times 2^3 \times 2^3$

$= 2^{3+3+3+3}$ (using the first rule)

$= 2^{3 \times 4}$

$= 2^{12}$

The third rule is $(a^m)^n = a^{mn}$.

Exercise

1. Use the first rule of indices to simplify:
   a) $8^3 \times 8^2$
   b) $4^4 \times 4^4$
   c) $6^3 \times 6^2 \times 6^7$
   d) $5^3 \times 5^8 \times 5$.
2. Use the second rule of indices to simplify:
   a) $3^8 \div 3^2$
   b) $2^{20} \div 2^5$
   c) $4^6 \div 4^5$
   d) $7^7 \div 7$.
3. Use the third rule of indices to simplify:
   a) $(5^2)^4$
   b) $(5^4)^2$
   c) $(9^8)^3$
   d) $[(4^2)^3]^5$.
4. Use the rules of indices to express each of the following in the form a^n:
   a) $2^6 \times 2^3 \div 2^4$
   b) $6^3 \times 6^5 \div 6^7$
   c) $(3^5 \div 3^3)^4$
   d) $(8^3)^3 \div 8^4$.

The rules of indices in algebra

Remember:
- the index or power only applies to the base to which it is attached. This means that in ab^2 only the b is squared, and in $(ab)^2$ a and b are both squared – this can also be written as a^2b^2
- the rules of indices can only be applied to powers that have the same bases, so $a^3 \times a^4 = a^7$ but $a^3 \times b^4 = a^3b^4$.

■ Examples

1. Simplify.

$2e^4 \times 5e^{10}$

$= 10e^{4+10}$

$= 10e^{14}$

$5y^5 \times 3y^2$

$= 15y^{5+2}$

$= 15y^7$

$4p^3 \times 9p^{-4}$

$= 36p^{3+(-4)}$

$= 36p^{-1}$

2. Simplify.

$12p^4 \div 4p^3$

$= 3p^{4-3}$

$= 3p$

$8y^2 \div y^4$

$= 8y^{2-4}$

$= 8y^{-2}$

$6g^4h^3 \div 2g^3h$ (a base with no power $= a^1$)

$= 3g^{4-3} h^{3-1}$

$= 3gh^2$

3. Remove the brackets.

$(3h^3)^2$

$= 9h^6$

$(p^3q)^4$

$= p^{12}q^4$

$y^2(y^3 + 5y - 1)$

$= y^2.y^3 + y^2.5y - y^2.1$

$= y^5 + 5y^3 - y^2$

Exercise

1. Given that $g = 3$ and $h = 4$, calculate the value of:
   a) $5g^4$ b) g^2h^3
   c) $2g^3 + h^3$ d) $3h^2 + 2h - 5$.

2. Use the first rule of indices to simplify:
   a) $4f^3 \times 3f^4$ b) $5y^2 \times y^6$
   c) $3e \times 2e^4$ d) $7pq^2 \times 6p^2q$.

3. Use the second rule of indices to simplify:
   a) $8p^8 \div 4p^4$ b) $7q^5 \div q^4$
   c) $9y^3 \div 3y$ d) $8p^3q^2 \div 4pq$.

4. Simplify:
   a) $(2x^2)^2$ b) $5y^2 \times (-3x^2)$
   c) $-2x^2 + 4x^2$ d) $-2y^2 (ay)$.

5. Remove the brackets in the following expressions:
   a) $(4k^2)^3$ b) $(3p^4)^2$
   c) $q(3q^2 - 5q - 1)$ d) $5y^4(y^2 + 2y - 7)$.

Zero and negative indices

The rules of indices apply in all situations. This means that indices that are negative and indices that are 0 all obey the same rules.

Consider:
$a^3 \div a^3$
You know any number divided by itself is 1.
So $a^3 \div a^3 = a^{3-3} = a^0 = 1$

$$a^2 \div a^5 = \frac{a \times a}{a \times a \times a \times a \times a} \text{ (cancel)}$$

$$= \frac{1}{a \times a \times a} = \frac{1}{a^3}$$

So $a^2 \div a^5 = a^{2-5} = a^{-3} = \dfrac{1}{a^3}$

a^{-3} is the same as $\dfrac{1}{a^3}$. In other words, a^{-3} is the reciprocal of a^3.

Exercise

1. Find the value of the following:
   a) $(5)^{-2}$
   b) $(0.0006)^0$
   c) $\left(\frac{2}{5}\right)^{-3}$
   d) $4^{-1} \times 4^0 \times 4^3$
   e) 2^{-3}
   f) 21^0
   g) $\left(\frac{2}{3}\right)^{-4}$
   h) $3^0 \times 3^2 \times 3^4$.

2. Simplify:
   a) $y^6 \times y^{-2}$
   b) $k^5 \times k \times k^{-6}$
   c) $p^3 \div p^6$
   d) $q^4 \div q^{-2}$
   e) $p^3 \times p^5 \times p^{-2}$
   f) $q^{-4} \times q \times q^2$
   g) $e^2 \div e^5$
   h) $f^{-2} \times f^2$.

3. Simplify:
   a) $e^6 \div e^2$
   b) $e^6 \div e^{-2}$
   c) $(2^{-1})^{-1}$
   d) $k \div k^{-3}$
   e) $\dfrac{y^5}{y^3}$
   f) $\dfrac{y^5}{y^{-3}}$
   g) $\left(\left(\frac{3}{4}\right)^{-1}\right)^{-1}$
   h) $\dfrac{k^5}{k^{-2}}$.

4. Express with positive indices and simplify, if possible:
   a) x^{-y}
   b) 4^{-1}
   c) $\dfrac{1}{x^{-a}}$
   d) 4^{-2}
   e) 11^{-1}
   f) $\dfrac{a^{-3}}{b^{-4}}$
   g) xy^{-1}
   h) $-(5^{-2})$.

5. Simplify:
   a) x^0 if $x \neq 0$
   b) $(3^2)^0$
   c) $\left(\frac{1}{4}\right)^{-1}$
   d) 0^1
   e) 1^0
   f) $(107)^0$.

Fractional indices

You now know the meaning of a^n when the index n is a positive whole number, a negative whole number or zero. You will now consider expressions such as $9^{\frac{1}{2}}$, $8^{-\frac{2}{3}}$ and $(32)^{0.4}$.

As before, our definitions of such expressions must fit in with the rules of indices.

Let us first consider $a^{\frac{1}{2}}$. If we multiply it by itself using the first rule of indices, we get $a^{\frac{1}{2}} \times a^{\frac{1}{2}} = a^{\frac{1}{2}+\frac{1}{2}} = a^1 = a$. In other words, the square of $a^{\frac{1}{2}}$ is a, so $a^{\frac{1}{2}}$ must be the square root of a $\left[(a^{\frac{1}{2}})^2 = a, \text{ so } a^{\frac{1}{2}} = \sqrt{a} \right]$.

Similarly, $a^{\frac{1}{3}} \times a^{\frac{1}{3}} \times a^{\frac{1}{3}} = a^{\frac{1}{3}+\frac{1}{3}+\frac{1}{3}} = a^1 = a$ $\left[(a^{\frac{1}{3}})^3 = a, \text{ so } a^{\frac{1}{3}} = \sqrt[3]{a} \right]$

so the cube of $a^{\frac{1}{3}}$ is a. So $a^{\frac{1}{3}}$ is the cube root of a.

$a^{\frac{1}{n}}$ is the nth root of a or $a^{\frac{1}{n}} = \sqrt[n]{a}$.

To interpret other fractional powers, we use the third rule of indices.

$$a^{\frac{m}{n}} = a^{m \times \frac{(1)}{n}} = [a^m]^{\frac{1}{n}} = \sqrt[n]{a^m}$$

Alternatively, $a^{\frac{m}{n}} = a^{\frac{1}{n} \times m} = [a^{\frac{1}{n}}]^m = (\sqrt[n]{a})^m$.

If the index is a decimal, we must change it to a fraction in order to interpret it. For example,

$$a^{0.4} = a^{\frac{4}{10}} = a^{\frac{2}{5}} = \sqrt[5]{a^2}.$$

If the index is negative, we must remember that a^{-1} stands for $\frac{1}{a}$, that is, the reciprocal of a.

Exercise

1. Work out the value of:
   a) $16^{\frac{1}{2}}$ b) $27^{\frac{2}{3}}$

   c) $25^{\frac{3}{2}}$ d) $10\,000^{0.75}$.

2. Work out the value of:
   a) $(\frac{25}{4})^{\frac{3}{2}}$ b) $8^{\frac{1}{4}} \times 8^{\frac{1}{12}}$

   c) $9^{\frac{3}{4}} \div 9^{\frac{1}{4}}$ d) $32^{0.6}$.

3. Simplify:
   a) $27^{-\frac{2}{3}}$ b) $(\frac{9}{25})^{-\frac{3}{2}}$

   c) $e^{-\frac{1}{2}} \div e^{-\frac{3}{2}}$ d) $(f^{\frac{1}{2}})^6$.

Algebraic manipulation

Most algebraic manipulation combines the rules that you have already learnt. You will be expected to remove brackets, and apply the rules for directed numbers as well as the laws of indices and BODMAS all at the same time. This requires clear thinking and a great deal of practice. Revise what you have learnt so far by completing the exercise below.

Exercise

1. Remove the brackets and simplify.
   a) $4(y + 3) - 3(2y - 1)$ b) $2(3s - 4t) - 5(s + 2t)$
   c) $-3(2z - 1) + 7(3z - 2)$ d) $6(p + q) - (q - p)$
   e) $3(y + 7) - 2(4y - 5)$ f) $4(3s - 2t) - (s + 3t)$
   g) $-5(6z - 1) + 6(2z - 3)$ h) $2(p + q) - (p - q)$
2. Remove the brackets and simplify.
   a) $3(2e + f) - 4f + 5(e - f)$ b) $t(2t + 1) - (4t - 3) - 5$
   c) $2(e + 3f) - 3e + 4(e - f)$ d) $t(3t + 2) - (5t - 4) + 7$

Multiplying two brackets

You have removed brackets in cases where there is one multiplier in front of the bracket. However, in some cases, you may need to work with expressions in the form of $(a + b)(c + d)$. Such an operation is called 'expanding the brackets'. To do this, you multiply all the terms in the second bracket by all the terms in the first bracket:

$$(a + b)(c + d) = a(c + d) + b(c + d)$$
$$= ac + ad + bc + bd$$

If there are like terms after expanding the brackets, you add them together.

■ Examples
Remove the brackets and simplify.

Remember

$pq = qp$

$(p + q)(p + 2q)$
$= p(p + 2q) + q(p + 2q)$
$= p^2 + 2pq + qp + 2q^2$
$= p^2 + 3pq + 2q^2$

$(t - 2)(t + 3)$
$= t(t + 3) - 2(t + 3)$
$= t^2 + 3t - 2t - 6$
$= t^2 + t - 6$

$(5p - 3)^2$
$= (5p - 3)(5p - 3)$
$= 5p(5p - 3) - 3(5p - 3)$
$= 25p^2 - 15p - 15p + q$
$= 25p^2 - 30p + 9$

$(s - 3)(s^2 - s - 4)$
$= s(s^2 - s - 4) - 3(s^2 - s - 4)$
$= s^3 - s^2 - 4s - 3s^2 + 3s + 12$
$= s^3 - 4s^2 - s + 12$

Some important expansions

The sum of two squares:

$$(a + b)^2 = (a + b)(a + b)$$
$$= a^2 + ab + ba + b^2 \qquad (ab = ba)$$
$$= a^2 + 2ab + b^2$$

The square of the difference between two numbers:

$$(a - b)^2 = (a - b)(a - b)$$
$$= a^2 - ab - ba + b^2 \qquad (ab = ba)$$
$$= a^2 - 2ab + b^2$$

The difference of two squares:

$$(a - b)(a + b) = a(a + b) - b(a + b)$$
$$= a^2 + ab - ba - b^2 \qquad (ba = ab)$$
$$= a^2 - b^2$$

Exercise

1. Remove the brackets and simplify.
    a) $(t - 5)(t - 4)$ 　　　　　b) $(2p - 1)(3p + 1)$
    c) $(4s + 3t)(3s - 4t)$

2. Expand the brackets and simplify.
    a) $(3p + 2)^2$ 　　　　　　b) $(4y - 3)(4y + 3)$
    c) $(s - 2)(s^2 + 2s - 3)$

3. Expand and simplify.
    a) $(m + n)(x - y)$ 　　　　b) $(3a - b)(2a - b)$
    c) $2(3a + b)(a + 2b)$ 　　　d) $(a - b)(3a - 2b + 3c)$
    e) $(2x + 4)(2x - 4)$ 　　　　f) $(3x - 5)^2$
    g) $(x + 2y)^2$ 　　　　　　h) $-2(x - y)^2$
    i) $(x + 2)(x - 3) - 2(x + 4)$ 　　j) $(x + 2)(x - 7) - 2(x + 4)(x - 3)$
    k) $(3x + 2)^2 - (4x - 2)^2$ 　　l) $(x + 7)^2 - (x + 2)(4 - 3x)$

Factorising

In algebra, factorising is the same as in arithmetic. When you factorise an expression, you write it as the product of its factors. Factorisation can be seen as the reverse of removing brackets.

Finding common factors

The first step in factorising is to find any common factors in all the terms. For example, in the expression $18p + 4q$:
- there are no variables in common
- the HCF of 18 and 4 is 2.

So the expression $18p + 4q$ can be written as $2(9p + 2q)$.

■ Examples

Factorise:

$18s + 12t + 24u$	$9pq - 6qr$	$t^2 - 2t$
HCF = 6	HCF = $3q$	HCF = t
$\therefore$ $6(3s + 2t + 4u)$	$\therefore$ $3q(3p - 2r)$	$\therefore$ $t(t - 2)$

Exercise

1. Factorise:
   a) $10s + 15t + 20u$ b) $12pq - 9qr$ c) $16yz - 4y$.
2. Factorise:
   a) $t^2 + 5t$ b) $9y^2 - 6yz$ c) $6p^2q + 14pq^2 - 10pq$.

Factorising by grouping

When an expression has an even number of terms, such as four, you may find that there is no common factor for all four terms. In such cases, you group the terms in pairs and find the common factor of each pair.

■ Examples

Factorise:

$ap + aq - bp - bq$

$= a(p + q) - b(p + q)$

$= (a - b)(p + q)$

$6b^2 + 4bd + 3bc + 2cd$

$= 2b(3b + 2d) + c(3b + 2d)$

$= (3b + 2d)(2b + c)$

$3y + 4pq - 3p - 4yq$

$= 3(y - p) + 4q(p - y)$

$= 3(y - p) - 4q(y - p)$

$= (y - p)(3 - 4q)$

$6ab - 3bc + 2ad - cd + 8a - 4c$

$= 3b(2a - c) + d(2a - c) + 4(2a - c)$

$= (2a - c)(3b + d + 4)$

Remember

$(p - y) = -(y - p)$

Factorising quadratic expressions

Expressions in the form of $ax^2 + bx + c$ (where a, b and c are numbers) are called quadratic expressions. When you factorise such expressions, you have to remember that the middle term bx is, in fact, a combination of two terms. Consider: $(2x - 3)(x + 4)$ can be multiplied out to give:

$2x(x + 4) - 3(x + 4) = 2x^2 + 8x - 3x - 12 = 2x^2 + \mathbf{5x} - 12$

When you try to factorise the expression $2x^2 + 5x - 12$, you have to decide which combination of terms gave $5x$. You can do this by trial and error.

■ Examples

Factorise: $x^2 + 2x - 15$

$= (x - 3)(x + 5)$

You know the first term: $x \times x$.
Find 2 numbers that add up to
+2 and have a product −15:
$-3 \times 5 = -15$ $-3 + 5 = 2$

Factorising $ax^2 + bx + c$

When the co-efficient of x^2 in a quadratic equation is not 1, you have to find the combination of factors that give ac and, when added together, give b. You can work this out by trial and error.

■ Examples

1. Factorise $3x^2 + 2x - 8$.

Find two numbers which, when multiplied, give $(3)(-8) = -24$ and, when added, give $+2$.

-1 and $+24$; no good – these add up to $+23$
-2 and $+12$; no good – these add up to $+10$
-3 and $+8$; no good – these add up to $+5$
-4 and $+6$; success! – these add up to $+2$

$$3x^2 + 2x - 8 = 3x^2 - 4x + 6x - 8$$
$$= x(3x - 4) + 2(3x - 4)$$
$$= (3x - 4)(x + 2)$$

2. Factorise $6x^2 - 7x + 2$.

Find two numbers which, when multiplied, give $(6)(+2) = +12$ and, when added, give -7. must give – when added   must give + when multiplied

$\therefore$ both numbers will be negative

-1 and -12; no good – these add up to -13
-2 and -6; no good – these add up to -8
-3 and -4; success! – these add up to -7

$$6x^2 - 7x + 2 = 6x^2 - 3x - 4x + 2$$
$$= 3x(2x - 1) - 2(2x - 1)$$
$$= (2x - 1)(3x - 2)$$

Factorising the difference of two squares

Remember that $(a - b)(a + b) = a^2 - b^2$.

If you work the reverse of this, then the factors of $a^2 - b^2 = (a - b)(a + b)$. You can use this rule to factorise the difference of two squares.

■ Examples

$9x^2 - 16y^2$
$= (3x + 4y)(3x - 4y)$

$t^2 - 25$
$= (t + 5)(t - 5)$

$18x^2 - 8 = 2(9x^2 - 4)$
$\qquad\quad = 2(3x + 2)(3x - 2)$

Sometimes you need to find the HCF before you get a difference of two squares.

Exercise

1. Factorise.

   a) $3y + 9z + by + 3bz$

   b) $4p + 6r - 2pq - 3qr$

   c) $3x^2 - 4x + 6x - 8$

   d) $ac - bc - ad + bd$

2. Factorise.
   a) $x^2 + 14x + 24$ b) $x^2 + x - 6$
   c) $x^2 - 8x + 15$
3. Factorise.
   a) $3x^2 + 11x + 6$ b) $5x^2 - 9x - 2$
   c) $4x^2 - 8x + 3$
4. Work out.
   a) $(10\ 001)^2 - (10\ 000)^2$ b) $(51)^2 - (49)^2$
   c) $(6\frac{1}{4})^2 - (5\frac{3}{4})^2$
5. Factorise.
   a) $4x^2 - 9y^2$ b) $36a^2 - 25b^2$
   c) $16c^2 - 1$ d) $27x^2 - 12y^2$

Algebraic fractions

Algebraic fractions can be simplified, added, subtracted, multiplied and divided in the same ways as arithmetic fractions. However, in order to perform these operations, you may need to find the factors of the numerator and denominator.

■ Examples

Remember

You can only cancel complete terms. You *cannot* cancel like this:

$\frac{\cancel{y}}{\cancel{y}+1}$ ✗

1. Simplify.

$$\frac{6pq}{8p^2}$$

$$= \frac{{}^3\cancel{6} \times \cancel{p} \times q}{{}^4\cancel{8} \times \cancel{p} \times p}$$

$$= \frac{3q}{4p}$$

$$\frac{y^2 + 5y}{y^2 + 6y + 5}$$

$$= \frac{y(\cancel{y+5})}{(y+1)(\cancel{y+5})}$$

$$= \frac{y}{y+1}$$

$$\frac{x^2 - 4}{x^2 - 5x + 6}$$

$$= \frac{(x+2)(\cancel{x-2})}{(x-3)(\cancel{x-2})}$$

$$= \frac{(x+2)}{(x-3)}$$

2. Express as a single fraction in simplest form.
   a) $\dfrac{3d}{4} + \dfrac{9d}{10}$ so $\dfrac{3d}{4} = \dfrac{15d}{20}$ and $\dfrac{9d}{10} = \dfrac{18d}{20}$

   $\text{LCM} = 20$ $\therefore \dfrac{3d}{4} + \dfrac{9d}{10} = \dfrac{15d}{20} + \dfrac{18d}{20} = \dfrac{33d}{20}$

   b) $\dfrac{t}{3} - \dfrac{t-4}{2}$

   6 is the LCM of 3 and 2.

   $\dfrac{t}{3} = \dfrac{2t}{6}$ and $\dfrac{t-4}{2} = \dfrac{3(t-4)}{6}$ (the brackets are essential!)

   so $\dfrac{t}{3} - \dfrac{t-4}{2} = \dfrac{2t}{6} - \dfrac{3(t-4)}{6} = \dfrac{2t - 3(t-4)}{6}$

   $$= \dfrac{2t - 3t + 12}{6}$$

   $$= \dfrac{12 - t}{6} \text{ (be careful with the signs!)}$$

   c) $\dfrac{1}{x-2} - \dfrac{2}{x-3}$

   Neither $x - 2$ nor $x - 3$ can be factorised, so the simplest expression which is a multiple of each of them is $(x-2)(x-3)$.

$$\frac{1}{x-2} = \frac{(x-3)}{(x-2)(x-3)} \quad [\times \text{ numerator and denominator by } (x-3)]$$

$$\text{and } \frac{2}{x-3} = \frac{2(x-2)}{(x-3)(x-2)} \quad [\times \text{ numerator and denominator by } (x-2)]$$

$$\text{so } \frac{1}{x-2} - \frac{2}{x-3} = \frac{(x-3)}{(x-2)(x-3)} - \frac{2(x-2)}{(x-3)(x-2)}$$

$$= \frac{(x-3) - 2(x-2)}{(x-2)(x-3)}$$

$$= \frac{x-3-2x+4}{(x-2)(x-3)}$$

$$= \frac{1-x}{(x-2)(x-3)} \quad \begin{array}{l}\text{[it is usual to leave the denominator}\\ \text{in factorised form]}\end{array}$$

3. Simplify: $\dfrac{3x+4}{x^2+x-6} - \dfrac{1}{x+3}$

$$\frac{3x+4}{x^2+x-6} - \frac{1}{x+3}$$

$$= \frac{3x+4}{(x+3)(x-2)} - \frac{1}{x+3}$$

$$= \frac{3x+4}{(x+3)(x-2)} - \frac{(x-2)}{(x+3)(x-2)} \quad [\text{LCM of denominator is } (x+3)(x-2)]$$

$$= \frac{3x+4-(x-2)}{(x+3)(x-2)}$$

$$= \frac{3x+4-x+2}{(x+3)(x-2)}$$

$$= \frac{2x+6}{(x+3)(x-2)} \quad [\text{factorise the numerator}]$$

$$= \frac{2\cancel{(x+3)}}{\cancel{(x+3)}(x-2)}$$

$$= \frac{2}{(x-2)}$$

Exercise

1. Simplify.

   a) $\dfrac{4p^2q}{6pq^2}$

   b) $\dfrac{y^2+3y}{y^2+8y+15}$

   c) $\dfrac{x^2-2x}{x^2-7x+10}$

2. Express as a single fraction in its simplest form.

   a) $\dfrac{3a}{4b} \times \dfrac{5ab}{3}$

   b) $\dfrac{x^2}{x^2-4} \times \dfrac{x^2+3x+2}{2x}$

   c) $\dfrac{3p-6}{p^2-p-6} \div \dfrac{p-2}{p+2}$

3. Express as a single fraction in its simplest form.

   a) $\dfrac{3a}{4} - \dfrac{7a}{10}$

   b) $\dfrac{2x}{3} - \dfrac{3(x-5)}{2}$

   c) $\dfrac{2}{t+1} + \dfrac{3}{t+2}$

4. Simplify: $\dfrac{2(x+2)}{x^2+4x-5} - \dfrac{1}{x+5}$.

Transforming more complicated formulae

The skills you have learnt in algebraic manipulation can be applied to formulae containing fractions, negative numbers and brackets. However, you may need to apply a number of rules or principles at the same time.

These steps are useful in manipulating a formula:
- clear the formula of any fractions
- eliminate any square or other roots
- expand any brackets
- rearrange the terms so that those containing the new subject are isolated on one side of the formula
- factorise this side so that it has the format (new subject) × (expression)
- divide both sides by (expression) so the formula becomes (new subject) = …

■ Examples

1. Given that $ab + c = d(b + 2)$, express b in terms of a, c and d.
   (This means 'make b the subject of the formula'.)
   $$ab + c = d(b + 2) \text{ [expand the brackets]}$$
   $$ab + c = db + 2d \text{ [collect terms containing } b \text{ on one side]}$$
   $$ab - db = 2d - c \text{ [factorise as } b \times \text{(expression)]}$$
   $$b(a - d) = 2d - c \text{ [divide both sides by (expression)]}$$
   $$b = \frac{2d - c}{(a - d)}$$

2. Make u the subject of the formula $\dfrac{1}{v} + \dfrac{1}{u} = \dfrac{2}{R}$.
   $$\frac{vuR}{v} + \frac{vuR}{u} = \frac{2vuR}{R} \text{ [multiply by } vuR \text{ (to clear the fractions)]}$$
   $$uR + vR = 2vu \text{ (collect terms containing } u \text{ on one side)}$$
   $$vR = 2vu - uR \text{ [factorise as } u \times \text{(expression)]}$$
   $$vR = u(2v - R) \text{ [divide both sides by (expression)]}$$
   $$\frac{vR}{(2v - R)} = u$$
   The required formula is $u = \dfrac{vR}{(2v - R)}$.

3. Make b the subject of the formula $r = \dfrac{a - b}{a + b}$.
   $$r(a + b) = a - b \text{ [multiply both side by } a + b \text{ (to clear the fraction)]}$$
   $$ra + rb = a - b \text{ [expand the brackets]}$$
   $$rb + b = a - ra \text{ [collect terms containing } b \text{ on one side]}$$
   $$b(r + 1) = a - ra \text{ [factorise as } b \times \text{(expression)]}$$
   $$b = \frac{a - ra}{(r + 1)} \text{ [divide both sides by (expression)]}$$

   This could be written as $b = \dfrac{a(1 - r)}{(1 + r)}$.

4. Given that $F = \dfrac{R}{R+r}$, express R in terms of F and r.

$$\begin{aligned}
F(R+r) &= R \text{ [multiply both sides by } R+r \text{ (to clear the fraction)]} \\
FR + Fr &= R \text{ [expand the bracket]} \\
Fr &= R - FR \text{ [collect terms containing } R \text{ on one side]} \\
Fr &= R(1-F) \text{ [factorise as } R \times \text{(expression)]} \\
\dfrac{Fr}{(1-F)} &= R \text{ [divide both sides by (expression)]}
\end{aligned}$$

The required formula is $R = \dfrac{Fr}{(1-F)}$.

5. Make a the subject of the formula $T = 2\pi \sqrt{\dfrac{a}{g}}$.

Divide both sides by 2π (to isolate the square root).

$$\dfrac{T}{2\pi} = \sqrt{\dfrac{a}{g}}$$

$$\dfrac{T^2}{4\pi^2} = \dfrac{a}{g} \text{ [square both sides]}$$

$$\dfrac{gT^2}{4\pi^2} = a \text{ [multiply both sides by } g\text{]}$$

The required formula is $a = \dfrac{gT^2}{4\pi^2}$.

Exercise

1. Given that $p(x+q) = x + r$, express x in terms of p, q and r.

2. Make y the subject of the formula $\dfrac{x}{a} + \dfrac{y}{b} = 1$.

3. Change the subject of the formula $y = \dfrac{x+3}{x-2}$ to x.

4. Make f the subject of the formula $s = ut + \dfrac{1}{2}ft^2$.

5. Given that $p = 3\sqrt{q} - 4$, express q in terms of p.

6. Make m the subject of the formula $E = mc^2$.

7. Express in terms of x: $y = (3x+2)^2$.
   a) Solve for x if $y = -3$.

   b) Solve for x if $y = 1\dfrac{1}{2}$.

8. What is r if $v = \dfrac{1}{3}\pi r^2 h$?

9. Make A, P and r the subject of this formula: $\dfrac{A}{P} = (1+r)^n$.

10. Make v the subject of the formula: $\dfrac{1}{v} + \dfrac{1}{u} = \dfrac{2}{R}$.

Solving simultaneous equations

Jan and Ashraf visit a café. Jan pays $11 for a bun and two cups of coffee. This is represented as:

$b + 2c = 11$.

Ashraf buys two buns and one cup of coffee for $10. This is represented as:

$2b + c = 10$.

Both of these equations have two variables, so you cannot solve them using the methods you have already learnt. However, there is only one value of b and one value of c that will make both equations true. Such equations are called simultaneous equations and they have to be solved at the same time. You will learn more about these in Module 3 when you work with graphs.

There are two methods of solving simultaneous equations algebraically:

- substitution method – used when one variable has a constant of 1 or –1
- method of equal co-efficients – used when none of the variables has a constant of 1 or –1.

Substitution

In this method, you change the subject of the equation in one equation to get a value for one of the variables. You then substitute the value in the other to get an equation with only one unknown.

■ Examples

1. Solve simultaneously: $b + 2c = 11$
   $2b + c = 10$.

   $b + 2c = 11$ ...① ⎤ Write above each other
   $2b + c = 10$ ...② ⎦ and name ① and ②

   Find b: $b = 11 - 2c$ ...③] Call the new equation ③

   Substitute ③ into ②: $2(11 - 2c) + c = 10$
       $\therefore 22 - 4c + c = 10$
       $\therefore -3c = -12$
       $\therefore c = 4$

   Substitute in ③: $b = 11 - 2(4)$
       $\therefore b = 3$

2. Find the value of x and y in these equations: $x - 3y = 13$
   $3x + 2y = 6$.

   $x - 3y = 13$ ... ①
   $3x + 2y = 6$ ... ②
   $x = 13 + 3y$ ... ③

   Substitute ③ into ②: $3(13 + 3y) + 2y = 6$
       $\therefore 39 + 9y + 2y = 6$
       $\therefore 11y = -33$
       $\therefore y = -3$

   Substitute in ③ : $x = 13 + 3(-3)$
       $\therefore x = 4$

Method of equal co-efficients

In this method, you transform both equations to make one of the variables (say x or y) equal in both equations. You then add or subtract the transformed equations to eliminate one unknown and solve for the other.

■ Examples

1. Solve simultaneously: $3x + 2y = 12$
$$5x - 3y = 1.$$

$3x + 2y = 12$ ...①⎤ Make y equal:
$5x - 3y = 1$ ...②⎦ ①×3 ②×2

① × 3: $\quad 9x + 6y = 36$ ... ③
② × 2: $\quad \underline{10x - 6y = 2}$ ... ④
③ + ④: $19x \quad\quad = 38$
$\therefore x \quad\quad = 2$

Substitute x in ②: $5(2) - 3y = 1$
$$\therefore 9 = 3y$$
$$y = 3$$

2. Solve the simultaneous equations: $4x = y + 7$
$$3x + 4y + 9 = 0.$$

$4x - y = 7$ ...①⎤
$3x + 4y = -9$ ...②⎦ Rearrange

① × 4: $\quad \underline{16x - 4y = 28}$ ... ③
② + ③: $19x \quad\quad = 19$
$x \quad\quad = 1$

Substitute x in ①: $4(1) - y = 7$
$$\therefore -y = 3$$
$$\therefore y = -3$$

Exercise

1. Solve the simultaneous equations.
   a) $3x + 2y = 10 \quad 4x - y = 6$
   b) $p + 2q = 7 \quad 3p - 2q = -3$
   c) $2u + v = 7 \quad 3u - 2v = 7$
2. Solve the simultaneous equations.
   a) $4s = 5t + 5,\ 2s = 3t + 2$
   b) $6f - 6g = 5,\ 3f - 4g = 1$
   c) $2x = 3y + 14,\ 3x + 2y + 5 = 0$
3. Three notebooks and five pencils cost $10. One notebook and ten pencils cost $10. Taking the cost of a notebook to be n dollars and the cost of a pencil to be p dollars, write down two simultaneous equations. Solve the equations and state the cost of a notebook and the cost of a pencil.
4. The variables x and y are related by $y = mx + c$, where m and c are constants. It is given that $y = 12$ when $x = 2$ and that $y = 4$ when $x = 6$. Find the value of m and the value of c.

Solving quadratic equations

A quadratic equation is one that has no power greater than x^2. Normally quadratic equations have two solutions (also called roots). When you are asked to solve a quadratic equation, you give the solution as: $x = $ (1st value) or $x = $ (2nd value).

Using factors to solve quadratic equations

You already know how to factorise algebraic expressions. Factorising a quadratic equation relies on a property of 0:

> When two numbers are multiplied to give 0, then one of those numbers must be equal to 0. If $xy = 0$, then $x = 0$ or $y = 0$.

■ Examples

1. Solve the equation $x^2 - 5x = 0$.
    Factorise the left-hand side: $x(x - 5) = 0$
    So, $x = 0$ or $(x - 5) = 0$
    ∴ $x = 0$ or $x = 5$
2. Solve for y: $y^2 + 2y - 15 = 0$.
    Factorise the left-hand side: $(y - 3)(y + 5) = 0$
    So, $(y - 3) = 0$ or $(y + 5) = 0$
    ∴ $y = 3$ or $y = -5$
3. Solve $25t^2 - 16 = 0$.
    Factorise using difference of squares: $(5t - 4)(5t + 4) = 0$
    So, $(5t - 4) = 0$ or $(5t + 4) = 0$
    $t = \frac{4}{5}$ or $t = \frac{-4}{5}$

Exercise

1. Solve the quadratic equations.
    a) $x^2 + 7x = 0$
    b) $y^2 - 16 = 0$
    c) $(2t - 5)(t + 3) = 0$
2. Solve the quadratic equations.
    a) $x^2 - 7x + 12 = 0$
    b) $p^2 - 5p - 6 = 0$
    c) $n^2 + 8n + 16 = 0$
3. Solve the quadratic equations.
    a) $x^2 + 36 = 13x$
    b) $2y^2 = 3y + 2$
    c) $t(t + 4) = 12$
4. The square of a boy's present age in years is equal to 9 times his age 2 years ago. Calculate his present age.

Using a formula to solve quadratic equations

The formula for solving quadratic equations is called the quadratic formula and it is written as follows:

If $ax^2 + bx + c = 0$, then $x = \dfrac{-b \pm \sqrt{b^2 - 4ac}}{2a}$.

Note that there are two solutions to the quadratic equation. The symbol $\pm$ means that one answer is obtained using the + sign, the other by using the – sign.

■ Examples

1. Solve the equation $2x^2 + 7x + 3 = 0$.

   Comparing with $ax^2 + bx + c = 0$, we see that $a = 2$, $b = 7$, $c = 3$.
   It is useful to work out the value of ac here: $ac = 2 \times 3 = 6$.

   Using the formula: $x = \dfrac{-b \pm \sqrt{b^2 - 4ac}}{2a}$

   $$x = \frac{-7 \pm \sqrt{49 - 4(6)}}{4}$$

   $$= \frac{-7 \pm \sqrt{49 - 24}}{4}$$

   $$x = \frac{-7 \pm \sqrt{25}}{4}$$

   $$x = \frac{-7 + 5}{4} \text{ or } \frac{-7 - 5}{4}$$

   $$x = \frac{-2}{4} \text{ or } \frac{-12}{4}$$

   $$x = \frac{-1}{2} \text{ or } -3$$

2. Solve the equation $y^2 = 2y + 1$.
   First rearrange: $y^2 - 2y - 1 = 0$

   Using the formula: $y = \dfrac{-b \pm \sqrt{b^2 - 4ac}}{2a} = \dfrac{-(-2) \pm \sqrt{4 - 4(-1)}}{2}$

   $$= \frac{+2 \pm \sqrt{4 + 4}}{2}$$

   $$= \frac{+2 \pm \sqrt{8}}{2}$$

   $$= \frac{+2 + 2.828}{2} \text{ or } \frac{+2 - 2.828}{2}$$

   $$= \frac{4.828}{2} \text{ or } \frac{-0.828}{2}$$

   $$= 2.41 \text{ or } -0.41$$
   (to 2 decimal places)

Hint

Using the calculator:
2 ⊞ 8 √ ⊟ ⊡ 2 ⊟
or if you have a DAL
calculator:
2 ⊞ √ 8 ⊟ ⊡ 2 ⊟

Exercise

1. Solve the quadratic equations using the formula.
   a) $5x^2 + 9x + 2 = 0$ b) $y^2 = 4y + 1$ c) $t^2 + t - 3 = 0$
2. Use the formula to solve the equations.
   a) $n^2 = 5n - 3$ b) $x^2 - 6x - 21 = 0$ c) $3m^2 + m - 1 = 0$
3. A picture has a length of L cm and a breadth of B cm. For the picture to be a pleasing shape, L and B should be related by $B(B + L) = L^2$. If the breadth of the picture is 30 cm, calculate the required length.

Linear inequalities

In equations, one side is equal to the other side. This is indicated by the presence of an equal sign.

In an inequality, one side is bigger than (or equal to) or smaller than (or equal to) the other side. This is indicated by the signs $<, >, \leq, \geq$.

For example: $x < 21$; $v \leq 100$; $5 < y < 10$.

Solving linear inequalities

The rules for solving inequalities are similar to those for solving linear equations. However, there are important exceptions to the rules for multiplication and division.

To solve an inequality, you can:
- add the same number or expression to both sides
- subtract the same number or expression from both sides
- multiply both sides by the same positive number or expression
- divide both sides by the same positive number or expression.

Consider the inequality $10 > 5$.
Multiply both sides of this inequality by –1 and note what happens:
   $-10 > -5$.

When you multiply by a negative, the inequality is no longer true. For this reason, there is an additional rule for multiplying and dividing inequalities by negative numbers:

> If you multiply or divide both sides of an inequality by a negative number, the direction of the inequality sign changes (from $<$ to $>$ or from $>$ to $<$).

■ Examples

1. Solve for x: $2x + 1 > 15$.
   (−1) $2x > 14$
   (÷2) $x > 7$
2. Solve the inequality $\dfrac{2y - 1}{3} \leq 4$.

   (×3) $2y - 1 \leq 12$
   (+1) $2y \quad \leq 13$
   (÷2) $y \quad \leq 6.5$

3. Solve the inequality $2t + 3 > 7t + 11$.
$(-7t)\quad -5t + 3 > 11$
$(-3)\qquad -5t > 8$
$(\div -5)\qquad t < \frac{-8}{5}$ (Remember to change the sign)

4. Solve $-5 \leq 2x + 1 < 5$.

$-5 \leq 2x + 1$ ... ① $\qquad$ $2x + 1 < 5$ ... ②
$-6 \leq 2x$ $\qquad\qquad\qquad$ $2x < 4$
$-3 \leq x$ $\qquad\qquad\qquad\qquad$ $x < 2$

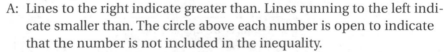

combine
$-3 \leq x < 2$

5. Find the two integer values of t which satisfy all of these conditions:
$2t + 6 \geq 0$, $t \neq -2$, $t < 0$.
$2t + 6 \geq 0$ gives $2t \geq -6$ $\quad \therefore\; t \geq -3$
$t < 0$, $-3 \leq t < 0$

But t is an integer, $t \neq -2$, so values of t are -3 and -1.

Showing inequalities on a number line
Inequalities can be shown on a number line. The examples on the left show you how to do this. Note the difference between a closed and an open circle above a number. An arrow indicates that the line continues to infinity in the direction of the arrow.

■ **Examples**

A: Lines to the right indicate greater than. Lines running to the left indicate smaller than. The circle above each number is open to indicate that the number is not included in the inequality.

B: In this case, the circle above the number is shaded to show that it is included in the inequality.

C: These lines show a range of values between two units. The same conventions apply but the line length is limited.

Hint

There are two inequalities in this example.

A
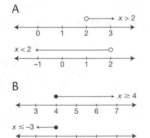

B

C

Exercise

1. Solve the inequalities:

   a) $3x + 7 > 1$ $\qquad$ b) $\dfrac{5y - 3}{4} < 8$ $\qquad$ c) $3 - 5t \geq 18$.

2. Solve the inequality $-3 \leq 2x + 3 < 3$ and illustrate your solution on a number line.

3. Given that n is a positive integer, list the values of n for which $-8 \leq 4n < 10$.

4. Given that x is a positive integer, $x \leq 5$ and $x \neq 3$, list the possible values of x.

Check your progress

1. At a party, there are c children, w women and m men. Write each of the following statements in terms of algebra, using appropriate mathematical symbols.
   a) There are 14 people at the party.
   b) There are more than three children.
   c) The number of men is different from the number of children.
   d) There are two more women than men.

2. Find the value of $\sqrt{\dfrac{a^2 + b^2}{6}}$ when $a = 5.3$ and $b = 4.8$.

3. $S = 13 + 5R$.
   a) If $R = 2.8$, find the value of S.
   b) If $S = 48$, find the value of R.
   c) Make R the subject of the formula.

4. Solve the following equations.
   a) $4x = 34$
   b) $\dfrac{y}{5} = \dfrac{3}{4}$
   c) $5(z + 2) = 3(z - 1) + 23$

5.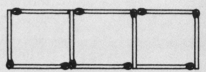

   The diagrams above show how matchsticks are used to make lines of squares.
   a) Complete the following table.

Number of squares in a line	1	2	3	4	5	6	7	8
Number of matchsticks needed	4	7	10					

   b) How many matchsticks are needed to make 6 squares in a line?
   c) Find the formula for the number of matches needed to make n squares in a line.

6. A builder has to deliver 20 tonnes of sand to a customer. He has a large truck which holds 3 tonnes of sand and a small truck which holds 2 tonnes. His driver makes x journeys with the large truck and y journeys with the small truck, each time fully loaded.
   a) Write down an equation connecting x to y.
   b) State why x and y must be whole numbers.
   c) Write down all the possible pairs of values of x and y.

7. y is inversely proportional to x^2, and $y = 1.5$ when $x = 12$.
   Calculate the value of y when $x = 3$.

8. Work out the value of:
   a) $4^3 \times 5^2$ b) $2^6 + 2^3 + 2^0$
   c) $5^{-1} + 5^{-2}$ d) $(6 \times 10^8) \div (2 \times 10^6)$.

9. a) Complete the statement below by putting the symbol $<$ or $>$ between the numbers.
   $33.8 \times 10^4 \ \square \ 2.7 \times 10^6$
   b) Which of the above two numbers is written in standard form?
   c) Rewrite the other number in standard form.

10. Simplify:
   a) $33p^2 \div 11p^{-4}$
   b) $(5q^{12})^2$
   c) $2y^2 \times 3y^{-5}$.

11. A heart beats once every second. How many times will the heart beat in a lifetime of 70 years? Take a year to be 365.25 days and give your answer in standard form.

12. A square centimetre of high-resolution photographic film holds 1.5×10^8 bits of information. A large encyclopedia holds 4×10^{10} bits of information. What area of high-resolution film is needed to hold all the information from the encyclopedia? Give your answer correct to 3 significant figures.

13. a) Simplify:
    (i) $(16e^{10})^{-\frac{1}{2}}$
    (ii) $2p^{\frac{1}{2}} \times 3p^{-\frac{5}{2}}$.
   b) Find the value of x which satisfies:
    (i) $3^x = 81$
    (ii) $3^x = \frac{1}{9}$
    (iii) $3^x = 1$.

14. Rearrange the following expressions in numerical order, putting the smallest one first:
   $(-2)^4$, $2(-5)^2$, $(-0.9)^3$, $(-3)^2 - 4(+2)(-6)$, $(+9) \div (-\frac{1}{3})$

15. a) Find the value of $2t^2 - 3t - 1$ when $t = -4$.
   b) Find the value of $(t - 3)(2t + 1)$ when $t = -2$.

16. a) Expand $2y(y^2 - 3y - 5)$.
   b) Expand $(s - 4t) - (3s - 2t)$ and factorise your answer.

17. a) Factorise $15a - 9b + 12$.
   b) Factorise completely $9x^2y - 12xy^2 + 15xy$.

18. a) Find the product of $(x + 3)$ and $(3x - 2)$, expressing your answer in its simplest form.
   b) Factorise $2t^2 - 3t - 2$.

19. a) Express $\frac{2}{v} - \frac{1}{v+1}$ as a single fraction, in its simplest form.
   b) Given that $V = \frac{1}{3}\pi r^2 h + \frac{2}{3}\pi r^3$, find a formula for h in terms of π, r and V. Give your answer as a single fraction, in its simplest form.

20. Solve the simultaneous equations: $3x - y = 11$
   $4x + y = 10$.

21. The pressure P at a depth x metres below the surface of the ocean is given by the formula $P = ax + b$, where a and b are constants.
   When $x = 0$, $P = 15$.
   When $x = 60$, $P = 45$.

a) Find the values of a and b.

b) Hence calculate the pressure at a depth of 100 m.

22. Consider these statements about two-digit numbers:

| 3 | 7 | $= 10 \times 3 + 7$

| 5 | 3 | $= 10 \times 5 + 3$

a) Fill in the blanks in this statement:

| 2 | 9 | $= \ldots + \ldots$

b) If | p | q | represents a two-digit number (like 37) then

| p | q | $= 10p + q$.

   (i) Complete the statement | q | p | $= \ldots + \ldots$

   (ii) Show that | p | q | $-$ | q | p | $= 9p - 9q$.

c) A two-digit number | p | q | is such that

| p | q | $-$ | q | p | $= 18$.

   (i) Use the results in part b)(ii) to write down an equation in p and q.

   (ii) It is also given that $p + q = 14$. Solve these simultaneous equations to find p and q.

   (iii) Write down the two-digit number.

23. Solve the quadratic equation $(2y + 5)^2 = 49$.

24. The formula connecting the variables x and y is $y = \frac{a}{x} + bx$.

   It is known that $y = 2$ when $x = 1$ and that $y = -5$ when $x = 2$.

   a) Find the value of a and the value of b.

   b) Show that when $y = 16$, the formula $y = \frac{a}{x} + bx$ becomes $2x^2 + 8x - 3 = 0$.

   c) Solve the equation $2x^2 + 8x - 3 = 0$, giving your answer correct to 2 decimal places.

25. Solve the inequality $7 < 3 - 2x \le 13$ and represent your answer on a number line.

26. It is given that $-5 \le x \le -3$ and $-1 \le y \le 2$.

   Find the largest possible value of:

   a) $x + y$

   b) xy

   c) x^2y.

Graphs and functions

In the seventeenth century, mathematicians began to combine the mathematical fields of algebra and geometry. Credit for this development is given to Blaise Pascal and René Descartes. The branch of mathematics that deals with the interaction of algebra and geometry is called coordinate geometry or Cartesian geometry (after Descartes).

In Cartesian geometry, plotting points and drawing graphs on a special grid shows the relationship between variables. This grid is called the Cartesian plane.

The Cartesian plane

In developing Cartesian geometry, Descartes made use of the fact that position of a point on a flat surface (plane) can be specified using two numbers. These numbers are called the coordinates of the point.

To develop the Cartesian plane, Descartes began with two perpendicular number lines. These lines form the axes of the Cartesian plane, which is also called a grid.

The horizontal line is called the x-axis.

The vertical line is called the y-axis.

Note that the axes extend beyond 0 in both directions (negative and positive) to infinity.

The position of any point on the plane can be given as an ordered pair of numbers, written in brackets. For example, P is the point (2, 3). The first number is read off the x-axis and is called the x-coordinate of the point. The second value is read off the y-axis and is called the y-coordinate. The point is the intersection of these two coordinates.

The order of the points is important. The x-coordinate is always given first. Remember this as in the alphabet: x comes before y.

The point where the axes cross each other is called the origin. It is usually given the letter O. The coordinates of the origin are (0, 0).

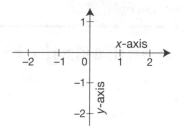

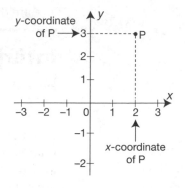

■ Examples

1.

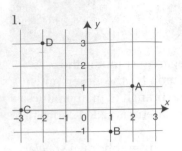

In the diagram:
point A has coordinates (2, 1)
point B has coordinates (1, −1)
point C has coordinates (−3, 0)
point D has coordinates (−2, 3)

2.

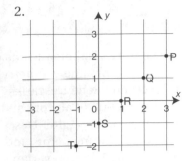

In the diagram:
point P has coordinates (3, 2)
point Q has coordinates (2, 1)
point R has coordinates (1, 0)
point S has coordinates (0, −1)
point T has coordinates (−1, −2)
Notice that P,Q,R,S,T are in a
straight line.

3.

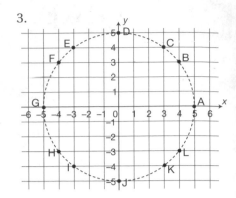

In the diagram, the following points are
plotted:

A (5, 0),   B (4, 3),   C (3, 4),
D (0, 5),   E (−3, 4),   F (−4, 3),
G (−5, 0),   H (−4, −3),   I (−3, −4),
J (0, −5),   K (3, −4),   L (4, −3).

Exercise

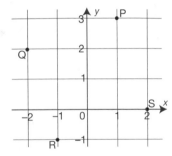

1. a) Write down the coordinates of the points P, Q, R and S shown in
     the diagram.
   b) If the lines PQ, QR, RS and SP are drawn, what shape is formed?
2. a) On a pair of axes, plot the following: point A (4, 3); point B (2, 2);
     point C (−4, −1).
   b) Points A, B and C are in a straight line. Find the coordinates of
     the points where this straight line crosses the x- and y-axes.
3. a) On a pair of axes, plot the following points: point H (3, 1);
     point J (−3, 0); point I (−1, 3); point K (3, −3).
   b) What do you notice about the lines IH and JK?
   c) If the lines HI, IJ, JK and KH are drawn, what shape is formed?
4. The points P, Q, R, S are the corner points of a rectangle. The
   coordinates of the points are: P (1, 2), Q (1, −1), R (−1, −1). What
   are the coordinates of point S?

Interpreting and using graphs

You have probably seen and used graphs from the media. Graphs are used
to present complicated information and figures in an accessible and easy-
to-understand format. In this section, you are going to learn to interpret
and draw graphs on a Cartesian plane. The graphs will illustrate a
mathematical relationship between two quantities – an x-quantity and
a y-quantity. Once the graph has been drawn, you can normally find a
great deal of information.

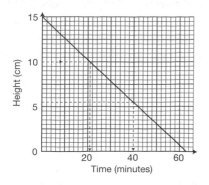

■ Examples

1. This graphs shows the change in height of a candle as it burns. Notice that the axes are labelled to show both what they represent and the units that are being used. Without these labels, the graph would be meaningless.

   In order to understand graphs, you have to work out what each small square, or interval, represents. On this graph, each small square on the horizontal axis is equal to $\frac{1}{10}$ of 20 minutes, or 2 minutes. On the vertical axis, each small square is $\frac{1}{10}$ of 5 cm, so each one represents 0.5 cm.

   If you look at the graph, you can see that:
   - the unlit candle was 15 cm high
   - it takes 21 minutes for the candle to burn down to a height of 10 cm
   - after 40 minutes, the candle is 5.5 cm high
   - it takes 63 minutes for the candle to burn out completely.

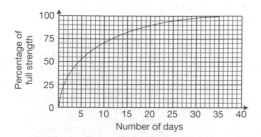

2. Some graphs are curved. This graph represents the increasing strength of concrete after it has been laid. Look at it carefully to check the answers to these questions:
   - What percentage of full strength does the concrete have after three weeks? (90%)
   - How many days does the concrete take to reach half strength (50%)? ($4\frac{1}{2}$ days)
   - How long does it take for the strength to increase from 70% to 80%? (4 days)

3. This graph shows the concentration of vaccine in the blood for 7 hours after a person has been vaccinated.

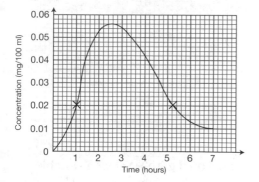

   a) After what time is the concentration of the vaccine 0.02 mg/100 ml?

      If you take a horizontal line across from this concentration, you can see that it crosses the graph in two places. This means that the concentration was 0.02mg/100 ml after 1 hour and again after 5 hours and 12 minutes (each small square $= \frac{1}{5} \times 60 = 12$).

   b) What is the concentration of vaccine in the blood after 5 hours?

      Take a vertical line from 5 hours to where it meets the graph. Read the corresponding value from the vertical scale. The concentration is 0.024 mg/100 ml (each small square is $0.02 \times \frac{1}{5} = 0.002$ mg/100 ml).

4. The graph below left shows the cost of making compact disks. Note that the graph starts at 50 on the vertical scale. This means there is a fixed cost plus an additional amount per disk.

   Disks are sold at $10. This information can be plotted in a table like this:

No. of disks sold (thousands)	0	5	10	15	20
Amount received ($)	0	50	100	150	200

This information can be added to the graph as shown below right. It can then be used to answer questions such as:

a) How many disks must be made and sold before there is a profit? This is the point at which the two graphs meet. In other words, 12 500 disks must be sold before costs are covered and a profit is made.

b) If 18 000 disks are made and sold, how much profit is made?
   Amount received = $180 000
   Cost of goods = $158 000
   Profit = amount received – cost of goods = $22 000

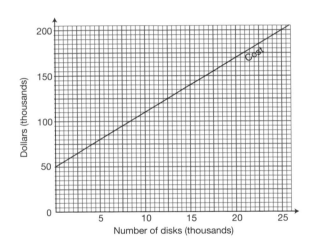

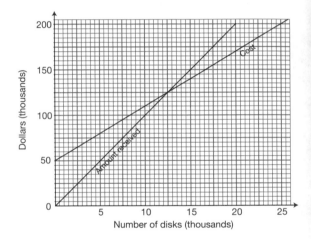

Exercise

1. Represent the information in the table on a graph. Select a suitable scale. Join the points you have plotted with a smooth curve. Use your graph to answer the questions that follow.

Year	1830	1890	1927	1960	1974	1987	2000
World population (thousands of millions)	1	1.5	2	3	4	5	6

a) In what year was the population of the world 2 500 million?
b) What was the population of the world in 1980?
c) By how much did the world population increase from 1830 to 1850?
d) Describe how the population of the world is changing.

2. Meena invested $3 000 in a special savings account. The interest rate was variable. The graph shows the amount in the savings account over the first 5 years.

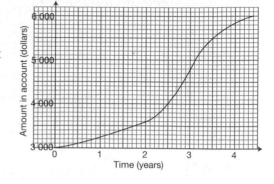

a) What does one small square on the horizontal axis represent?
b) What does one small square on the vertical axis represent?
c) What was the amount in the account after 3 years?
d) How long did it take for the amount in the account to reach $5 500?
e) How long did it take for the amount in the account to rise from $4 000 to $6 000?
f) After how long was the amount in the account rising most rapidly?

Conversion graphs

Conversion graphs are used to convert from one measurement to another, for example, from miles to kilometres or from dollars to pounds (or any other currency). If a conversion graph passes through the origin, the two variables are directly proportional.

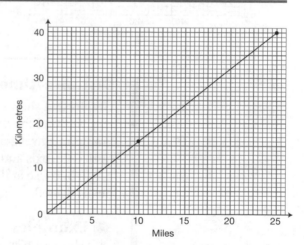

■ Example

5 miles is more or less the same as 8 kilometres. Draw a graph to convert miles to kilometres.

To obtain the graph, plot two points using the information given:
    0 miles = 0 kilometres
    5 miles = 8 kilometres.

The line has been extended so that we can read higher values from the graph. For example, using the graph, we can work out that:
    20 km = 12.5 miles   10 miles = 16 km
    27 km = 17 miles   12 miles = 19 km.

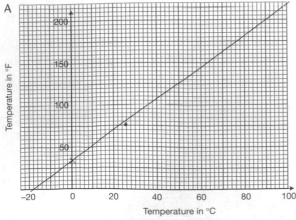

A

Temperature in °F

Temperature in °C

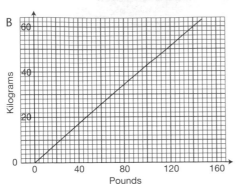

B

Kilograms

Pounds

Exercise

1. Graph A shows the relationship between temperature in degrees Celsius (°C) and degrees Fahrenheit (°F). Use the graph to convert:
   a) 60 °C to °F   b) 16 °C to °F
   c) 0 °F to °C   d) 100 °F to °C.

2. Graph B is a conversion graph for kilograms and pounds. Use the graph to answer the questions below:
   a) What does one small square on the horizontal axis represent?
   b) What does one small square on the vertical axis represent?
   c) Change 80 pounds to kilograms.
   d) The minimum weight to qualify as an amateur lightweight boxer is 57 kg. What is this in pounds?
   e) Which of the following conversions are incorrect? What should they be?
      (i)   30 kg = 66 pounds
      (ii) 18 pounds = 40 kg
      (iii) 60 pounds = 37 kg
      (iv) 20 pounds = 9 kg

Distance–time graphs

Graphs that show the connection between the distance an object has travelled and the time taken to travel that distance are called distance–time graphs or travel graphs. On such graphs, time is normally shown along the horizontal axis and distance on the vertical. The graphs normally start at the origin because at the beginning, no time has elapsed and no distance has been covered.

■ Examples

1. This graph shows the following journey:
   • a cycle for 4 minutes from home to a bus stop 1 km away
   • a 2-minute wait for the bus
   • a 7-km journey on the bus that takes 10 minutes.

Note that the graph line remains horizontal while the person is not moving (waiting for the bus) because no distance is being travelled at this time.

The steeper the line, the faster the person is travelling.

Distance travelled (km)

Time (minutes)

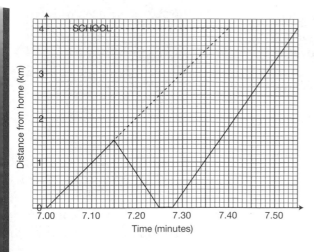

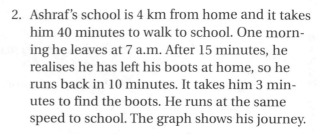

2. Ashraf's school is 4 km from home and it takes him 40 minutes to walk to school. One morning he leaves at 7 a.m. After 15 minutes, he realises he has left his boots at home, so he runs back in 10 minutes. It takes him 3 minutes to find the boots. He runs at the same speed to school. The graph shows his journey.

   Note that:
   - Ashraf walked 1.5 km before he remembered his boots
   - the graph goes downwards (back towards 0 km) as he goes home
   - the horizontal part of the graph corresponds with the 3 minutes at home
   - running to school at the same speed means 1.5 km in 10 minutes, so the graph is extended to 4 km.

Exercise

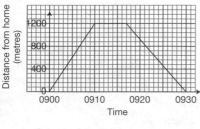

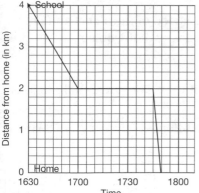

1. This distance–time graph represents Monica's journey from home to a supermarket and back again.
   a) How far was Monica from home at 0906 hours?
   b) How many minutes did she spend at the supermarket?
   c) At what times was Monica 800 m from home?
   d) On which part of the journey did Monica travel faster – going to the supermarket or returning home?

2. Omar left school at 1630. On his way home, he stopped at a friend's house before going home on his bicycle. The graph shows this information.
   a) How long did he stay at his friend's house?
   b) At what time did Omar arrive home?
   c) Omar's brother left school at 1645 and walked home using the same route as Omar. Work out at what time the brother passed Omar's friend's house.

3. A swimming pool is 25 m long. Jasmine swims from one end to the other in 20 seconds. She rests for 10 seconds and then swims back to the starting point. It takes her 30 seconds to swim the second length.
   a) Draw a distance–time graph for Jasmine's swim.
   b) How far was Jasmine from her starting point after 12 seconds?
   c) How far was Jasmine from her starting point after 54 seconds?

Speed in distance–time graphs

The graphs on pages 92 and 93 give some indication of speed by their steepness.

- A straight line graph indicates a constant speed.
- The steeper the graph, the greater the speed.
- An upward slope and a downward slope represent movement in opposite directions.

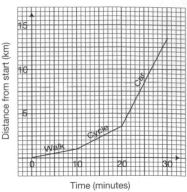

Time (minutes)

The distance–time graph above is for a person who walks, cycles and then drives for three equal periods of time. For each period, speed is given by the formula: $\text{speed} = \dfrac{\text{distance travelled}}{\text{time taken}}$.

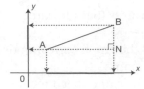

In this diagram, the steepness of the line AB is measured by $\dfrac{\text{increase in } y\text{-coordinate}}{\text{increase in } x\text{-coordinate}}$.

In mathematics, it is necessary to be more precise about what is meant by steepness of a line. In the diagram on the left, the steepness of line AB is measured by $\dfrac{\text{increase in } y\text{-coordinate}}{\text{increase in } x\text{-coordinate}}$.

This is the same as $\dfrac{\text{NB}}{\text{AN}}$ and is called the gradient (slope) of the line.

For a distance–time graph, a positive gradient indicates the object is moving in the direction of y increasing, a zero gradient indicates the object is not moving and a negative gradient indicates the object is moving in the direction of y decreasing. For a distance–time graph:

$$\frac{\text{change in } y\text{-coordinate (distance)}}{\text{change in } x\text{-coordinate (time)}} = \frac{\text{time travelled}}{\text{time taken}} = \text{speed.}$$

Thus, the gradient of the graph gives us the speed of the object and its direction of motion. This is known as the velocity of the object.

■ Examples

The travel graph below represents a car journey. The horizontal sections have zero gradient.

For section AB, the gradient is positive. Gradient $= \dfrac{\text{NB}}{\text{AN}} = \dfrac{40}{0.5} = 80$ km/h.

For section CD, the gradient is negative.

Gradient $= \dfrac{20-40}{0.25} = -80$

∴ the velocity was 80 km/h in the direction towards home.

For the section EF, the gradient is negative. Gradient $= \dfrac{0-20}{0.35} = -57.1$

∴ the velocity was 57.1 km/h in the direction towards home.

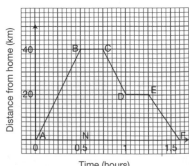

Time (hours)

Exercise

1. The travel graph shows Andile's daily run.
   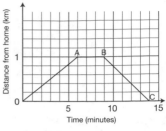
   a) For how many minutes does he run before taking a rest?
   b) Calculate the speed in km/h at which he runs before taking a rest.
   c) For how many minutes does he rest?
   d) Calculate the speed in m/s at which he runs back home.

Speed–time graphs

In certain cases, the speed (or velocity) of an object may change. An increase in speed is called acceleration, a decrease in speed is called deceleration. A speed–time graph shows speed (rather than distance) on the vertical axis.

■ Example

This graph shows a train travelling between two stations.
- The train starts at zero speed.
- The speed increases steadily, reaching 18 m/s after 15 seconds.
- The train travels at a constant speed (horizontal section) of 18 m/s for 25 seconds.
- It then slows down at a steady rate and then stops.
- The entire journey took 60 seconds.

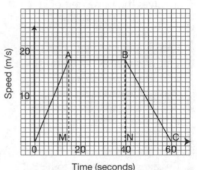

Consider the first part of the journey.
The speed increased by 18 m/s in 15 seconds. This is a rate of 1.2 m/s every second. This is the rate of acceleration and is written as 1.2 m/s^2 (or m/s/s). $\dfrac{18 \text{ m/s}}{15 \text{ seconds}}$ is the gradient of the line representing the first part of the journey. This gives us a particular result:

For a speed–time graph, gradient = acceleration.
Positive gradient (acceleration) is an increase in speed.
Negative gradient (deceleration) is a decrease in speed.

Distance travelled

Distance = speed × time.

Hint

You will find the formulae for calculating the area of shapes on page 148.

On a graph, these calculations correspond with the area of shapes under the sections of the graph. We can thus use the graph to work out the distance travelled.

For a speed–time graph, area under the graph = distance travelled.

■ Examples

1. This speed–time graph represents the motion of a particle over a period of 5 seconds.

   a) During which periods of time was the particle accelerating?

   b) Calculate the particle's acceleration 3 seconds after the start.

   c) Calculate the distance travelled by the particle in the 5 seconds.

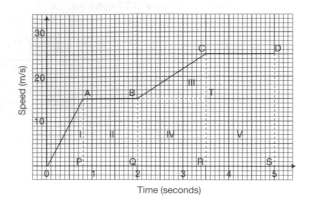

Time (seconds)

   a) The particle was accelerating in the period 0 to 0.8 seconds (section OA) and in the period 2 to 3.5 seconds (section BC).

   b) The acceleration was constant in the period 2 to 3.5 seconds, so the acceleration 3 seconds after the start

   $$= \frac{25 - 15 \text{ (m/s)}}{3.5 - 2 \text{ (seconds)}} = \frac{10 \text{ (m/s)}}{1.5 \text{ (s)}} = 6\frac{2}{3} \text{ m/s}^2.$$

   Acceleration $= \dfrac{\text{speed}}{\text{time}}$.

   c) Distance travelled = area under graph

   $= \text{area I} + \text{area II} + \text{area III} + \text{area IV} + \text{area V}$

   $= \frac{1}{2}(0.8 \times 15) + (1.2 \times 15) + \frac{1}{2}(1.5 \times 10) + (1.5 \times 15)$

   $\quad + (1.5 \times 25)$

   $= 6 + 18 + 7.5 + 22.5 + 37.5$

   $= 91.5 \text{ m}$

2. The diagram on the left is the distance–time graph for a short car journey. The greatest speed reached is 60 km/h. The acceleration in the first 2 minutes and the deceleration in the last 2 minutes are constant.

   a) Draw the speed–time graph of this journey.

   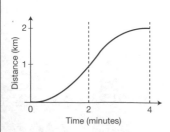

   Since the acceleration and deceleration are both constant, the speed–time graph consists of straight lines. The greatest speed is 60 km/h so the graph is drawn as in the second diagram on the left.

   b) Calculate the average speed, in km/h, for the journey.

   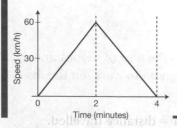

   $$\text{Average speed} = \frac{2 \text{ km}}{4 \text{ minutes}} = \frac{2 \times 15}{60 \text{ minutes}} = 30 \text{ km/h}.$$

Exercise

1. The distance–time graph below represents Ibrahim's journey from home to school one morning.
   a) How far was Ibrahim from home at 0830 hours?
   b) How fast, in m/s, was Ibrahim travelling during the first 10 minutes?
   c) Describe the stage of Ibrahim's journey represented by the line BC.
   d) How fast, in m/s, was Ibrahim travelling during the last 20 minutes?

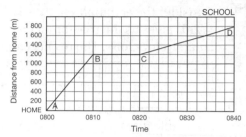

2. The graph below shows the speed, in m/s, of a car as it comes to rest from a speed of 10 m/s.
   a) Calculate the rate at which the car is slowing down during the first 3 seconds.
   b) Calculate the distance travelled during the 10-second period shown on the graph.
   c) Calculate the average speed of the car for this 10-second period.

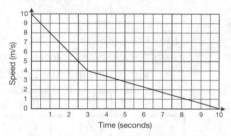

3. The diagram below is the speed–time graph for a car journey.
   a) Calculate the acceleration during the first 20 seconds of the journey.
   b) Calculate the distance travelled in the last 10 seconds of the journey.
   c) Calculate the average speed for the whole journey.

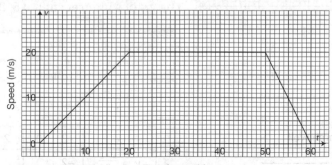

Drawing and using algebraic graphs

In algebra, the relationship between quantities is generalised and a rule is formed. In these rules, the letters x and y are used instead of time, distance, kilometres or any other quantities. Rules are written in the form of equations that give the coordinates on a Cartesian plane.

Straight line graphs

Consider the equation $y = 2x + 1$.

This gives us the rule for taking values of x and using them to calculate values for y. The results can be shown on a table of values:

x	−3	−2	−1	0	1	2	3	4
$y = 2x + 1$	−5	−3	−1	1	3	5	7	9

This table gives us a set of coordinates for points in the form (x, y). These ordered pairs can be plotted on a graph.

■ Examples

1. Make a table of values for the relation $y = \frac{1}{2}x - 1$ for values between −3 and 3. Draw the Cartesian graph for this relation. Substitute the values of x to find the y-values.

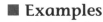

x	−3	−2	−1	0	1	2	3
$y = \frac{1}{2}x - 1$	$-2\frac{1}{2}$	−2	$-1\frac{1}{2}$	−1	$-\frac{1}{2}$	0	$\frac{1}{2}$

2. Draw the graphs of $y = 3x + 4$ and $y = 2 - x$ on the same diagram, for values of x from −2 to 2. Write down the coordinates of the point where the two graphs intersect.

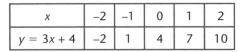

x	−2	−1	0	1	2
$y = 3x + 4$	−2	1	4	7	10

x	−2	−1	0	1	2
$y = 2 - x$	4	3	2	1	0

The graphs intersect at (−0.5, 2.5).

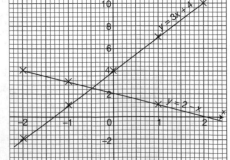

Exercise

1. Complete the following table of values for the relation $y = \dfrac{x-2}{2}$ and draw the graph of the relation.

x	−2	0	2	4	6
$y = \frac{x-2}{2}$	−2				2

2. Make a table of x-values from –3 to 3 for each of the following equations. Plot the coordinates on separate grids and draw the graphs.
   a) $y = 3x - 2$ b) $y = 5 - 3x$
   c) $y = x + 2$ d) $y = 6 - x$
   e) $y = -2x + \frac{1}{2}$ f) $y = x$

3. On a set of axes, draw the graphs $y = x - 1$ and $y = 4 - x$ for $-1 < x < 4$. Write down the coordinates of the point where the graphs intersect.

4. On the same set of axes, graph the relations:
   a) $y = 2x$ b) $y = 2x + 2$
   c) $y = 2x - 1$ d) $y = 2x + 4$.
   What do you notice about these graphs?

Recognising straight line graphs

Any relation in the form $y = mx + c$ (where m and c are numbers) will give a straight line graph.

If the relation is not given in this form, you need to change the subject of the relation to $y = \dots$

Remember that the value of c may be 0, as in the case of $y = 2x$.

Lines parallel to the axes

Graphs A and B show lines parallel to the axes.

In graph A, note that:
• every point on this line has a y-coordinate that is equal to c
• the equation of the line is $y = c$
• the x-axis itself has an equation $y = 0$ (every point on it has a y-coordinate of 0).

In graph B, note that:
• every point on the line has an x-coordinate that is equal to d
• the equation of the line is $x = d$
• the y-axis itself has an equation of $x = 0$ (every point on it has an x-coordinate of 0).

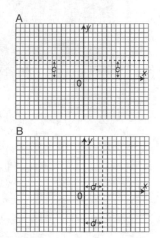

The gradient of a straight line

You have already learnt how to work out the gradient of a distance–time graph.

$$\text{Gradient} = \frac{\text{increase in } y\text{-coordinate from A to B}}{\text{increase in } x\text{-coordinate from A to B}}$$

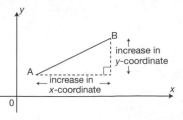

On a straight line graph, you can choose any two points as A and B to find the gradient. This is because a straight line rises or falls at a constant rate.

When the equation of a straight line is known, it is possible to find the gradient without drawing the graph. Consider an equation in the form of $y = mx + c$. You will see that an increase of 1 in the value of x will result in an increase of m in the value of y. This means that:

> The gradient of the line $y = mx + c$ is m.

For example, the gradient of the line $y = 2x - 1$ is 2; the gradient of the line $y = 10 - \frac{1}{2}x$ is $-\frac{1}{2}$.

When the equation is not in the form $y = mx + c$ (such as the one above), you need to change it to this form in order to work out the gradient.

Parallel lines have the same gradient.

The lines $y = 5x$ and $y = 5x + 4$ both have a gradient of 5 (m), so they are parallel.

■ Examples

1. Draw the line $3y - 2x = 12$ and find its gradient.

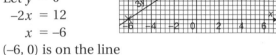

Let $x = 0$	Let $y = 0$
$3y = 12$	$-2x = 12$
$y = 4$	$x = -6$
(0, 4) is on the line	(–6, 0) is on the line

To find the gradient, take any two points on the line. For example,

A (0, 4) and B (3, 6). Gradient $= \dfrac{\text{increase in } y}{\text{increase in } x} = \dfrac{6 - 4}{3 - 0} = \dfrac{2}{3}$

2. Find the gradient of the line $5x + 4y - 8 = 0$.
   Do not draw the graph!

   $4y = -5x + 8$

   $y = \dfrac{-5x}{4} + 2$

   $\therefore$ the gradient is $-\dfrac{5}{4}$

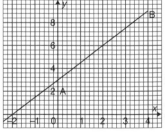

3. Find the gradient of the straight line shown in this diagram.
   Find any two points on the line, for example A and B.

   Gradient $= \dfrac{\text{increase in } y}{\text{increase in } x} = \dfrac{9 - 3}{4 - 0} = \dfrac{6}{4} = \dfrac{3}{2}$.

4. The coordinates of points P and Q are (–3, –1) and (2, 1) respectively. Find the gradient of line PQ.
   Draw a rough sketch as shown on the left.

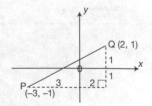

   Gradient $= \dfrac{\text{increase in } y \text{ from P to Q}}{\text{increase in } x \text{ from P to Q}} = \dfrac{1 - (-1)}{2 - (-3)} = \dfrac{1 + 1}{2 + 3} = \dfrac{2}{5}$.

Exercise

1. Draw the straight line $4x + 5y = 20$ and find its gradient.
2. Find the gradient of each of the straight lines:
   a) $3x - 4y = 24$ b) $4x + 5y + 6 = 0$
   c) $x = 2y + 3$ d) $y = \frac{3}{4}x$.
3. Show that the lines $2x = 4y + 3$ and $6y - 3x + 5 = 0$ are parallel.
4. Given the points P $(0, 4)$ and Q $(2, -1)$, calculate the gradient of line PQ.
5. Find the gradient of the graph on the right.

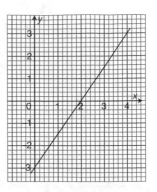

Interpreting the equation $y = mx + c$

The greater the value of m, the steeper the line.

If m is positive, the line slopes upwards from left to right.

If m is zero, the line is horizontal (it is parallel to the x-axis)

If m is negative, the line slopes downwards from left to right.

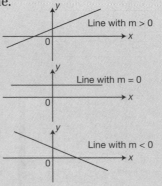

What does the value of c (the constant in the equation $y = mx + c$) tell us about the straight line?

Putting $x = 0$ in $y = mx + c$ gives $y = c$, so the point $(0, c)$ is on the line. This means that the line crosses the y-axis at a point which is c units above or below the origin (depending on whether c is a positive or negative number).

The letter c refers to the *intercept on the y-axis* of the line $y = mx + c$. This is sometimes shortened to *y-intercept*.

■ Examples

1. For the line $2y - 3x = 12$, find the gradient and where it cuts the y-axis.
   $$2y = 3x + 12$$
   $$y = \frac{3}{2}x + 6$$
   Gradient $= \frac{3}{2}$, cuts y-axis at 6.

2. Write down and simplify the equation of a line that has a gradient of $-\frac{1}{2}$ and a y-intercept of 3.
   $$y = mx + c \qquad m = -\frac{1}{2} \qquad c = 3 \qquad \therefore\ y = -\frac{1}{2}x + 3$$

3. Find the gradient and the y-intercept on the diagram and use them to find the equation of the line.

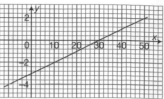

y-intercept $= -3$

Gradient $= \frac{3}{30} = \frac{1}{10}$, so $y = \frac{x}{10} - 3$

or $10y = x - 30$.

Exercise

1. For each of the following lines, find the gradient and the intercept on the y-axis:
   a) $y = 4 - 3x$ b) $2y - 4 = x$ c) $x + y = 3$.
2. Write down and simplify the equation of the line that has:
   a) a gradient of $\frac{3}{5}$ and a y-intercept of -2
   b) a gradient of $-\frac{1}{2}$ and a y-intercept of $\frac{3}{4}$.
3. Find the gradient and the intercept on the y-axis for the line drawn in diagrams A to C and hence find the equation of each line.
4. Find the equation of the line through the points P (5, 2) and Q (7, 8).

A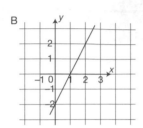

B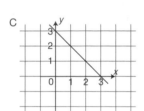

The length of a line segment

You can use Pythagoras' theorem to find the length of a straight line segment. This theorem states that in a right-angled triangle $c^2 = a^2 + b^2$. In other words, the square of the longest side equals the sum of the squares of the other two sides. The examples show you how this works in straight line graphs.

C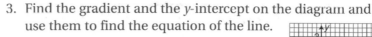

■ Examples

1. Given the points P(–2, 5) and Q(4, –3), calculate the length of line segment PQ.

   Show the points and their distances from the axes on a rough sketch.
   Apply Pythagoras' theorem:
   $$(PQ)^2 = (PN)^2 + (NQ)^2$$
   $$= 8^2 + 6^2$$
   $$= 64 + 36$$
   $$= 100$$
   Length PQ $= \sqrt{100} = 10$

2. Given the points S(–3, –2) and T(4, 1), calculate the length of ST.
   $$(ST)^2 = (SN)^2 + (NT)^2$$
   $$= (3 + 4)^2 + (2 + 1)^2$$
   $$= 49 + 9$$
   $$= 58$$
   Length ST $= \sqrt{58}$
   $$= 7.62 \text{ to 3 significant figures}$$

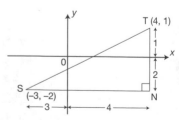

Exercise

1. In $\triangle DEF$, $\hat{D}$ is a right angle, side DE = 8 and side DF = 15. Calculate the length of side EF.
2. Given the points A (–4, 2) and B (5, 14), calculate the length of the line segment AB.
3. Given the points C (–4, –10) and D (6, 14), calculate the length of the line segment CD.
4. Given the points G (–2, 3) and H (4, 12), calculate the length of the line segment GH.

Curved graphs

Many of the relations you will work with for IGCSE are of the type known as functions. (Revise your work on functions from Module 2.)

Quadratic functions are those in which the index of x is no greater than 2. The graphs of such functions are not straight lines – they are all fairly similar curves. However, they can be drawn on a set of axes in much the same way as straight line graphs.

■ Examples

1. Draw the graph of the function $y = x^2$ for the values of x from –3 to 3. Start by drawing up a table of values:

A

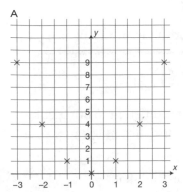

x	–3	–2	–1	0	1	2	3
$y = x^2$	9	4	1	0	1	4	9

Plot the points on a set of axes as shown in diagram A. Note that they do not lie on a straight line.

To obtain a better idea of the shape, you can find some fractional or decimal values of x.

x	–2.5	–1.5	–0.5	0.5	1.5	2.5
$y = x^2$	6.25	2.25	0.25	0.25	2.25	6.25

When these are plotted on the graph (see diagram B), the shape becomes clearer.

B

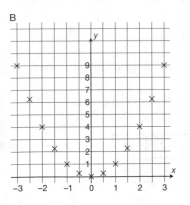

$y = x^2$ is the simplest of the quadratic functions. The graph of this function is a curve called a parabola.

2. Draw the graph of the function $y = x^2 + 2x - 1$ for values of x from –4 to 2.

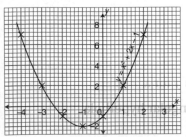

x	–4	–3	–2	–1	0	1	2
x^2	16	9	4	1	0	1	4
$+2x$	–8	–6	–4	–2	0	+2	+4
-1	–1	–1	–1	–1	–1	–1	–1
$y = x^2 + 2x - 1$	7	2	–1	–2	–1	2	7

Notice that the lowest point on the graph is where $x = -1$. If you work out the value of y at this point, you will find it is –2.

3. Draw the graph of the function $y = 6 + x - x^2$ for values of x from -3 to 4.

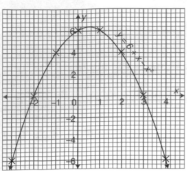

x	-3	-2	-1	0	1	2	3	4
6	6	6	6	6	6	6	6	6
$+x$	-3	-2	-1	0	$+1$	$+2$	$+3$	$+4$
$-x^2$	-9	-4	-1	0	-1	-4	9	-16
$y = 6 + x - x^2$	-6	0	4	6	6	4	0	-6

Notice that this graph is still a parabola but it is 'upside down' because the term containing x^2 is negative.

Exercise

1. Make a table of values and draw the graph of the function $y = x^2 + 2x$ for values of x from -4 to 2.
2. Make a table of values and draw the graph of the function $y = x^2 - 5x - 4$ for values of x from -2 to 7.
3. Draw the graph of the function $y = 4 - x^2$ for values of x from -3 to 3.

Graphs of reciprocal functions ($y = \frac{a}{x}$)

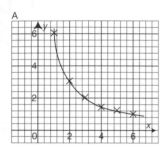

A

The graph of a reciprocal function such as $y = \dfrac{6}{x}$ is different to other curves because there is no value of y that corresponds with $x = 0$ (division by 0 is meaningless).

If you draw a table of values for this function, you find that:

x	1	2	3	4	5	6
$y = \frac{6}{x}$	6	3	2	1.5	1.2	1

When you plot the points (see diagram A), you can see that they do not lie on a straight line.

However, this is only part of the graph because no negative values of x have been included.

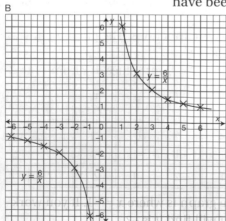

B

x	-6	-5	-4	-3	-2	-1
$y = \frac{6}{x}$	-1	-1.2	-1.5	-2	-3	-6

This table now gives a more complete picture of the graph. The curve is, in fact, two separate pieces (see diagram B). The graph gets closer and closer to the x-axis as the value of x increases, but it never meets the axis. Each piece of the curve also gets closer to the y-axis as x gets closer to 0 but it never meets because there is no value of y for $x = 0$.

This type of curve is called a rectangular hyperbola.

■ Example

Draw the graph of the function $y = \frac{-12}{x}$ $(x \neq 0)$ for $-12 \leqslant x \leqslant 12$.

Make a table of values for positive and negative numbers:

x	−12	−11	−10	−9	−8	−7	−6	−5	−4	−3	−2	−1
$y = -\frac{12}{x}$	1	1.1	1.2	1.3	1.5	1.7	2	2.4	3	4	6	12

x	1	2	3	4	5	6	7	8	9	10	11	12
$y = -\frac{12}{x}$	−12	−6	−4	−3	−2.4	−2	−1.7	−1.5	−1.3	−1.2	−1.1	−1

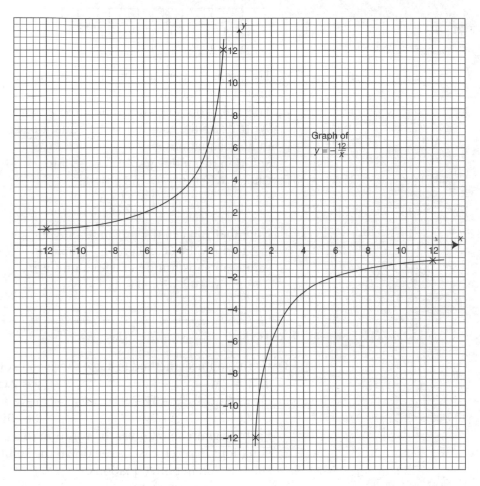

Graph of $y = -\frac{12}{x}$

Notice that the graph of $y = -\frac{12}{x}$ is the same shape as the graph of $y = \frac{6}{x}$ but the two pieces of $y = -\frac{12}{x}$ are in the top left and bottom right quarters of the graph paper, whereas the two pieces of $y = \frac{6}{x}$ are in the bottom left and top right quarters. This is because $y = -\frac{12}{x}$ has a negative value of a and $y = \frac{6}{x}$ has a positive value of a in the general function $y = \frac{a}{x}$.

Using graphs to solve equations

The point at which two straight line graphs intersect each other is common to both lines. This means that the x- and y-values of this point satisfy the equations of both graphs. This fact allows mathematicians to use graphs to find the solution to simultaneous linear equations.

■ **Examples**

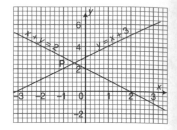

1. Use a graphical method to solve the simultaneous equations $y = x + 3$ and $x + y = 2$.

    $y = x + 3$

x	-4	0	4
y	-1	3	7

    $x + y = 2$

x	-4	0	4
y	6	2	-2

    The point P where the lines cross has the value $(-0.5, 2.5)$. This is the only point that falls on both lines $\therefore x = -0.5$ and $y = 2.5$.

2. This graph has been drawn to solve a pair of simultaneous equations. Write down the solution of the equations and check it by substitution in both equations.

    Solution (from the graph): $x = -3$ and $y = -1$.

    Check: When $x = -3$ and $y = -1$,

    $$x + 3y + 6 = -3 - 3 + 6 = 0$$
    $$x + 2 = -3 + 2 = -1 = y$$

    Note that you can use any method you like to draw the actual graphs – table of values, or substitution of 0 for x and y to find the x- and y-intercepts.

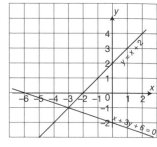

Exercise

1. Use a graphical method to solve these pairs of simultaneous equations:
    a) $y = 2x + 4$ $x + y = 1$ b) $y = 2x - 5$ $2y = x + 2$
    c) $x + 3y = 0$ $x - 3y = 6$ d) $x - y = 1$ $x + 2y = 7$
    e) $4x - 2y = 7$ $x + 3y = 7$ f) $2x - 5y = 8$ $3x - 7y = 11$.

2. Each diagram below gives the solution to a pair of simultaneous equations. In each case, write down the solution and check it by substitution in both equations.

    a)
    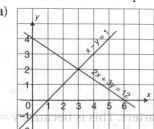

    b)
    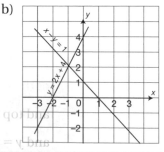

Solving quadratic and other equations graphically

Suppose you were asked to solve the equation $x^2 - 3x - 1 = 0$.

In other words, you need to find the value or values that make $x^2 - 3x - 1$ equal to 0.

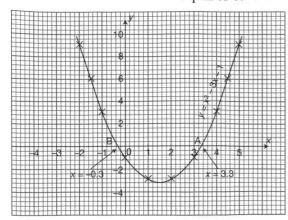

You could try to do this by trial and error, but you will soon find that the value of x that you need is not a whole number; in fact, it lies between the values of 3 and 4.

It is quicker and more efficient to draw the graph of this equation. To solve the equation, you need to find the points where the graph crosses the x-axis – in other words, the points (x, y) where $y = 0$.

If you look at the graph, you will see that the curve crosses the x-axis at two points. The x values of these points are 3.3 and –0.3 respectively.

■ Examples

1. Use the graph of $y = x^2 - 2x - 7$ to solve the equations:
   a) $x^2 - 2x - 7 = 0$
   b) $x^2 - 2x - 7 = 3$
   c) $x^2 - 2x = 1$.

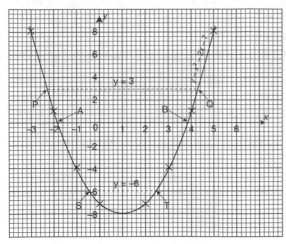

a) You have to find the points on the curve that have a y-coordinate of 0. There are two such points, marked A and B on the graph. The x-coordinates of these points are –1.8 and 3.8, so the solutions of the equation $x^2 - 2x - 7 = 0$ are $x = -1.8$ and $x = 3.8$.

b) You have to find the points on the curve that have a y-coordinate of 3. There are two such points, marked P and Q, on the graph. The x-coordinates of these points are –2.3 and 4.3, so the solutions of the equation $x^2 - 2x - 7 = 3$ are $x = -2.3$ and $x = 4.3$.

c) You must first rearrange the equation $x^2 - 2x = 1$ so that the left-hand side matches the function whose graph you are using. Taking 7 from both sides, you get $x^2 - 2x - 7 = 1 - 7$, that is $x^2 - 2x - 7 = -6$.

You can now proceed as you did in parts a) and b). You find the points on the curve that have a y-coordinate of –6. They are marked S and T on the graph. Their x-coordinates are –0.4 and 2.4. The solutions of the equation $x^2 - 2x = 1$ are $x = -0.4$ and $x = 2.4$.

2. a) Given $x > 0$, use the graph of $y = \dfrac{8}{x}$ to solve the equation $\dfrac{8}{x} = 5.7$.

   b) On the diagram, draw a graph of $y = x$ and hence find a value of x such that $\dfrac{8}{x} = x$.

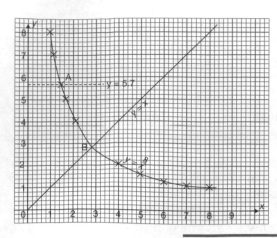

a) You have to find a point on the curve that has a y-coordinate of 5.7. The point is marked A on the diagram. Its x-coordinate is 1.4, so the solution of the equation $\frac{8}{x} = 5.7$ is $x = 1.4$.

b) The straight line $y = x$ crosses the curve $y = \frac{8}{x}$ at the point B, whose x-coordinate is 2.8. Hence, a value of x such that $\frac{8}{x} = x$ is 2.8.

Exercise

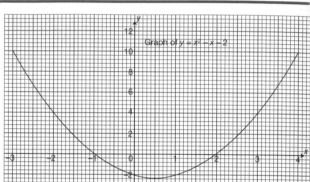

1. Use this graph of the function $y = x^2 - x - 2$ to solve the following equations:
   a) $x^2 - x - 2 = 0$
   b) $x^2 - x - 2 = 6$
   c) $x^2 - x = 6$.

2. A person makes a journey of 240 km. The average speed is x km/h and the time the journey takes is y hours.
   a) Complete this table of corresponding values for x and y:

x	20	40	60	80	100	120
y	12		4			2

   b) On a set of axes, draw a graph to represent the relationship between x and y.
   c) Write down the relation between x and y in its algebraic form.

Other algebraic graphs

It is possible to apply the rules you have already learnt for drawing graphs to functions in different forms. The table of values may take longer to work out because the expressions are more complicated and you may need to apply your previous knowledge of graphing and of the rules of algebra.

Cubic curves

When the highest power of x is x^3, the function is said to be a cubic function. All graphs of this function have more or less the same shape. Their shape is different to the parabolic and hyperbolic curves.

■ Examples

1. Draw the graph of the function $y = x^3 - 6x$ for $-3 \leq x \leq 3$. Draw tables of values for whole numbers and 'half values' of x:

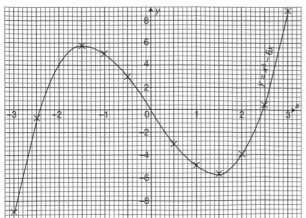

x	-3	-2	-1	0	1	2	3
x^3	-27	-8	-1	0	1	8	27
$-6x$	$+18$	$+12$	$+6$	0	-6	-12	-18
$y = x^3 - 6x$	-9	$+4$	$+5$	0	-5	-4	$+9$

x	-2.5	-1.5	-0.5	$+0.5$	$+1.5$	$+2.5$
x^3	-15.625	-3.375	-0.125	$+0.125$	$+3.375$	$+15.625$
$-6x$	$+15$	$+9$	$+3$	-3	-9	-15
$y = x^3 - 6x$	-0.625	$+5.625$	$+2.875$	-2.875	-5.625	$+0.625$

Plot the points on the axes and join them with a smooth curve.

2. Draw the graph of the function $y = x^3 - 2x^2 - 1$ for $-1 \leq x \leq 3$.
   Use the graph to solve the equations:
   a) $x^3 - 2x^2 - 1 = 0$ b) $x^3 - 2x^2 = -1$ c) $x^3 - 2x^2 - 5 = 0$.

   Draw a table of values for whole and half values:

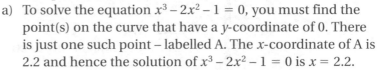

x	-1	-0.5	0	0.5	1	1.5	2	2.5	3
x^3	-1	-0.125	0	0.125	1	3.375	8	15.625	27
$-2x^2$	-2	-0.5	0	-0.5	-2	-4.5	-8	-12.5	-18
-1	-1	-1	-1	-1	-1	-1	-1	-1	-1
$y = x^3 - 2x^2 - 1$	-4	-1.625	-1	-1.375	-2	-2.125	-1	$+2.125$	$+8$

Plot the points on the axes.

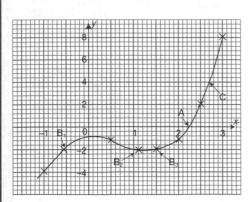

   a) To solve the equation $x^3 - 2x^2 - 1 = 0$, you must find the point(s) on the curve that have a y-coordinate of 0. There is just one such point – labelled A. The x-coordinate of A is 2.2 and hence the solution of $x^3 - 2x^2 - 1 = 0$ is $x = 2.2$.

   b) To solve the equation $x^3 - 2x^2 = -1$, you must first rearrange it so that the left-hand side is the function y. Subtracting 1 from both sides, the equation becomes $x^3 - 2x^2 - 1 = -2$. You have to find the point(s) on the curve that have a y-coordinate of -2. There are three such points – they are labelled B_1, B_2 and B_3. The x-coordinates of these points are the solutions of the equation. Hence, the solutions of $x^3 - 2x^2 = -1$ are $x = -0.6$, 1 and 1.6.

   c) As in part b), you need to rearrange the equation $x^3 - 2x^2 - 5 = 0$ if you want to use the graph of $y = x^3 - 2x^2 - 1$ to solve it. Adding 4 to both sides of the equation, you get $x^3 - 2x^2 - 1 = 4$ and so you must find the point(s) on the curve that have a y-coordinate of 4. There is only one point (C) where this is the case. At this point, the x-coordinate is 2.7. The approximate solution is therefore $x = 2.7$.

Exercise

1. a) Complete the table of values for the function $y = x^3 - 6x^2 + 8x$.

x	−1	−0.5	0	0.5	1	1.5	2	2.5	3	3.5	4	4.5	5
$y = x^3 - 6x^2 + 8x$	−15	−5.6		2.6		1.9		−1.9		−2.6		5.6	15

   b) On a set of axes, draw the graph of the function
      $y = x^3 - 6x^2 + 8x$ for $-1 \le x \le 5$.
   c) Use the graph to solve the equations:
      (i) $x^3 - 6x^2 + 8x = 0$
      (ii) $x^3 - 6x^2 + 8x = 3$.
2. a) Draw the graphs of $y = \frac{x^3}{10}$ and $y = 6x - x^2$.
   b) Hence solve the equation $\frac{x^3}{10} + x^2 - 6x = 0$.

Other curves

In the IGCSE examination, you will be expected to be able to draw and use graphs for functions of the type $y = ax^n$ (where $n = 2, -1, 0, 1, 2$ or 3) and sums of such functions. This means graphs of functions such as

$$y = 2x^2 - x - 1, \quad y = x^2 + \frac{36}{x} \quad \text{and} \quad y = \frac{x^3}{12} - \frac{6}{x}.$$

There will never be more than three terms in these functions.

   The last type of function you will have to deal with is $y = a^x$ (where a is a positive whole number). This is sometimes called the *growth function* or the *exponential function*. It is useful in describing, mathematically, any situation where the rate of increase of a quantity at any time is proportional to the size of the quantity at that time. This occurs, for example, in the growth of a population (human beings or animals).

■ Examples

1. Draw the graph of the function $y = x^2 + \frac{36}{x}$ for $0.5 \le x \le 8$.

   Use the graph to estimate the smallest possible value of $x^2 + \frac{36}{x}$ for values of x in the range $0.5 \le x \le 8$.
   A table of corresponding values of x and y is shown below.

x	0.5	1	2	3	4	5	6	7	8
$y = x^2 + \frac{36}{x}$	72.25	37	22	21	25	32.2	42	54.1	68.5

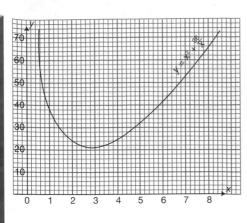

The shape of the graph is clear. There is no need to plot more than nine points.

The lowest point on the curve will give the smallest possible value of $x^2 + \dfrac{36}{x}$.

The lowest value of y is about 20.5 and the corresponding value of x is 2.6.

It follows that the smallest possible value of $x^2 + \dfrac{36}{x}$ for $0.5 \leq x \leq 8$ is 20.5.

It is worth checking this by substituting $x = 2.6$ in $x^2 + \dfrac{36}{x}$. You will find that this gives 20.6.

2. On the same axes, draw the graphs of $y = x + 10$ and $y = \dfrac{36}{x^2}$ for $-4 \leq x \leq 4$.

Use the graphs to find two solutions of the equation $\dfrac{36}{x^2} = x + 10$. Does this equation have more than two solutions?

The graph of $y = x + 10$ is a straight line. It passes through the points $(-4, 6)$, $(0, 10)$ and $(4, 14)$.

Remember that $\dfrac{36}{x^2}$ has no value when $x = 0$, so the graph of $y = \dfrac{36}{x^2}$ has a break in it at $x = 0$.

A table of values is shown below.

x	-4	-3	-2	-1	0	1	2	3	4
$y = \frac{36}{x^2}$	2.25	4	9	36	–	36	9	4	2.25

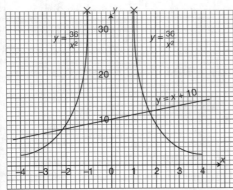

To find the solutions of the equation $\dfrac{36}{x^2} = x + 10$, read off the x-coordinates of the points where the graphs cross. The two solutions are $x = -2.2$ and $x = 1.7$.

Consider whether the graphs will cross again if they are extended outside the range $-4 \leq x \leq 4$.

It is clear that the value of $x + 10$ increases as x increases beyond $x = 4$, but the value of $\dfrac{36}{x^2}$ decreases.

Hence, the graphs will not cross when $x > 4$.

At $x = -4$, the graph of $y = x + 10$ is above the curve $y = \dfrac{36}{x^2}$. The graph of $y = x + 10$ crosses the x-axis where $x = -10$, but $\dfrac{36}{x^2}$ is always positive, so the curve always stays above the x-axis. It follows that the graphs must cross at some point for which $x < -4$.

Hence, the equation $\dfrac{36}{x^2} = x + 10$ has more than two solutions.

Hint

The equation has three solutions.

---

3. Draw the graph of the function $y = 2^x$ for $-2 \leq x \leq 4$. Use your graph to find the value of $2^{2.5}$ and check your result by using the fact that $2^{2.5} = 2^{\frac{5}{2}} = \sqrt{2^5}$.

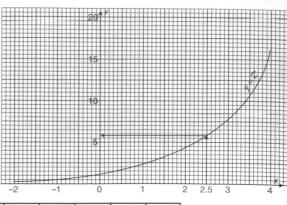

The table of values is as follows:

x	-2	-1	0	1	2	3	4
$y = 2^x$	0.25	0.5	1	2	4	8	16

From the graph shown above, you can see that when $x = 2.5$ the value of y is 5.75.

Thus the value of $2^{2.5}$ is approximately 5.75.

Check: $2^{2.5} = 2^{\frac{5}{2}} = \sqrt{2^5} = \sqrt{32} = 5.66$.

Exercise

1. a) Complete this table of values for the function $y = 3x - \dfrac{12}{x}$.

x	1	1.5	2	2.5	3	3.5	4	4.5	5
$y = 3x - \frac{12}{x}$		3.5	0		5	7.1		10.8	

b) Draw the graph of $y = 3x - \dfrac{12}{x}$ for $1 \leq x \leq 5$.

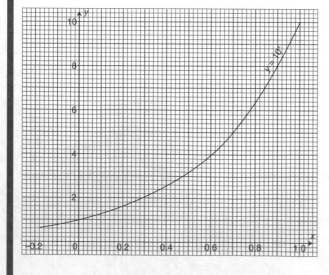

2. The graph of $y = 10^x$ for $-0.2 \leq x \leq 1.0$ is shown here.
   a) Use the graph to find the value of:
      (i) $10^{0.3}$
      (ii) $10^{-0.1}$.
   b) In the diagram, draw a straight line graph that will enable you to solve the equation $10^x = 8 - 5x$. Write down the solution of this equation.

3. a) Draw the graph of $y = x^2 - 2x - 4$ for values of x from -3 to 5.
   b) Use your graph to find approximate solutions to the equations:
      i) $x^2 - 2x - 4 = 0$
      ii) $x^2 - 2x - 4 = 3$
      iii) $x^2 - 2x - 4 = -1$.

The gradient of a curve

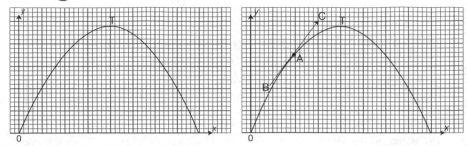

Look at the left-hand graph, which represents the path of a cricket ball. The x-axis represents the horizontal ground and the y-axis represents the vertical line through the point 0 from which the ball was thrown. The path of the ball is quite steep at first but, by the time the ball reaches its highest point, T, it is moving horizontally – the steepness has reduced to zero.

After the ball has passed the highest point, its path gets steeper and steeper in a downwards direction.

For a straight line graph, we defined the gradient to be the measure of its steepness. Since a straight line rises (or falls) at a constant rate, its gradient is constant. It is clear that the steepness of a curve is not constant and so we cannot talk about *the* gradient of a curve – the steepness (and therefore the gradient) changes from point to point.

Consider the situation when the cricket ball is at point A in the right-hand graph. The direction in which the ball is travelling is shown by the line BAC. (If gravity suddenly ceased to exist, this is the direction in which the ball would move). We shall define the gradient of the curve at point A to be the gradient of the straight line BAC.

The line BAC just *touches* the curve at A – it does not cross it – and we say that it is the *tangent* to the curve at A. We can now give the following definition:

> The gradient of a curve at a point is the gradient of the tangent to the curve at that point.

Drawing a tangent to a curve

When you find the gradient of a curve by drawing and measuring, the accuracy of your result depends on the accuracy of the curve you have drawn and the accuracy of the tangent you draw. It is particularly important that you position the tangent correctly.

Let A be the point at which you want to draw the tangent.

Turn the page around until A is the point on the curve nearest to you. Place your ruler below the curve and move it upwards until it touches the curve. If the point of contact is not A, rotate the ruler (so that it rolls along the curve) until the point of contact is A.

Use a pencil to draw the tangent.

There should be no gap between the tangent and the curve. The tangent must pass through A and not through any other point on the curve near A.

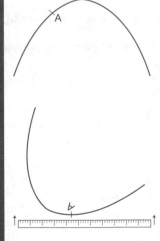

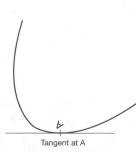

Tangent at A

Calculating the gradient of the tangent

Mark two points, P and Q, on the tangent. Try to make the horizontal distance between P and Q a whole number of units (measured on the *x*-axis scale).

Draw a horizontal line through P and a vertical line through Q to form a right-angled triangle PNQ.

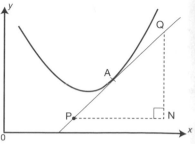

Gradient of the curve at A = Gradient of the tangent PAQ.

$$= \frac{\text{distance NQ (measured on the } y\text{-axis scale})}{\text{distance PN (measured on } x\text{-axis scale})}$$

■ Examples

1. The graph of the function $y = 5x - x^2$ is shown in the diagram below left. Find the gradient of the graph:

   a) at the point (1, 4)   b) at the point (3, 6).

   a) At the point A(1, 4), gradient $= \dfrac{\text{NQ}}{\text{PN}} = \dfrac{6}{2} = 3$.

   b) At the point B(3, 6), gradient $= \dfrac{\text{MR}}{\text{QM}} = -\dfrac{3}{3} = -1$.

2. The graph below right shows the height of a tree (*y* metres) plotted against the age of the tree (*x* years).

   Estimate the rate at which the tree was growing when it was 4 years old.

   The rate at which the tree was growing when it was 4 years old is equal to the gradient of the curve at the point where $x = 4$.

   The tangent at this point (A) has been drawn.

   Gradient of the curve at A $= \dfrac{\text{NQ}}{\text{PN}} = \dfrac{22.5}{8} = 2.8$.

   The tree was growing at a rate of 2.8 metres per year.

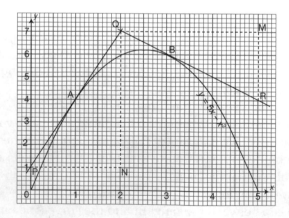

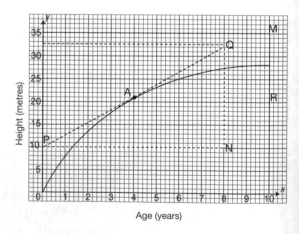

Age (years)

Exercise

1. The graph of the function $y = x^2$ is shown in the diagram.
    a) Find the gradient of the graph at the point:
        (i) (2, 4)
        (ii) (−1, 1).
    b) The gradient of the graph at the point (1.5, 2.25) is 3. Write down the coordinates of the point at which the gradient is −3.

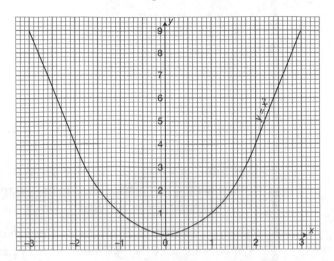

2. The graph shows how the population of a village has changed since 1930. Find the gradient of the graph at the point (1950, 170). What does this gradient represent?

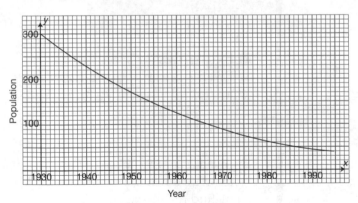

Functions

In Module 2 you worked with functions (see page 60). A function is the instruction for what to do with a number. If y is a function of x, the value of one variable (y) depends on another (x) and for each value of x there is only one possible value of y. If there is more than one possible value, then y is not a function of x.

Function notation

Consider the familiar $y = x + 2$.

This can also be written as $f(x) = x + 2$.

In function notation, the y is left out and the convention $f(x)$ is used to denote a function of x. This is read as 'f of x'.

If $f(x) = 6 - 3x$, then $f(5)$ refers to the value of the function when $x = 5$.

So, $f(5) = 6 - 15 = -9$

Similarly: $f(-2) = 6 - 3(-2) = 6 + 6 = 12$

$\qquad\qquad f(0.5) = 6 - 3(0.5) = 6 - 1.5 = 4.5$

Functions can also be written as $f{:}x \rightarrow 6 - 3x$ (you saw this in Module 2).

This is read as 'f is the function that maps x onto $6 - 3x$'.

The number $6 - 3x$ is called the image of x (under function f).

When there are two or more functions in the same problem, different symbols are used to represent them. For example:

$g(x) = x^2 - 2x - 3$ and $h(x) = 4x + 1$.

Bear in mind that the following are all the same function:

$x \rightarrow 2x + 3$

$t \rightarrow 2t + 3$

$y \rightarrow 2y + 3$.

■ Examples

1. Given $f(x) = x^2 - 3x$ and $g(x) = 4x - 6$, find the value of:

    a) $f(6)$ b) $f(-3)$ c) $g(\frac{1}{2})$ d) $g(6)$

           $= 36 - 18$ $= (-3) - 3(-3)$ $= 2 - 6$ $= 24 - 6$

           $= 18$ $= 9 + 9$ $= -4$ $= 18$

                                   $= 18$

2. Given $h{:}x \rightarrow 9 - x^2$,

    a) write down the expression for $h(x)$: $h(x) = 9 - x^2$

    b) find the image of:

        (i)   0   (ii) 3

             $h(0) = 9 - 0$ $h(3) = 9 - (3)^2$

                  $= 9$ $= 9 - 9$

                                            $= 0$

        (iii) 9   (iv) −9

             $h(9) = 9 - (9)^2$ $h(9) = 9 - (-9)^2$

                  $= 9 - 81$ $= 9 - (81)$

                  $= -72$ $= -72$

3. Given the functions $f(x) = x^2$ and $g(x) = x + 2$,
   a) solve the equation $f(x) = g(x)$.

$$f(x) = g(x)$$
$$\therefore x^2 = x + 2$$
$$\therefore x^2 - x - 2 = 0$$
$$\therefore (x - 2)(x + 1) = 0$$
$$\text{so } x = 2 \text{ and } x = -1$$

   b) solve the equation $4g(x) = g(x) - 3$.

$$4g(x) = g(x) - 3$$
$$\therefore 3g(x) = -3$$
$$\therefore g(x) = -1$$
$$\therefore x + 2 = -1$$
$$\text{so } x = -3$$

Exercise

1. $f(x) = 4x - 1$.
   Find a) $f(-1)$ b) $f(0)$ c) $f(0.5)$ d) $f(-4)$.
2. $f: x \rightarrow x^2 - 4$.
   Find a) $f(2)$ b) $f(0)$ c) $f(-3)$ d) $f(0.25)$.
3. Given the functions $f(x) = x^3 - 8$ and $g(x) = 3 - x$, find the values of:
      a) $f(2)$ b) $f(-1)$ c) $g(5)$ d) $g(-2)$.
4. Given the function $h: x \rightarrow 4x^2$, find the image of:
      a) 2   b) -2 c) $\frac{1}{2}$.
5. Given the functions $f(x) = x^2 - x$ and $g(x) = x^2 + 3x - 12$,
   a) solve the equation $f(x) = 6$
   b) solve the equation $f(x) = g(x)$.
6. $f: x \rightarrow 2x$.
   Find a) $f(1)$ b) $f(3)$ c) $f(a)$ d) $f(a + 2)$
        e) $f(4a)$ f) $4 f(a)$.
7. $f(x) = \frac{4 + x}{x}$ $(x \neq 0)$.
   a) Calculate $f(\frac{1}{2})$, simplifying your answer.
   b) Solve $f(x) = 3$.
8. $f(x) = (2x + 1)(x + 1)$.
   Find a) $f(2)$ b) $f(-2)$ c) $f(0)$.

Composite functions

You can think of a composite function as a function of a function. It is the result of applying one function to a number and then applying another function to the result.

Consider the two functions: $f(x) = 2x + 1$ and $g(x) = x^2$

$$f(4) = 9$$
$$g(9) = 81$$

This can be written as $g[f(4)] = 81$, but it is normally shortened to $gf(4)$.

Remember that $gf(x)$ stands for $g[f(x)]$.

Thus $gf(x)$ is a composite function in which f is applied first and g second.

■ Examples

Given the functions $f(x) = x^2 - 2x$ and $g(x) = 3 - x$, find the values of:

a) $gf(4) = g[f(4)] = g[16 - 8] = g[8] = 3 - 8 = -5$

b) $fg(4) = f[g(4)] = f[3 - 4] = f[-1] = (-1)^2 - 2(-1) = 1 + 2 = 3$

c) $ff(-1) = f[f(-1)] = f[1 + 2] = f[3] = 9 - 6 = 3$

d) $gg(100) = g[g(100)] = g[3 - 100] = g[-97] = 3 - (-97) = 3 + 97 = 100$

Exercise

Given the functions $g(x) = x^2 + 1$ and $h(x) = 2x + 3$, find the values of:

a) $gh(1)$ b) $hg(1)$ c) $gg(2)$ d) $hh(5)$.

Flow diagrams

The steps taken to work out the value of any function $f(x)$ can be shown on a flow diagram. For example, the function $f(x) = 2x + 5$ can be represented as:

$$x \rightarrow \boxed{\times 2} \xrightarrow{2x} \boxed{+5} \rightarrow 2x + 5$$

$g(x) = 2(x + 5)$ can be represented as:

$$x \rightarrow \boxed{+5} \xrightarrow{x+5} \boxed{\times 2} \rightarrow 2(x + 5)$$

Notice that these flow diagrams show the same operations but the order is different.

Inverse of a function

The inverse of a function is the function that will do the opposite of f, or, in other words, undo the effects of f. For example, if f maps 4 onto 13, then the inverse of f will map 13 onto 4.

In general, if f is applied to a number and the inverse of f is applied to the result, you will get back to the number you started with.

In simple cases, you can find the inverse of a function by inspection. For example, the inverse of $x \rightarrow x + 5$ must be $x \rightarrow x - 5$ because subtraction is the opposite of addition, and to undo +5 you have to subtract 5.

Similarly, the inverse of $x \rightarrow 2x$ is $x \rightarrow \dfrac{x}{2}$ because to undo $\times 2$ you have to divide by 2.

The inverse of the function f is denoted by f^{-1}.

Hence, if $f(x) = x + 5$, then $f^{-1}(x) = x - 5$.

And if $g(x) = 2x$, then $g^{-1}(x) = \dfrac{x}{2}$.

Not all functions have an inverse function.

Consider $x \to x^2$. This is a function because for every value of x there is only one value of x^2. However, the inverse is not a function because a positive number has two square roots (one negative and one positive).

Finding the inverse of a function

There are two methods of finding the inverse:

- **Method 1 – Using a flow diagram:**
  In this method, you draw a flow diagram for the function and then obtain the inverse by reversing the flow and 'undoing' the operations in the boxes.

■ Examples

1. Find the inverse of $f(x) = 3x - 4$.

   $f: \overset{\text{input}}{\longrightarrow} \boxed{\times 3} \to \boxed{-4} \to$ output

   $f^{-1}:$ output $\leftarrow \boxed{\div 3} \leftarrow \boxed{+4} \leftarrow$ input

   Let x be input:

   $\dfrac{x+4}{3} \leftarrow \boxed{\div 3} \overset{x+4}{\leftarrow} \boxed{+4} \leftarrow x$

   $\therefore f^{-1}(x) = \dfrac{x+4}{3}$

2. Given $g(x) = 5 - 2x$, find $g^{-1}(x)$.

   $g(x): \boxed{\times -2} \to \boxed{+5} \to$

   $g^{-1}(x):$ output $\leftarrow \boxed{\div -2} \leftarrow \boxed{-5} \leftarrow$ input

   Let x be input:

   $\dfrac{x-5}{-2} \leftarrow \boxed{\div -2} \overset{x-5}{\leftarrow} \boxed{-5} \leftarrow x$

   $\therefore g^{-1}(x) = \dfrac{x-5}{-2} = \dfrac{5-x}{2}$

- **Method 2 – Reversing the mapping:**
  In this method, you use the fact that if f maps x onto y, then f^{-1} maps y onto x. To find f^{-1}, you have to find a value of x that corresponds to a given value of y.

■ Examples

1. Find the inverse of the function $f(x) = 3x - 4$.

   Suppose f maps x onto y: Then $y = 3x - 4$.

   Make x the subject of the formula: $y + 4 = 3x$

   $$\text{Hence } \frac{y+4}{3} = x$$

   Now f^{-1} maps y onto x and so $f^{-1}: y \to \dfrac{y+4}{3}$

   This means that $f^{-1}(y) = \dfrac{y+4}{3}$.

   Thus we can say that, if $f(x) = 3x - 4$, then $f^{-1}(x) = \dfrac{x+4}{3}$.

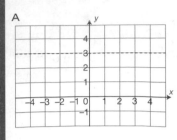

A

2. Given that $g(x) = 5 - 2x$, find $g^{-1}(x)$.

Let $y = 5 - 2x$, so we can say that g maps x onto y.
Make x the subject of the formula: $2x = 5 - y$

and so $x = \dfrac{5 - y}{2}$

g^{-1} maps y onto x and hence $g^{-1}(y) = \dfrac{5 - y}{2}$

It follows that $g^{-1}(x) = \dfrac{5 - x}{2}$.

Exercise

1. Find the inverse of the function $f(x) = 4x + 3$.
2. Given the function $g(x) = \dfrac{x}{3} - 4$, find $g^{-1}(x)$.
3. Given the function $h(x) = 2\,(x - 3)$, find the value of:
   a) $h^{-1}(10)$ b) $hh^{-1}(20)$ c) $h^{-1}h^{-1}(26)$.

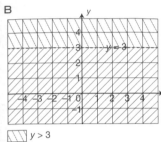

B

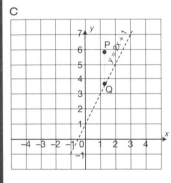

C

Inequalities and regions in a plane

In Module 2 you worked with inequalities and solved them using algebraic methods. In this section, you are going to represent these on the Cartesian plane.

In diagram A, the broken line is parallel to the x-axis. Every point on the line has a y-coordinate of 3. This means that the equation of the line is $y = 3$.

All of the points above the line $y = 3$ have y-coordinates that are > 3. The region above the line thus represents the inequality $y > 3$. Similarly, the region below the line represents the inequality $y < 3$. These regions are shown on diagram B.

In diagram C, the graph of $y = 2x + 1$ is shown as a broken line. Every point on the line has coordinates (x, y) which satisfy $y = 2x + 1$.

Point P has a y-coordinate that is greater than the y-coordinate of Q. P and Q have the same x-coordinate. This means that for any point P in the region above the line, $y > 2x + 1$.

The region above the line represents the inequality $y > 2x + 1$.

Similarly the region below the line represents the inequality $y < 2x + 1$. You can see this on diagram D.

If the equation of the line is in the form $y = mx + c$, then:
• the inequality $y > mx + c$ is above the line
• the inequality $y < mx + c$ is below the line.

If the equation is not in the form $y = mx + c$, you have to find a way to check which region represents which inequality.

D

$\boxed{||||}$ $y < 2x + 1$
$\boxed{\backslash\backslash}$ $y > 2x + 1$

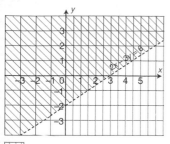

2x − 3y < 6

2x − 3y > 6

■ Examples

In a diagram, show the regions that represent the inequalities $2x - 3y < 6$ and $2x - 3y > 6$.

The boundary between the two required regions is the line $2x - 3y = 6$.

This line crosses the x-axis at (3, 0) and the y-axis at (0, −2). It is shown as a broken line in this diagram.

Consider any point in the region above the line. The easiest point to use is the origin (0, 0). When $x = 0$ and $y = 0$, $2x - 3y = 0$. Since 0 is less than 6, the region above the line represents the inequality $2x - 3y < 6$.

Rules about boundaries and shading of regions

Inequalities are not always < or >. They may also be ≤ or ≥. Graphical representations have to show the difference between these variations.

When the inequality includes equal to (≤ or ≥), the boundary line must be included in the graphical representation. It is therefore shown as a boundary line.

When the inequality does not include equal to (< or >), the boundary line is not included in the graphical representation, so it is shown as a broken line.

■ Examples

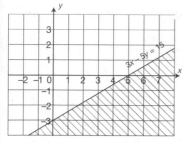

1. By shading the unwanted region, show the region that represents the inequality $3x - 5y \le 15$.

   The boundary line is $3x - 5y = 15$ and it is included in the region (because the inequality includes *equal to*).

   This line crosses the x-axis at (5, 0) and crosses the y-axis at (0, −3). It is shown as a solid line in this diagram.

   When $x = 0$ and $y = 0$, $3x - 5y = 0$. Since 0 is less than 15, the origin (0, 0) is in the required region. (Alternatively, rearrange $3x - 5y \le 15$ to get $y \ge \frac{3}{5}x - 3$ and deduce that the required region is above the line.)

   The unshaded region in this diagram represents the inequality $3x - 5y \le 15$.

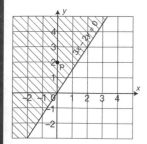

2. By shading the unwanted region, show the region that represents the inequality $3x - 2y \ge 0$.

   We cannot take the origin as the check-point because it lies on the boundary line. Take, instead, the point P (0, 2) which is *above* the line. When $x = 0$ and $y = 2$, $3x - 2y = -4$, which is less than 0. Hence P is *not* in the required region.

   The boundary line is $3x - 2y = 0$ and it is included in the region. It is shown as a solid line in this diagram.

3. Find the inequality that is represented by the unshaded region in this diagram. First find the equation of the boundary.

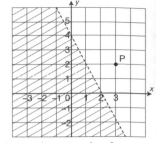

Its gradient $= -\frac{4}{2} = -2$ and its intercept on the y-axis $= 4$. Hence the boundary line is $y = -2x + 4$, that is $y + 2x = 4$. Take P $(3, 2)$ as the check-point: $2 + 6 = 8$. Note that 8 is greater than 4. Hence, the unshaded region represents $y + 2x > 4$. As the boundary is a broken line, it is not included and thus the sign is not $\geq$.

Exercise

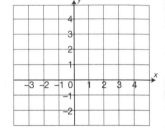

For questions 1 to 3, show your answers on a grid like the one on the right.

1. By shading the unwanted region, show the region that represents the inequality $2y - 3x \geq 6$.

2. By shading the unwanted region, show the region that represents the inequality $x + 2y < 4$.

3. By shading the unwanted region, show the region that represents the inequality $x - y \geq 0$.

4. For each of the following diagrams, find the inequality that is represented by the unshaded region.

a)

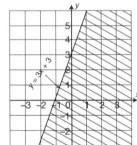

b)

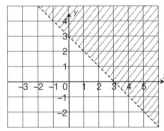

c)

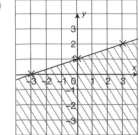

d)

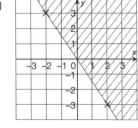

■ Example

By shading the unwanted regions, show the region defined by the set of inequalities $y < x + 2$, $y \leq 4$, $x \leq 3$.

The boundaries of the required region are $y = x + 2$ (broken line), $y = 4$ (solid line) and $x = 3$ (solid line).

The unshaded region in the diagram represents the set of inequalities $y < x + 2$, $y \leq 4$, $x \leq 3$.

Notice that this region does not have a finite area – it is not 'closed'.

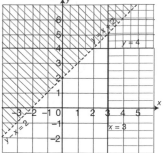

Exercise

1. By shading the unwanted regions, show the region defined by the set of inequalities $x + 2y \geq 6$, $y \leq x$, $x < 4$.

2. By shading the unwanted regions, show the region defined by the set of inequalities $x + y \geq 5$, $y \leq 2$, $y \geq 0$.

3. a) On a grid, draw the lines $x = 4$, $y = 3$ and $x + y = 5$.
   b) By shading the unwanted regions, show the region that satisfies all the inequalities $x \leq 4$, $y \leq 3$ and $x + y \geq 5$. Label the region R.

4. Write down the three inequalities that define the unshaded triangular region R in diagram A.

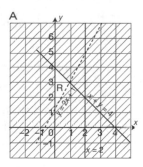
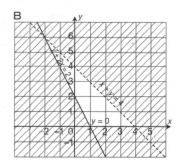

5. The unshaded region in diagram B represents the set of inequalities $y \geq 0$, $y + 2x \geq 2$, $x + y < 4$. Write down the pairs of integers (x, y) that satisfy all the inequalities.

6. Draw graphs to show the solution sets of these inequalities. Write down the coordinates (x, y) which satisfy all the inequalities in this case:
   $y \leq 4$, $y \geq x + 2$ and $3x + y \geq 4$.

Representing simultaneous inequalities

When two or more inequalities have to be satisfied at the same time, they are called simultaneous inequalities. These can also be represented graphically. On the diagram, the inequalities are represented by regions on the same diagram. The unwanted regions are shaded or crossed out. The unshaded region will contain all the coordinates (x, y) that satisfy all the inequalities simultaneously.

■ Examples

1. By shading the unwanted regions, show the region defined by the set of inequalities $3x + 2y \geq 6$, $2x - 3y > 6$, $x \leq 4$.

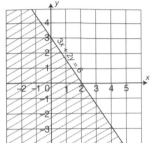 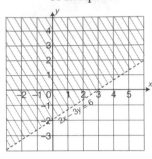 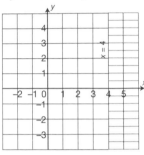 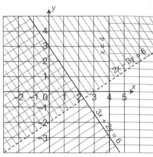

In the diagrams above, the unshaded regions represent the inequalities $3x + 2y \geq 6$, $2x - 3y > 6$ and $x \leq 4$ respectively.

Putting these regions on the same diagram, we obtain the diagram on the right. The unshaded triangular region in this diagram is defined by the set of inequalities $3x + 2y \geq 6$, $2x - 3y > 6$, $x \leq 4$.

2. Given that x and y are whole numbers, find the pairs of values (x, y) that satisfy all the inequalities $x + y \leq 4$, $y - 2x \leq 2$, $y > 0$.

The unshaded region in the left diagram below presents the set of inequalities $x + y \leq 4$, $y - 2x \leq 2$, $y > 0$.

We now have to use the fact that, in this question, x and y have to be whole numbers. This means that the only points in the unshaded region that we need are the grid points. These are marked by dots on the right.

Remember

Points on the broken line are not in the region.

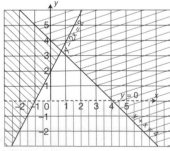 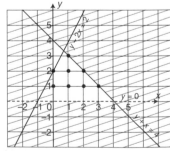

In the right-hand graph, we see that the pairs of whole numbers (x, y) that satisfy $x + y \leq 4$, $y - 2x \leq 2$ and $y > 0$ simultaneously are $(0, 1)$, $(0, 2)$, $(1, 1)$, $(1, 2)$, $(1, 3)$, $(2, 1)$, $(2, 2)$ and $(3, 1)$.

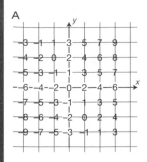

A

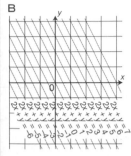

B

Linear programming

Many of the applications of mathematics in business and industry are concerned with obtaining the greatest profits or incurring the least cost, subject to restrictions such as the number of workers, machines available or capital available.

When these restrictions are expressed mathematically, they take the form of inequalities. When the inequalities are linear (such as $3x + 2y < 6$), the branch of mathematics you would use is called linear programming.

Greatest and least values

The expression $2x + y$ has a value for every point (x, y) in the Cartesian plane. Values of $2x + y$ at some grid points are shown in diagram A.

If points that give the same value of $2x + 7$ are joined, they result in a set of contour lines. These contour lines are straight lines – their equations are in the form $2x + y = k$ (k is the constant).

You can see that as k increases, the line $2x + y$ moves parallel to itself towards the top right-hand side of diagram B. As k decreases, the line moves parallel to itself towards the bottom left-hand side of the diagram.

The expression $2x + y$ has no greatest or least value if there are no restrictions on the values of x and y. When there are restrictions on the values, there is normally a greatest and/or a least value for the expression.

■ Examples

1. The numbers x and y satisfy all the inequalities $x + y \leq 4$, $y \leq 2x - 2$ and $y \geq x - 2$. Find the greatest and least possible values of the expression $2x + y$.

   You have to consider the values of $2x + y$ for points in the unshaded region. If $2x + y = k$, then $y = -2x + k$. Draw a line with gradient $= -2$. Use your pencil to do this – put your pencil on the line $2x + y = k$ and then move it parallel to the line.

   Draw a line with gradient $= -2$. As you move the line $2x + y = k$ parallel to itself towards the top right of the diagram, the value of k increases. The line is about to lose contact with the region when it passes through the point $(3, 1)$. $2x + y$ has its greatest value at this point. Substitute $x = 3$ and $y = 1$ in $2x + y$. Hence, the greatest possible value of $2x + y$ is 7.

   As you move the line $2x + y = k$ towards the bottom left of the diagram, the value of k decreases. The line is about to lose contact with the region when it passes through the point $(0, -2)$. $2x + y$ has its least value at this point. Substitute $x = 0$ and $y = -2$ in $2x + y$. Hence, the least possible value of $2x + y$ is -2.

2. The numbers x and y satisfy all the inequalities $x \geq 0$, $x + 2y \geq 7$, $2x + y \leq 8$ and $7x + 6y \leq 42$. Find the greatest and least possible values of the expression $3x + 2y$.

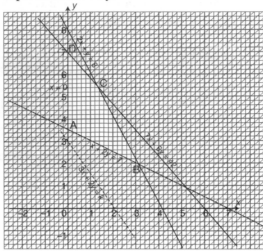

The unshaded region ABCD in this diagram consists of the point (x, y) which will satisfy all the given inequalities.

Consider the line $3x + 2y = k$. Find the gradient by making y the subject of the formula: $y = -\dfrac{3}{2}x + \dfrac{k}{2}$. The gradient is $-\dfrac{3}{2}$, so draw a line with this gradient anywhere on the Cartesian plane. If you move this line parallel to itself towards the top right of the diagram, it will first come into contact with the region at the point A(0, 3.5). If you continue to move the line parallel to itself, its last point of contact with the region is the point C(1.2, 5.6).

It follows that the least possible value of $3x + 2y = 0 + 7 = 7$ and the greatest possible value of $3x + 2y = 3.6 + 11.2 = 14.8$.

Exercise

1. In the diagram, the unshaded region represents the set of inequalities $x \leq 6$, $0 \leq y \leq 6$, $x + y \geq 4$.

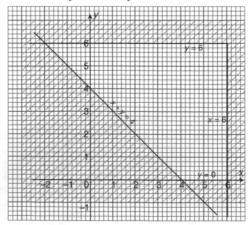

Find the greatest and least possible values of $3x + 2y$ subject to these inequalities.

2. a) On a grid, shade the unwanted regions to indicate the region satisfying all the inequalities $y \leq x$, $x + y \leq 6$, $y \geq 0$.
    b) What is the greatest possible value of $2x + y$ if x and y satisfy all these inequalities?
3. The whole numbers x and y satisfy all the inequalities $y \geq 1$, $y \leq x + 3$ and $3x + y \leq 6$. Find the greatest and least possible values of the expression $x + y$.

Practical applications

Computers are used to solve most linear programming problems in the real world. However, the example below shows you how inequalities and the greatest and least values can be used to solve simple problems. All problems have to be translated into mathematical language before they can be solved.

■ Example

A farmer keeps x cows and y sheep, where $x \geq 4$ and $y \geq 10$.
a) On graph paper, draw axes from 0 to 60, using a scale of 2 cm to represent 10 units on each axis.
    Draw and label the lines $x = 4$ and $y = 10$.
b) The total number of cows and sheep must not be more than 49. Write this as an inequality and draw the appropriate line on your graph.
c) Shade the *unwanted* regions of your graph.
d) The farmer makes $100 profit per cow and $50 per sheep. What is his maximum profit?

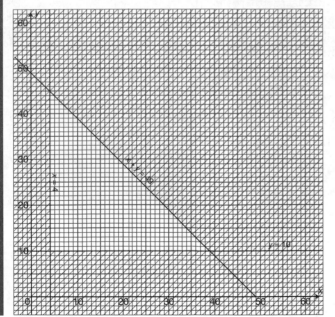

$x + y \leq 49$
The vertices of the unshaded region are (4, 10), (4, 45) and (39, 10).
The profit on x cows is $100x$ and the profit on y sheep is $50y$.

At (4, 10), the total profit
= $[(4 \times 100) + (10 \times 50)]$
= $900.

At (4, 45), the total profit
= $[(4 \times 100) + (45 \times 50)]$
= $2\,600.

At (39, 10), the total profit
= $[(39 \times 100) + (10 \times 50)]$
= $4\,500.

Hence, the maximum profit is $4\,500.

Check your progress

1. The graph shows the cost of repair work carried out by an electrician.
   a) What is the cost of work which takes $1\frac{3}{4}$ hours to complete?
   b) If the cost of work was $85, how long did the electrician take to do the work?
   c) The cost is made up of a fixed charge, p dollars, together with a cost of q dollars per hour. Find the value of p and the value of q.

2. The graph below shows the changes that take place when a substance is heated.
   a) How many degrees are represented by one small square on the temperature axis?
   b) Write down the temperature after 9 minutes.
   c) After how many minutes was the temperature 130 °C?
   d) Calculate the average rate of increase in temperature over the whole 16-minute period.
   e) In which state (solid, liquid or gas) is the temperature increasing most quickly? Explain how the graph helps to answer this.

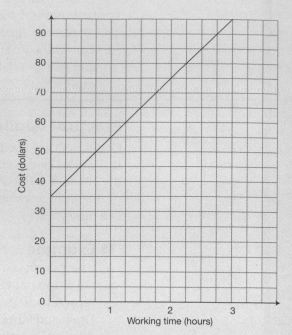

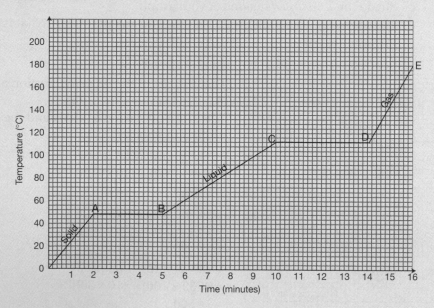

3. A lorry driver who is delivering goods to North Africa finds that fuel is sold in US gallons. He usually buys fuel in litres.
   a) Taking 1 US gallon to be 3.8 litres, draw a conversion graph for US gallons and litres.
   b) The driver knows the fuel tank will hold 90 litres. How many gallons is this?
   c) The lorry will travel 5 km on a litre of fuel. How far will it travel on 1 gallon of fuel?

4. The speed–time graph on the right represents the journey of a train between two stations. The train slowed down and stopped after 15 minutes because of engineering work on the railway line.

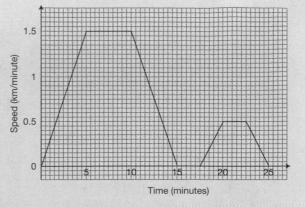

a) Calculate the greatest speed, in km/h, which the train reached.

b) Calculate the deceleration of the train as it approached the place where there was engineering work.

c) Calculate the distance the train travelled in the first 15 minutes.

d) For how long was the train stopped at the place where there was engineering work?

e) What was the speed of the train after 19 minutes?

f) Calculate the distance between the two stations.

5. a) (i) On a grid, plot the points $(-6, -2)$, $(-4, 0)$, $(-2, 2)$ and $(0, 4)$. Draw a straight line through the four points.

   (ii) Find the gradient of the straight line that you have drawn.

   b) (i) If $y = \frac{1}{3}x + 2$, fill in the blanks in the table of values below.

x	−6	−3		3
y	0		2	

   (ii) Draw the graph of $y = \frac{1}{3}x + 2$ on the same grid.

   c) Write down the coordinates of the point at which the two graphs meet.

6. The graph of $y = x^2$ is drawn on the grid below.

   a) The table shows some corresponding values of x and y for the function $y = x^2 + 3$. Work out the missing values of y, and put them in the table.

x	−2	−1.5	−1	−0.5	0	0.5	1	1.5	2
y		5.25	4	3.25	3		4	5.25	7

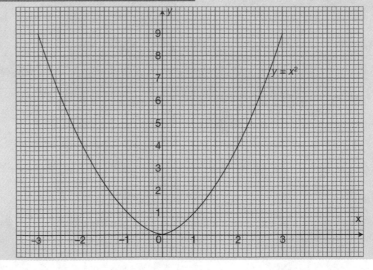

   b) Plot the points on a copy of the grid on the right and join them up with a smooth curve.

   c) Will the two curves ever meet? Explain your answer.

   d) By drawing a suitable straight line on the same grid, solve the equations:

   (i) $x^2 = 6$

   (ii) $x^2 + 3 = 6$.

7. The graph of $x + 2y = 6$ is drawn on the grid.

   a) Find the gradient of the straight line $x + 2y = 6$.

   b) (i) If $y = \frac{3}{2}x - 3$, complete the table below.

x	0	2	4
y			

   (ii) Draw the graph of $y = \frac{3}{2}x - 3$.

   c) Solve the simultaneous equations:

   $x + 2y = 6,$

   $y = \frac{3}{2}x - 3.$

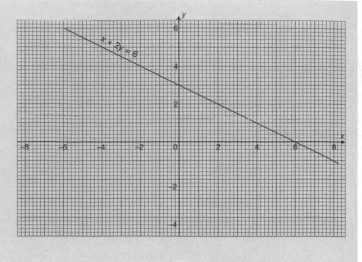

8. Answer the whole of this question on a sheet of graph paper.

x	0.6	1	1.5	2	2.5	3	3.5	4	4.5	5
y	p	−5.9	−3.7	−2.3	−1.1	0.3	1.9	3.8	q	r

Some of the values for the function $y = \dfrac{x^3}{12} - \dfrac{6}{x}$ are shown in the table above.

Values of y are given correct to one decimal place.

a) Find the values of p, q and r.

b) Using a scale of 2 cm to represent 1 unit on the x-axis, and 1 cm to represent 1 unit on the y-axis, draw the graph of $y = \dfrac{x^3}{12} - \dfrac{6}{x}$ for $0.6 \le x \le 5$.

c) Find, from your graph, correct to 1 decimal place, the value of x for which $\dfrac{x^3}{12} - \dfrac{6}{x} = 0$.

d) Draw the tangent to the curve at the point where $x = 1$, and hence estimate the gradient of the curve at that point.

(i)
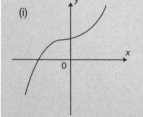

(ii)
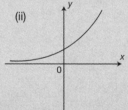

(iii)
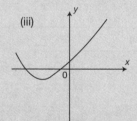

(iv)
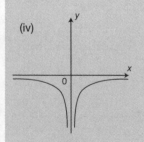

(v)
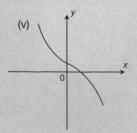

(vi)
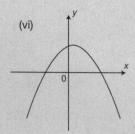

9. Look at the sketch graphs from (i) to (vi) on the left.

   a) Which one could be the graph of $y = 1 + x - 2x^2$?

   b) Which one could be the graph of $y = 3^x$?

   c) Which one could be the graph of $y = x^3 + x^2 + 1$?

   d) Which one could be the graph of $y = -\dfrac{16}{x^2}$?

10. a) In a chemical reaction, the mass, M grams, of a chemical is given by the formula $M = 160 \times 2^{-t}$, where t is the time, in minutes, after the start. A table of values for t and M is given below.

t	0	1	2	3	4	5	6	7
M	p	80	40	20	q	5	r	1.25

   (i) Find the values of p, q and r.
   (ii) Draw the graph of M against t for $0 \le t \le 7$. Use a scale of 2 cm to represent 1 minute on the horizontal t-axis and 1 cm to represent 10 grams on the vertical M-axis.
   (iii) Draw a suitable tangent to your graph and use it to estimate the rate of change of mass when $t = 2$.

   b) The other chemical in the same reaction has mass m grams, which is given by $m = 160 - M$. For what value of t do the two chemicals have equal mass?

11. f and g are the functions $f: x \to x - 5$ and $g: x \to 5 - x$. Which of the following are true and which are false?
    a) $f^{-1} = g$
    b) $g^{-1}: x \to 5 - x$
    c) $fg: x \to -x$
    d) $fg = gf$

12. $f(x) = 3x^2 - 2x - 4$ and $g(x) = 4 - 3x$.
    a) State the value of $f(-2)$.
    b) Solve the equation $f(x) = -3$.
    c) Solve the equation $f(x) = 0$, giving your answers correct to 2 decimal places.
    d) Solve the equation $g(x) = 2g(x) - 1$.
    e) Find $g^{-1}(x)$.

13. a) By shading the unwanted regions on a diagram, show the region that satisfies all the inequalities $y \ge \frac{1}{2}x + 1$, $5x + 6y \le 30$ and $y \le x$.

    b) Given that x and y satisfy these three inequalities, find the greatest possible value of $x + 2y$.

14. There are two popular electronic games, 'Cluedo' and 'Fantasy'.
    A retailer decides to order 60 Cluedo games and 40 Fantasy games.
    There are two suppliers, S and T.
    a) If the retailer orders x Cluedo games and y Fantasy games from S, how many of each type must he order from T? Write down the inequalities that express the fact that each of his orders must be a positive number or zero.
    b) The total number of games the retailer orders from S must not be more than 80. The total number he orders from T must not be more than 55. Write down two inequalities, in terms of x and y, that represent these restrictions.

c) On graph paper, show the region containing the points whose coordinates (x, y) satisfy all the inequalities you have obtained in parts a) and b). Shade the unwanted regions.

d) Supplier S charges $90 for each Cluedo game and $180 for each Fantasy game. Supplier T charge $120 for each Cluedo game and $160 for each Fantasy game. Show that the total cost of the retailer's order is $10 $(1\ 360 - 3x + 2y)$, and find the values for x and y which make this cost as small as possible.

15. Arnie and Bernie are tailors. They make x jackets and y suits each week. Arnie does all the cutting and Bernie does all the sewing.

To make a jacket takes 5 hours of cutting and 4 hours of sewing.
To make a suit takes 6 hours of cutting and 10 hours of sewing.
Neither tailor works for more than 60 hours a week.

a) For the sewing, show that $2x + 5y \le 30$.

b) Write down another inequality in x and y for the cutting.

c) They make at least 8 jackets a week. Write down another inequality.

d) (i) Draw axes from 0 to 16, using 1 cm to represent 1 unit on each axis.
   (ii) On your grid, show the information in parts a), b) and c). Shade the *unwanted* regions.

e) The profit on a jacket is $30 and on a suit is $100. Calculate the maximum profit that Arnie and Bernie can make in a week.

Geometry

Angles and shapes

Geometry allows us to work out the relationships between shapes, forms and spaces. The following definitions will help you work with geometry.

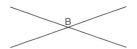

A point is where two lines meet. Here, the point of intersection is called B.

A line is the shortest distance between two points. Line AB connects the points A and B.

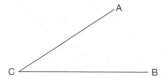

An angle is formed when two line segments meet at a point. The meeting point is called the vertex of the angle, and the line segments form the arms of the angle. Here C is the vertex, and line segments AC and BC are the arms of AĈB or BĈA.

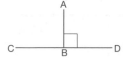

When two lines meet at a right angle (90°), they are perpendicular. AB is perpendicular to CD: AB ⊥ CD. A right angle equals 90°: ⌐ = 90°.

When two lines are the same perpendicular distance apart at all points, they are parallel. ST is parallel to QR: ST ∥ QR. We show parallel lines by drawing an arrow on each line, pointing in the same direction.

An acute angle is less than 90°: AĈB < 90°.

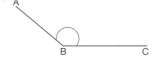

An obtuse angle is greater than 90° but less than 180°: 90° < AĈB < 180°.

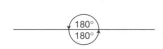

A straight line can also be seen as two angles at 180°.

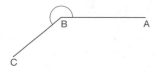

A reflex angle is greater than 180° but less than 360°: 180° < AĈB < 360°.

Complementary angles add up to 90°. You can form a complementary angle by intersecting a right angle: AB̂D + DB̂C = 90°.

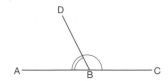

Supplementary angles add up to 180°. You can form a supplementary angle by intersecting a straight line: AB̂D + DB̂C = 180°.

The angles around a point always add up to 360°, which is a full turn or revolution.

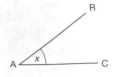

Naming and measuring angles

The angle in the diagram can be named as angle A, or Â or ∠A. It could also be called x, BÂC or CÂB. The vertex of the angle is always the middle letter in the angle's name.

Exercise

Using three different ways of naming angles, state which of the angles marked with a small letter in the following diagrams are acute, which are obtuse, which are right angles and which are reflex.

a)
b)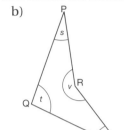
c)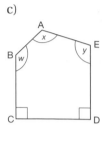

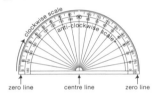
Using a protractor to measure an angle < 180°

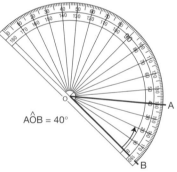

AÔB = 40°

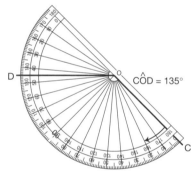

CÔD = 135°

- Estimate the size of the angle. Is it < 90° or > 90°?
- Place the protractor over the angle so that the centre point is on the vertex of the angle, and the zero line is on one arm of the angle. If necessary, extend the arms of the angle so that they reach the outer edge of the protractor.
- Start from the zero mark that is on top of the arm of the angle. Notice which scale you are using, and stick to that one.
- Go around the scale until you reach the other arm of the angle, and count the degrees. Does your reading sound right according to your estimate?

Using a protractor to measure an angle > 180°

Method 1: To measure the reflex angle x in this diagram, you could extend BA to point D.

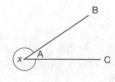

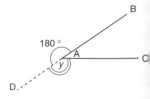

With your protractor, measure the obtuse DÂC.
The reflex angle $x = 180° + $ DÂC.

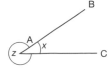

Method 2: Measure the acute angle and subtract it from 360°.

$$x + z = 360° \qquad x = 360° - z$$

Drawing angles

To draw an angle, you need a protractor, a ruler and a *sharp* pencil. Suppose you have a straight line segment AB and you want to draw an angle of 125° with its vertex at A and with AB as one of its arms. Follow these steps:

- Start with a straight line and mark A and B.
- Place the centre point of your protractor on top of the vertex (A) of the angle, and the zero line of the protractor on top of the arm AB.
- Find the 0 (zero) mark on the protractor that is on top of the arm AB and count around its scale: 0, 10, 20, …, 110, 120, then the extra degrees 1, 2, 3, 4, 5.
- Put a dot (C) next to the mark 125.
- Remove your protractor and use your ruler to draw a straight line from A through C.
- BÂC is the angle you wanted. Check that it looks about the right size.

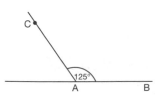

Exercise

1. Use your protractor to measure the reflex angles marked in these diagrams.

    a) 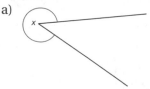   b)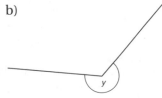

2. This is a sketch of triangle XYZ.
    a) Make an accurate drawing of the triangle.
    b) Measure the size of $\hat{Z}$ of the triangle.

    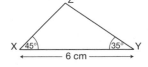

3. Here is another sketch.
    a) Make an accurate drawing of the diagram.
       (Remember that ∟ = 90°).
    b) Measure the size of BĈD.
    c) Measure the length of CD.

    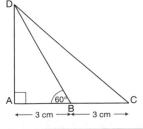

Different kinds of angles

Angles at a point

Remember

Angles at a point add up to 360°.

In this sketch, the four angles together make one complete
turn, so they must add up to 360°: $x + 90° + 110° + 115° = 360°$

$$x + 315° = 360°$$
$$\text{so } x = 45°$$

Angles on one side of a straight line

Remember

Angles on one side of a straight line add up to 180°.

In this sketch, the three adjacent angles together make a straight line,
so they must add up to 180°: $23° + x + 48° = 180°$

$$x + 71° = 180°$$
$$\text{so } x = 109°$$

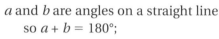

Vertically opposite angles

Remember

Vertically opposite angles are equal.

When two straight lines cross one another, they form four angles
   (a, b, c, d).
Angles a and c are the same size and angles b and
   d are the same size.
We can show this without doing any measuring,
   as follows:

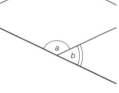

a and b are angles on a straight line
   so $a + b = 180°$;
b and c are angles on a straight line
   so $b + c = 180°$.
It follows that $a + b = b + c$ and hence $a = c$.

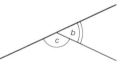

You should be able to show that $b = d$ in the
   same way.
d and b are vertically opposite angles. So are a and c.

Exercise

1. Find the size of x in each diagram.

a)

b)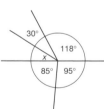

c)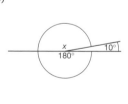

2. Measure the unnamed angles and find x in each diagram.

a)

b)

c)

Angles associated with parallel lines

Parallel lines are straight lines that are always the same distance apart. The lines are in the same direction. In diagrams, you mark parallel lines with matching arrow heads.

  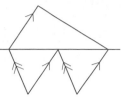

A *transversal* is a straight line that crosses parallel lines.

Corresponding angles (or F angles)

The transversal and the parallel lines form pairs of *corresponding angles*, as shown in the diagrams below.

 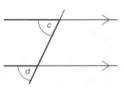 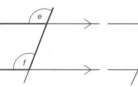

This means that, in the diagrams, $a = b$, $c = d$, $e = f$ and $g = h$.

Alternate angles (or Z angles)

The transversal and the parallel lines also form pairs of *alternate angles*. In these diagrams, w and x are alternate angles, and so are y and 115°.
So $w = x$ and
   $y = 115°$.

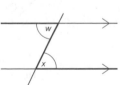 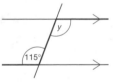

Co-interior angles (or C angles)

In the diagrams below, angles r and s are a pair of *co-interior angles* and so are angles t and u.

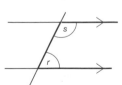 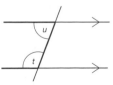

Using alternate angles and angles on a straight line, you should be able to prove that: $r + s = 180°$ and $u + t = 180°$.

Exercise

Calculate the size of the lettered angles in these sketches.

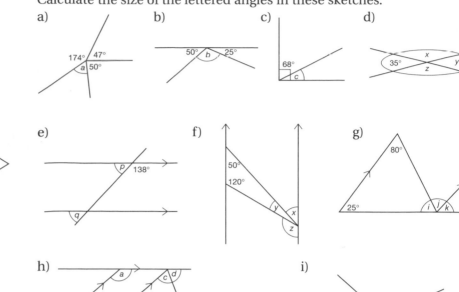

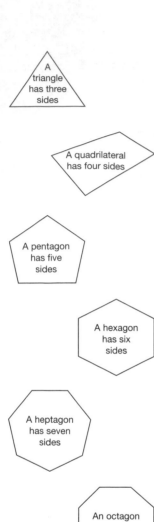

A triangle has three sides

A quadrilateral has four sides

A pentagon has five sides

A hexagon has six sides

A heptagon has seven sides

An octagon has eight sides

A nonagon has nine sides

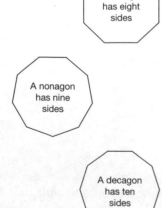

A decagon has ten sides

Properties of shapes (polygons)

A closed shape whose boundary consists of straight lines is called a polygon. The word 'polygon' comes from the Greek word meaning 'many angles'. The angles inside a polygon are called interior angles.

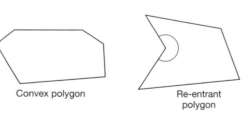

Convex polygon

Re-entrant polygon

If all the interior angles are less than 180°, the polygon is said to be *convex*. (All *regular* polygons are convex.) If one or more of the interior angles is reflex (greater than 180°), the polygon is said to be *re-entrant* or concave.

A polygon that has all its angles the same size *and* all its sides the same length is called a *regular polygon*.

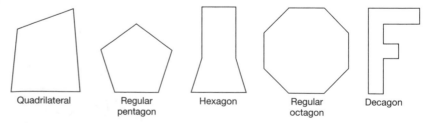

Quadrilateral   Regular pentagon   Hexagon   Regular octagon   Decagon

Look at the polygons above and on the left. You should be able to see easily which are convex, re-entrant or regular.

Triangles

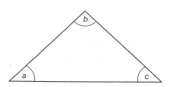  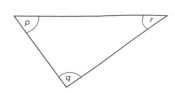

Use a protractor to measure the angles in each diagram. What do you find if you add up the three interior angles of each triangle?

The sum of the interior angles of a triangle is 180°.

The smallest angle is always opposite the smallest side, and the largest angle is always opposite the largest side.

AĈD = 180° – x (angles on a line = 180°).

BÂC + AB̂C = 180° – x (the sum of the interior angles of a triangle = 180°).

So AĈD = BÂC + AB̂C.

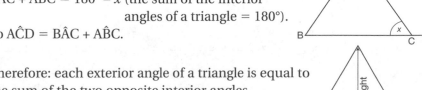

Therefore: each exterior angle of a triangle is equal to the sum of the two opposite interior angles.

The area of a triangle is A = $\frac{1}{2}$ × base × height.

Kinds of triangles

  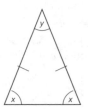

An equilateral triangle has all sides and all angles equal.

A scalene triangle has all sides and all angles different.

An isosceles triangle has two sides the same length and two angles the same size.

Exercise

Calculate the size of each lettered angle in these sketches.

a)

b)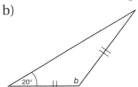

c)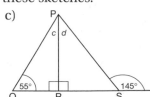

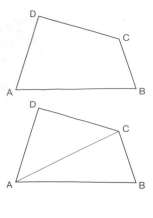

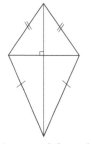

Quadrilaterals

A quadrilateral is a polygon with 4 sides, 4 vertices and 4 interior angles.

Here is quadrilateral ABCD. What is the sum of its 4 interior angles?
You can find the answer to this question by drawing the line from A to C.
(This is called a *diagonal* of the quadrilateral.)

Angle sum of $\triangle$ABC = 180°.
Angle sum of $\triangle$ADC = 180°.
Hence, angle sum of quadrilateral ABCD is 180° + 180° = 360°.

The angle sum of any quadrilateral is 360°.

Kinds of quadrilaterals

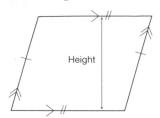

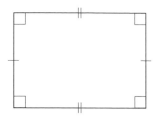

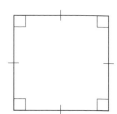

A kite is a quadrilateral with two pairs of equal sides adjacent to each other.
The area of a kite:
$A = \frac{1}{2} ab$

(where a and b are the lengths of the diagonals)

A parallelogram has two pairs of opposite sides which are parallel and equal.
The area of a parallelogram:
A = base × perpendicular height

A rectangle has four right angles.
The area of a rectangle:
A = base × height

A square is a rectangle with four equal sides.
The area of a square:
A = base × height, or
$A = s^2$

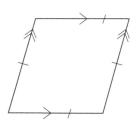

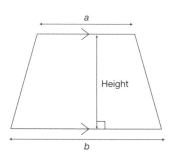

Note

The formulae for calculating area are given here. You will study perimeter and area more closely on pages 150–154.

A rhombus is a parallelogram with four equal sides.
The area of a rhombus:
A = base × perpendicular height

A trapezium is a quadrilateral with one pair of opposite sides parallel.
The area of a trapezium:
$A = \frac{1}{2} (a + b) h$

(where a and b are the lengths of the parallel sides and h is the height)

Interior angle sum of a polygon

All polygons can be split into triangles. The sum of the interior angles of a polygon can be found by calculating the sum of the angles of the triangles within the polygon.

A pentagon can be divided into 3 triangles.
Angle sum of pentagon = $3 \times 180° = 540°$.

A hexagon can be divided into 4 triangles.
Angle sum of a hexagon = $4 \times 180° = 720°$.

A heptagon can be divided into 5 triangles.
Angle sum of a heptagon = $5 \times 180° = 900°$.

The sum of the interior angles of a polygon with n sides is $(n - 2) \times 180°$.

Exterior angles

If you make a side of a convex polygon longer, the angle it makes with the next side is called an *exterior angle*.

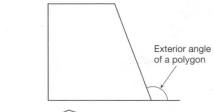

Exterior angle of a polygon

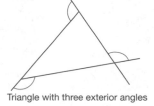

Triangle with three exterior angles

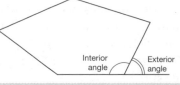

Interior angle   Exterior angle

An exterior angle and its adjacent interior angle always make a straight line.

An exterior angle + adjacent interior angle = 180°.

$a + b + c + d = 360°$

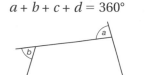

$p + q + r + s + t = 360°$

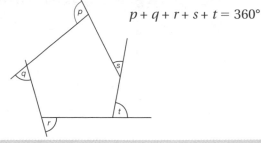

The exterior angles of a convex polygon add up to 360°.

Exercise

1. Calculate the size of each lettered angle in these sketches.

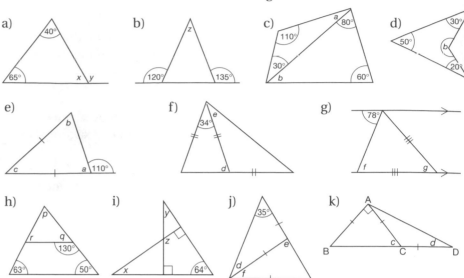

a)   b)   c)   d)

e)   f)   g)

h)   i)   j)   k)

2. Calculate the size of each lettered angle in these sketches.

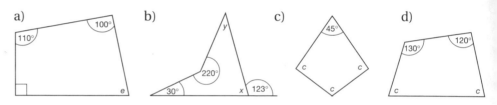

a)   b)   c)   d)

3. A 7-sided polygon has 2 equal angles of 120° and 5 equal angles of $a°$.
   Find the value of a.
4. Each of the interior angles of a regular polygon is 165°.
   Calculate the number of sides of the polygon.
5. Find the size of each interior angle of:
   a) a regular decagon   b) a regular 12-sided polygon.
6. Find the angle sum of:
   a) an 11-sided polygon   b) a polygon with 32 sides.
7. Work out the area of the following diagrams.

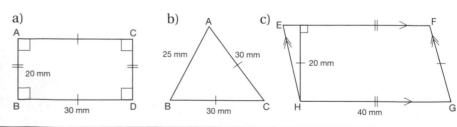

a)   b)   c)

8. Work out the area of each polygon and the size of each lettered angle.

a)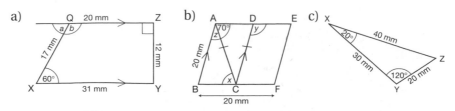

b)

c)

Circles

A circle is a shape enclosed by one continuous line called its *circumference*. All points on the circumference are at an equal distance from a point in the middle of the circle. This point is called the *centre* of the circle.

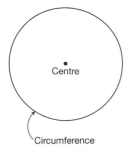

Here are some other terms and facts you need to know.

Terms and facts
- A line from the centre to the circumference is a *radius*.
- All radii (plural of radius) of a circle are equal in length.
- A line joining two points on the circumference is a *chord*.
- A chord which passes through the centre is a *diameter*.
- The length of a diameter is twice the length of a radius (D = 2r).
- Every diameter is a *line of symmetry* of the circle. (We shall cover lines of symmetry in more detail later.)
- A diameter divides the circle evenly into two equal parts. Each of these parts is called a semi-circle.
- A line that touches the circle (at one point only, and no matter how far it is extended) is a *tangent*.
- The point where the tangent touches the circle is called the *point of contact*.

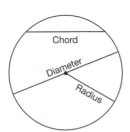

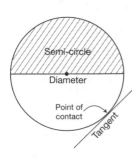

Exercise

1. State the correct names for the lettered parts of these circles.

a)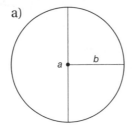

b)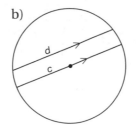

c)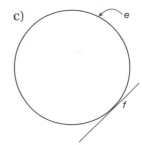

Arc, sector and segment

When describing angles and parts of circles, we use the term 'minor' to mean an angle $< 180°$. 'Major' means an angle $> 180°$.

Minor arc

Major arc

An arc is a part of the circumference (outline) of the circle.

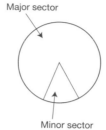

Major sector

Minor sector

A sector is the part of the circle formed by two radii and the arc between them.

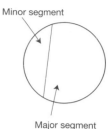

Minor segment

Major segment

A segment is formed when a chord divides the circle into parts.

Angles subtended by an arc

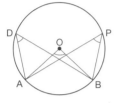

'The angle *subtended by an arc AB at the centre*' means AÔB. All three angles on the left are subtended by AB. AD̂B and AP̂B are subtended by AB at the circumference. We can also call them angles in the same segment.

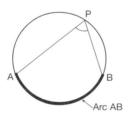

Arc AB

If AB is a minor arc, AP̂B is acute $(0 < AP̂B < 90°)$.

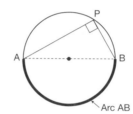

Arc AB

If AB is a semi-circular arc, AP̂B $= 90°$.

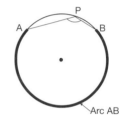

Arc AB

If AB is a major arc, AP̂B is obtuse.

Exercise

1. In this diagram, O is the centre of the circle. What can you say about the size of $\angle a$? Why?

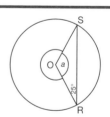

2. In the diagram, O is the centre of the circle, AC and BD are diameters. What is the size of AD̂C and BĈD? Why?

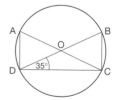

Constructions

Using geometrical instruments

You need the following equipment to take accurate measurements and construct geometrical diagrams:

- a ruler (to measure lines in cm and mm)
- a protractor (to measure angles)
- a compass with a sharp HB pencil (to draw circles)
- a set square (to draw parallel and perpendicular lines)
- another sharp HB pencil
- an eraser.

Drawing a circle

Using a compass

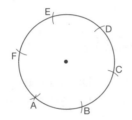

To draw a circle, you need to know the radius of the circle and the position of its centre. Make the distance between the sharp point of the compass and the pencil point equal to the length of the radius. Place the sharp point of the compass where the centre of the circle is to be. Now draw the circle, keeping the point of the compass steady.

Exercise

1. Draw a circle with a radius of 3 cm and mark a point A on its circumference. Using point A as centre, draw an arc with radius 3 cm to cut the circle at B. Using point B as centre, draw an arc with radius 3 cm to cut the circle at C. Repeat this process to obtain the points D, E and F.

When you use point F as centre and draw an arc with radius 3 cm, you will find that the arc passes through the point A where you started. Label the centre of the original circle O. Join A to B, B to O and O to A.
   a) What special type of triangle is △OAB?
   b) What is the size of AÔB?
   c) Explain why the arc with centre F and radius 3 cm passes through the point A.

2. Draw a circle with a radius of 3 cm and obtain the points A, B, C, D, E, F as in question 1. Join A to B, C to D, D to E, E to F and F to A.
   a) What can you say about the length of the sides of the figure ABCDEF?
   b) What is the size of AB̂C?
   c) Explain why it is this size.
   d) What is the size of BĈD, CD̂E, DÊF, EF̂A and FÂB?
   e) What kind of figure is ABCDEF?

Note

Sometimes you only need to draw part of a circle. This is called an arc. If the arc is more than half the circumference, it is a major arc. If it is less than half the circumference, it is a minor arc. If the arc is exactly half the circumference, it is a semi-circle.

Drawing an angle

Sometimes you may need to draw an angle without using a protractor.

■ Example

Draw an angle of 60° without using a protractor.

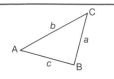

Draw a straight line segment with an end point.
With centre O, draw an arc of a circle to cut the line segment (choose any radius you like). In the diagram, the arc is PQ and it cuts the line segment at P.
Now draw an arc with centre P and the same radius as before.
Join O to R, the point where the two arcs cross.
Measure PÔR.
You should find that it is 60°.

Drawing triangles

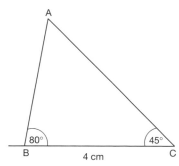

In triangle ABC, we refer to the side opposite Â by the letter a, the side opposite B̂ by the letter b and so on. Always leave your construction lines on your diagram.

In the last section, you used a compass to construct an angle of 60°. You also need to know how to draw triangles and various angles using a compass, ruler and protractor. The method you use to draw a triangle depends on how much information you have been given.

Given two angles and a side

■ Examples

1. Construct △ABC where BÂC = 55°, BĈA = 45° and a = 4 cm.

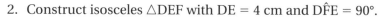

Draw a rough sketch. Draw BC = 4 cm. Use your protractor to draw BĈA at 45°.

Draw B̂ at 80° (∠s of a triangle equal 180°).

Label the point of intersection A.

2. Construct isosceles △DEF with DE = 4 cm and DF̂E = 90°.

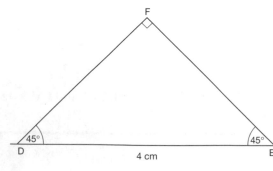

Draw a rough sketch.
Draw DE = 4 cm.
Calculate D̂ = Ê = 45° (∠ sum of △ = 180°, D̂ = Ê).

Measure D̂ and Ê and draw lines. Where they intersect is point F.

Given two sides and the included angle

When you are given more than one side, always draw the longest side first.

■ Example

Construct △XYZ with XY = 5 cm. YZ = 8 cm and XŶZ = 25°.

Draw a rough sketch. Draw the line YZ = 8 cm.

Draw an angle of 25° at Y. Construct the point X by using your compasses, with Y as the centre point and 5 cm as the radius.

The point of intersection between the line drawn from Y and the arc is the point X. Join X and Z.

Given three sides

Again, remember to draw the longest side first.

■ Example

Construct △PQR where $p = 5.5$ cm, $q = 4.2$ cm and $r = 3$ cm.

Draw a rough sketch. Draw the longest side. $p = 5.5$ cm.

Use Q as your centre and 3 cm as your radius to draw an arc 3 cm from Q.

Use R as your centre, and a radius of 4.2 cm to draw an arc 4.2 cm from R. The point of intersection of the arcs is P. Join QP and RP.

Note

When you study trigonometry in Module 5, you will learn different methods of drawing triangles.

Exercise

1. Draw △XYZ in which XY = 4 cm, XZ = 6 cm and X̂ = 70°.
2. Draw △PQR in which QR = 5 cm P̂ = 80° and Q̂ = 60°.
3. Draw △DEF in which DE = 5 cm, EF = 6 cm and FD = 7 cm.
4. Draw △JKL in which JK = 5 cm, KL = 6 cm and Ĵ = 90°.
5. Draw △ABC in which AB = 7 cm, Â = 60° and B̂ = 50°.
   Measure the lengths of the sides BC and CA.
6. Draw △PQR in which QR = 6 cm, P̂ = 70° and Q̂ = 30°.
   Measure the lengths of the sides PQ and PR.
7. Draw △XYZ in which XY = 5 cm, XZ = 6 cm and X̂ = 60°.
   Measure Ŷ, Ẑ and the side YZ.
8. Draw △DEF in which DE = 3 cm, EF = 4 cm and FD = 5 cm.
   Measure D̂, Ê and F̂.

Drawing parallels and perpendiculars

Drawing parallel lines using a set square

■ Example

Suppose you have a diagram with line AB and point P. You want to draw a line through P which is parallel to AB.

Place your set square so that one of its edges (not the longest edge) is along AB. Then place your ruler along the edge at the right of AB. Hold the ruler firmly. Slide the set square along the ruler until the edge touches P. On the paper, draw the line along the edge through P. This line is parallel to AB.

Drawing perpendicular lines using a set square

■ Example

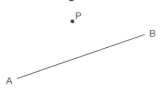

Suppose you want to draw a line through the point P, so that the new line is perpendicular to AB.

Place your ruler along AB. Place your set square so that one of its shorter edges is also along AB, against the ruler. Slide the set square along the ruler until the edge of the set square touches P. Draw the line along the edge, passing P and AB. This line is perpendicular to AB.

You can use the same method for drawing a perpendicular line from P when P is on the line AB.

Drawing the perpendicular bisector of a line using a compass

A line that cuts a figure or line segment into two equal parts is called a bisector. When a bisector forms an angle of 90° with the line segment through which it passes, it is called a perpendicular bisector.

■ Example

Construct the perpendicular bisector of AB.

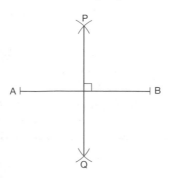

Set your compass to any radius greater than half the length of AB. Using A as the centre, draw arcs above and below the line.

Using the same radius, use B as the centre and draw arcs above and below. Join the points of intersection of the arcs (P and Q).

PQ is the perpendicular bisector of AB.

Using a compass to construct a perpendicular from a point on a straight line

You also need to know how to use a compass for constructing a perpendicular from a point on a line (as opposed to a line segment).

■ Example

Construct the perpendicular from the point x on a line.

Set your compass to a radius of 4 cm. Use x as your centre. Draw arcs to intersect the line on which x is situated. Label the points of intersection A and B.

Now construct the perpendicular bisector of AB as shown before.

Using a compass to construct a perpendicular from a point to a straight line

You can also use a compass to construct a perpendicular from a point to a straight line.

■ Example

Construct the perpendicular from a point P above the line.

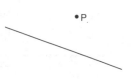

Set your compass to a suitable radius. Use P as your centre and draw arcs that intersect with the line. Label the points of intersection B and C. Set the compass to a radius greater than half the length of BC. Use C as the centre, and draw

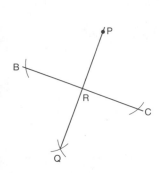

an arc on the opposite side of the line from P at Q. Keeping the same radius, use B as the centre and draw an arc on the opposite side of the line from P, to cut the arc drawn in the previous step at Q. Join PQ. Label the point of intersection of BC and PQ, R.
PR is the perpendicular bisector of the line BC.

Bisecting an angle

Work through the following example on how to bisect an angle.

■ Example

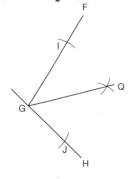

Bisect FĜH.

Set your compass to a radius of 2 cm. Use G (the vertex of the angle) as your centre. Draw arcs to intersect FG and GH, at I and J.

With the compass set at the same radius, use I as the centre. Draw an arc between F and H. Repeat this step, using J as the centre. Label the point where the arcs intersect as Q. Join GQ. GQ is the bisector of FĜH.

Perimeter and area of two-dimensional shapes

Polygons

The *perimeter of a polygon* is the sum of the length of its sides.

The *area of a polygon* is the size of its surface, and is always expressed in units squared, for example 15 m². You have already learnt how to work out the area of some polygons. The table below summarises some of the properties of polygons, as well as the formulae for the area of each polygon.

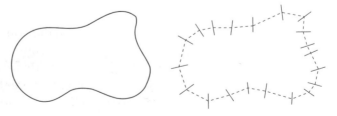

Remember

Height (h) always means perpendicular height. A means area, b means base.

Type of polygon	Area formula
Triangle	$A = \frac{1}{2}\text{base} \times \text{height}$ $(\frac{1}{2}bh)$
Trapezium (includes isosceles trapezium)	$A = \frac{1}{2}(\text{sum of parallel sides}) \times \text{height}$
Kite	A = sum of area of triangles formed by drawing a diagonal
Parallelogram Rhombus Rectangle Square (includes rhombus, rectangle and square)	$A = \text{base} \times \text{height}$ (bh)

Perimeter of a curved shape

One way to measure the perimeter of a curved shape is to lay a piece of string along the outline. Then measure the string on a ruler or tape measure.

Another way is to mark points on the outline and join them to form a polygon.

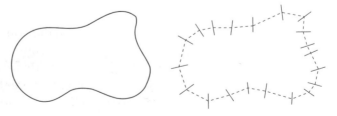

Exercise

1. Measure to find the perimeter of the following polygons.

a)

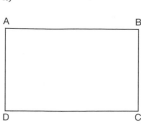

b)

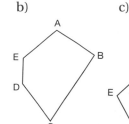

c)

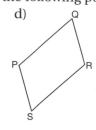

d)

e)

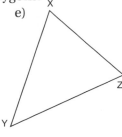

2. Calculate the area of each of the following shapes.

a)

b)

c)

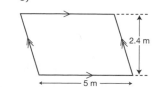

d)

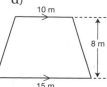

3. The shapes below are drawn on a grid. Each block represents $\frac{1}{2}$ cm². Calculate the area of each shape.

a) b) c) d)

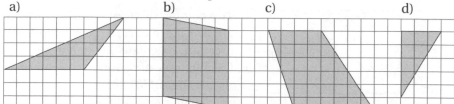

4. Calculate the area of the following shapes.

a)

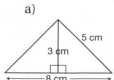

b)

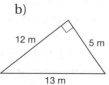

c)

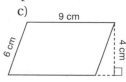

d)

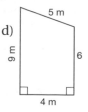

Circles

The *perimeter of a circle* is also known as its circumference.

Note

$\pi \approx \frac{22}{7}$

This is the fraction that comes closest to describing the relationship between the diameter and circumference of a circle. The circumference is slightly more than three times the diameter.

Circumference of circle = $\pi \times$ diameter
= πd or $2\pi r$
Also remember that $\pi \approx \frac{22}{7}$.

The formula for the *area of a circle* also uses π $(\frac{22}{7})$. The letter r stands for the radius of the circle. Remember that the radius is always equal to half the length of the diameter.

A = $\pi \times$ radius²
= πr^2

Exercise

1. Calculate the circumference of a circle with:
   a) a radius of 20 m
   b) a radius of 12.3 cm
   c) a diameter of 57 mm.
2. Taking the radius of the Earth to be 6 400 km, calculate the length of the equator.
3. Cotton is wound on a reel that has a diameter of 2.3 cm. There are 900 turns of cotton on the reel. What is the total length, in metres, of the cotton on the reel?
4. Calculate the area of the following figures. Round your answer off to 2 decimal places if necessary.

a)
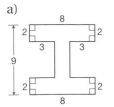

b)
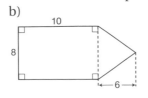

c)
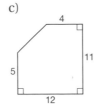

d)
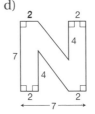

e)
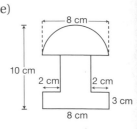

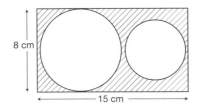

5. Two circular discs, of radii 4 cm and 3 cm, are cut from a rectangular piece of card measuring 15 cm by 8 cm, as shown in the diagram. Calculate the area of the remaining card.
6. Calculate the area and perimeter of a circle with:
   a) a radius of 5.2 cm
   b) a diameter of 1.56 m
   c) a radius of 23 mm.
7. A circus ring is a circle with diameter 15 m. Calculate its area.
8. A circular dance floor has a diameter of 10.4 m. The floor is to be sealed. One drum of sealant will cover 9 m². How many drums of sealant have to be bought?

Arcs and sectors

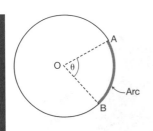

An *arc* is part of the circumference of a circle. The diagram shows an arc AB. The length of the arc depends on the angle θ, which it subtends at the centre O of the circle. The complete angle at O = 360°. If this angle were divided into 360 equal angles of 1°, each would be subtended by an arc $\frac{1}{360°}$ of the circumference.

∴ arc AB = $\frac{θ}{360°}$ of the circumference.

Hence, the formula is:

Length of an arc = $\frac{θ}{360°} \times 2\pi r$ where θ = angle subtended by the arc at the centre of the circle.

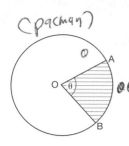

(pacman)

The region bounded by an arc AB of a circle (with centre O) and the two radii OA and OB is called a *sector* of the circle. AÔB is the angle of the sector.

If θ = 180°, the sector is a half circle – that is a semi-circle.
If θ = 90°, the sector is a quarter circle – that is a quadrant.
Using a similar argument as we used for the arc length, we find that the *area of a sector* is $\frac{\theta}{360°}$ of the area of the circle.

Hence, the formula is:

> Area of a sector = $\frac{\theta}{360°} \times \pi r^2$ where θ is the angle of the sector.

Exercise

1. A circle has a radius of 12 cm. For a sector of this circle with angle 50°, calculate:
   a) the perimeter
   b) the area.

2. A circle has a diameter of 14 cm. For a sector of this circle with angle 210°, calculate:
   a) the perimeter
   b) the area.

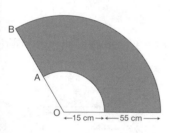

3. The diagram on the left shows the region of a windscreen that is wiped by a rubber blade AB which is 55 cm long. The end A moves along an arc of a radius 15 cm and centre O. The arc subtends an angle of 120° at O. Calculate the area of the region wiped by the blade.

4. A circle (shown on the left) has a radius of 10 cm. Calculate, correct to the nearest degree, the angle subtended at the centre by an arc which is 14 cm long.

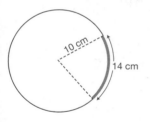

5. A circle has a diameter of 12 cm. Calculate, correct to the nearest degree, the angle subtended at the centre by an arc of length 20 cm.

6. The diagram on the right represents the floor of a room. OABC and OCDE are trapeziums. OAE is a sector of a circle with centre O, and AÔE = 106°.
   a) Calculate the perimeter of the floor.
   b) Calculate the area of the floor.

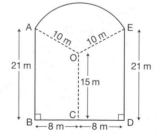

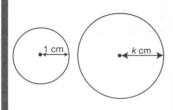

Perimeters and areas of similar figures

Look at the two squares on the left.

The sides, diagonals and perimeters are all lengths. These lengths are three times bigger in the large square than in the small square. The ratio of corresponding sides is called the *linear scale factor*. For these two squares, the linear scale factor is 3. The areas are in the ratio 9:1 and the *area scale factor* is 9. (Note that 9 = 3 *squared* and that area is measured in *square* units.)

Now look at the two circles.

The radii are 1 cm and k cm. The diameters are 2 cm and $2k$ cm. The circumferences are 2π cm and $2k\pi$ cm. You can see that the linear scale factor is k. However, the areas are π^2 and πk^2, so the *area scale factor* is k^2.

In fact, we can prove the following:

> If the *linear scale factor* for a pair of similar shapes is k, then the *perimeter scale factor* is k and the *area scale factor* is k^2.

Note

You can read about similar figures on page 162. You will need to apply the scale factors of similar figures to scale drawings on page 169.

Solids

The 3-dimensional equivalent of a polygon is a *polyhedron* – this is a solid whose surface consists of polygons.

The cube

The most common polyhedron is a *cube*. Its surface consists of 6 squares. Each of these squares is called a *face* of the solid. Where two faces meet is a straight line called an *edge* of the solid. A point where edges meet is called a *vertex* of the solid.

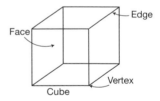

The cuboid

A polyhedron whose surface consists of 6 rectangles is called a *cuboid* or *rectangular block*. A cuboid has the same number of faces, edges and vertices as a cube. That is, 6 faces, 12 edges and 8 vertices.

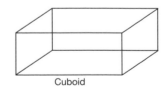

Triangular prism

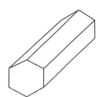

Hexagonal prism

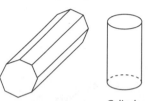

Octagonal prism

Cylinder

Sphere   Hemisphere

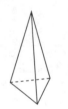

Triangular pyramid

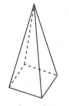

Square-based pyramid

Cone

The prism

A *prism* is a polyhedron which has 2 faces (the top and bottom faces) which are identical polygons, with their corresponding sides parallel. The other faces are usually rectangles, in which case we call the solid a *right prism*. However, the faces could be parallelograms, in which case the prism is 'leaning over'. The shape of the top and bottom faces is often used to describe the prism – for example, 'triangular prism', 'hexagonal prism' and 'pentagonal prism'.

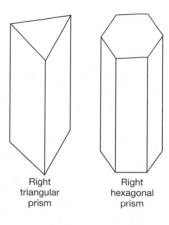

Right triangular prism

Right hexagonal prism

Nets of solids

A shape that can be cut out and folded to make the surface of a solid is called a net of the solid.

For example, the diagram below shows the net for a cube.

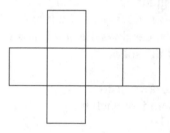

Here are two more nets for a cube. How many more can you think of?

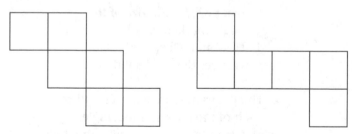

Exercise

1. Two similar solids have surface areas of 200 cm² and 450 cm². Find:
    a) the surface area scale factor
    b) the linear scale factor.

Exercise

1. The diagram represents a cuboid.
   a) How many edges does the cuboid have?
   b) Name two edges of the cuboid that are parallel to the edge BC.
   c) Name two edges of the cuboid that are perpendicular to the edge BC.

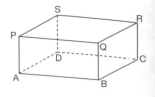

2. The diagram shows the net of a solid.
   a) What type of solid is this?
   b) How many faces does the solid have?
   c) How many edges does the solid have?
   d) In an accurately drawn net, which of the lengths PQ, QR, RS, ST must be equal?

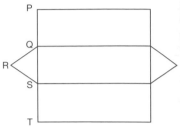

3. This diagram represents a solid with 8 faces. Each face is an equilateral triangle. (The solid is a regular octahedron.)
   a) How many edges does the solid have?
   b) How many vertices does the solid have?
   c) Draw a net of the solid.

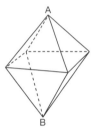

4. When the net is drawn accurately,
   a) why must PB and PA each be 50 mm long?
   b) what special type of triangle will ADS be?
   c) what will be the size of the angle marked x?
   d) How many faces, edges and vertices does the solid have?

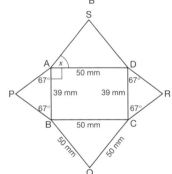

5. The diagram represents a small box. Each of the faces is a rectangle.
   a) Give the mathematical name for the shape of the box.
   b) How many faces, edges and vertices does the box have?
   c) Draw an accurate net for the box.
   d) How many of these boxes would fit into a carton 8 cm by 9 cm by 11 cm?

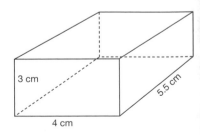

Note

You will learn more about solids in the next section, when you explore surface area and volume.

Surface area and volume of solids

A solid's boundary consists of surfaces (called faces) which may be planes (flat) or curved. The total area of the faces is the *surface area* of the solid.

Volume is the amount of space a solid or three-dimensional shape takes up, and is measured in cubic units. The units of volume are related to each other and to the units of length. The units and abbreviations you need to know are shown in the table.

Length	Volume	Units	Abbreviation
1 mm	1 mm³	cubic millimetre	mm³
1 cm = 10 mm	1 cm³ = 10 mm × 10 mm × 10 mm = 1 000 mm³	cubic centimetre	cm³
1 m = 100 cm	1 m³ = 100 cm × 100 cm × 100 cm = 1 000 000 cm³	cubic metre	m³

Cuboids

Surface area of a cuboid

A cuboid has six faces and each of them is a rectangle. If the length, breadth and height of the cuboid are l units, b units and h units, respectively, then the surface area is A square units, where A = $2lb + 2hb + 2hl$.

To find the *surface area* of any cuboid, you can calculate the sum of the areas of each face. Alternatively, find the area of the net.

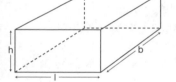

Volume of a cuboid

Suppose you have to find the *volume* of a cuboid which has a length of 4 cm, a breadth of 3 cm and a height of 2 cm.

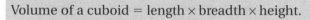

Volume of a cuboid = length × breadth × height.

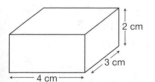

$$v = l \times b \times h$$
$$l = 4 \text{ cm} \qquad b = 3 \text{ cm} \qquad h = 2 \text{ cm}$$
$$\therefore v = 4 \text{ cm} \times 3 \text{ cm} \times 2 \text{ cm} = 24 \text{ cm}^3$$

Cylinders

Surface area of a cylinder

The surface of a cylinder consists of two flat circles (base and top) and a curved face. The curved face can be cut parallel to the axis of the cylinder and then flattened to make a rectangle. The breadth (b) of the rectangle is the same as the height (h) of the cylinder. The length of the rectangle is the same as the circumference of the base (or the top) of the cylinder, that is $2\pi r$.

$\therefore$ Surface area of the rectangle = length × breadth = $2\pi r \times h = 2\pi rh$.

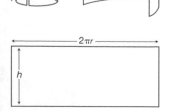

The curved surface area of a cylinder = $2\pi rh$.

Volume of a cylinder

A cylinder is a prism with a circular cross-section. The area of a circle (the cross-section of a cylinder) is πr^2. We take the height of the cylinder as h.

Volume of a cylinder $= \pi r^2 h$.

Prisms

Surface area of a prism

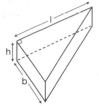

To find the *surface area* of a prism, first work out the surface area of one of its parallel end faces. Multiply this by two. Add the product to the sum of the surface area of each of its sides.

Volume of a prism

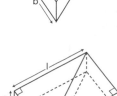

For a cuboid, length $\times$ breadth gives the area of the base, and so the formula for its *volume* could be written as follows:

volume = area of base $\times$ height.

This formula applies to some other solids too. Consider the upright prism shown in the diagram. Its cross-section is a right-angled triangle. Two of these prisms can be put together to form a cuboid.

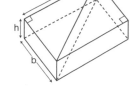

$$\text{volume of the cuboid} = l \times b \times h$$
$$\therefore \text{ volume of each prism} = \tfrac{1}{2} \times l \times b \times h$$
$$\text{but } \tfrac{1}{2} \times l \times b = \text{area of cross-section of prism}$$
$$\therefore \text{ volume of prism} = \text{area of cross-section} \times \text{height}$$

If the cross-section of a triangular prism is not right-angled, the prism can be divided into two prisms that do have right-angled cross-sections.

Volume of a prism $=$ area of cross-section $\times$ height.

Volumes and surface areas of similar solids

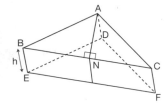

Look at the cubes on the left. The linear scale factor is k. The surface area of the larger cube is k^2 times the surface area of the unit cube ($6k^2:6$).

Hence, the surface area scale factor = (linear scale factor)^2.

The volume of the larger cube is k^3.

Hence, the volume scale factor = (linear scale factor)^3.

You can also use these formulae to work out the areas and volumes of other similar solids, including prisms, cylinders, pyramids and spheres.

For a pair of similar solids, surface area scale factor = (linear scale factor)^2
volume scale factor = (linear scale factor)^3.

Exercise

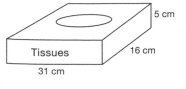

Tissues
31 cm
16 cm
5 cm

1. Find the surface area of a cuboid with:
   a) length = 56 mm, breath = 45 mm and height = 36 mm
   b) length = 1.2 m, breadth = 75 cm and height = 60 cm
2. A rectangular box of tissues has dimensions 31 cm by 16 cm by 5 cm. Calculate the surface area of the box.
3. Calculate the volume of:
   a) a prism with cross-sectional area 1.4 cm² and length 11 cm
   b) a cylindrical coin with diameter 22 mm and thickness 3 mm.

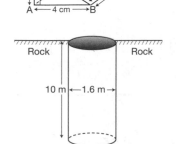

4. The diagram shows a triangular prism ABCDEF.
   CÂB = 90°, AB = 4 cm, BC = 5 cm, CA = 3 cm and BE = 16 cm.
   a) Calculate the surface area of the prism.
   b) Calculate the volume of the prism.
5. The diagram represents a cylindrical well. It is 10 m deep and has a diameter of 1.6 m.
   a) (i) Write down the radius of the well.
      (ii) Calculate the volume of the well.

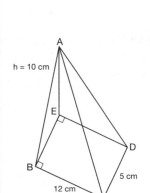

Rock   Rock
10 m   ←1.6 m→

   b) The well is dug through rock. Every cubic metre of this rock has a mass of 2.3 tonnes. Calculate the mass of the rock removed in digging the well.
6. A manufacturer is making a batch of 800 plastic knitting needles. Each needle is cylindrical in shape, with a diameter of 6 mm and a length of 32 cm. Calculate the total volume of plastic used in making these needles.

Pyramids

The base of a pyramid is a polygon and all the other faces are triangles. To find the *surface area* of a pyramid, you need to find the area of the base and the area of each of the triangles, and add them together. Alternatively, you could find the area of the net of the pyramid. To find the *volume*, use the following formula:

Volume of a pyramid $= \frac{1}{3} \times$ area of base $\times$ perpendicular height.

■ Example

Look at the pyramid on the left: BC = 12 cm, DC = 5 cm, h = 10 cm.

Volume of a pyramid $= \frac{1}{3}$ area of base $\times$ perpendicular height

Base $= 12 \times 5$

$\quad = 60$ cm²

$\therefore v = \frac{1}{3} \times 60$ cm² $\times 10$ cm

$\quad = 200$ cm³

Cones

A cone is a pyramid with a circular base. In this course, you will only consider right cones, which are upright and have their axis of symmetry perpendicular to the base. In the diagram, l represents slant height.

Curved surface area of a cone $= \pi r l$, where $l =$ slant height.

Volume of a cone $= \frac{1}{3}\pi r^2 h$, where $h =$ perpendicular height.

Spheres

A sphere has only one face, which is curved.

Surface area of a sphere $= 4\pi r^2$, where $r =$ radius.

Volume of a sphere $= \frac{4}{3}\pi r^3$.

A hemisphere is half a sphere, with one flat circular face and one curved face.

Surface area of a hemisphere $= \pi r^2 + 2\pi r^2 = 3\pi r^2$.

Volume of a hemisphere $= \frac{1}{2} \times \frac{4}{3}\pi r^3 = \frac{2}{3}\pi r^3$.

Exercise

1. An ice-cream scoop forms ice-cream in the shape of a sphere of diameter 5 cm.
   a) Taking π to be 3.14, find the volume of a sphere of ice-cream. Give your answer correct to one decimal place.
   b) How many spheres of ice-cream can be made from $1l$ of ice-cream?

2. The diagram shows the net of a solid. The straight lines ABC and DBE are perpendicular, EB = BC = 5 cm, AB = BD = 12 cm, AE = 13 cm.
   a) Of what special type is this solid?
   b) Write down the length of EF.
   c) Calculate the volume of the solid.

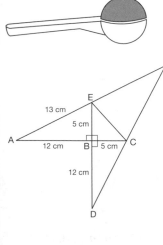

3. The diagram shows a heap of sand in the shape of a cone. The radius of the base of the cone is 1.2 m, its perpendicular height is 1.5 m. Its slant height is 1.92 m.
   a) Calculate the volume of the sand.
   b) Calculate the curved surface area of the cone.

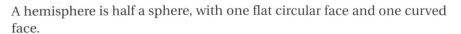

4. A glass sphere has a radius of 5 cm.
   a) Calculate the volume of the sphere.
   b) The sphere is tightly packed in a cylindrical gift box, touching the curved surface of the box and both the top and the bottom.
      (i) Calculate the total surface area of the interior of the box.
      (ii) Express the volume of the sphere as a fraction of the capacity of the box.

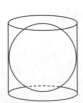

Congruent triangles

Triangles that are the same shape (all corresponding angles equal) *and* the same size (all corresponding sides equal) are said to be *congruent*.

AC = XY, AB = XZ and BC = YZ ∴ △ABC ≡ △XZY
△ABC ≡ △XZY means △ABC
and △XZY are congruent.

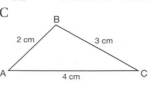

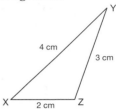

Two triangles are congruent if you can show one of the following conditions.

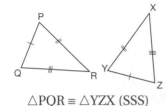

△PQR ≡ △YZX (SSS)

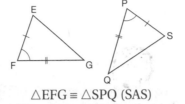

△EFG ≡ △SPQ (SAS)

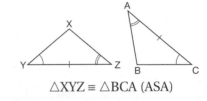

△XYZ ≡ △BCA (ASA)

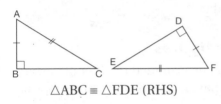

△ABC ≡ △FDE (RHS)

Side, Side, Side (SSS). If three sides of one triangle are equal to the three sides of another triangle, then the two triangles are congruent.

Side, Angle, Side (SAS). If two sides of one triangle are equal to two sides of another triangle and the angle between this pair of sides is the same in both triangles, then the two triangles are congruent.

Angle, Side, Angle (ASA). If two angles of one triangle are equal to two corresponding angles of another triangle and the side between each pair of angles is the same length in both triangles, then the two triangles are congruent.

Right Angle, Hypotenuse, Side (RHS). If two right-angled triangles have their longest sides equal in length and another side of the first triangle is equal to a side of the other triangle, then the two triangles are congruent.

Remember

The longest side of a right-angled triangle is always opposite the right angle. This side is known as the hypotenuse of the triangle. You will learn more about right-angled △s in Module 5.

Exercise

In each of the following cases, state whether the two triangles are congruent and, if they are, give a reason. (The diagrams are not to scale.)

a)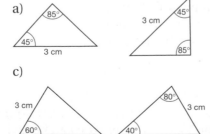

b)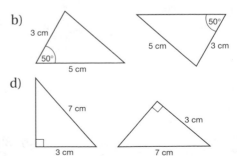

c)

d)

Similar triangles

Triangles that are the same shape but not the same size are said to be *similar*.

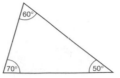 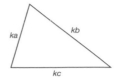

If two triangles are congruent, then they must also be similar. But if two triangles are similar, they are not necessarily congruent. Two triangles are similar if you can show one of the following conditions.

1. If two triangles have their corresponding angles equal, then they are similar (and their corresponding sides are in the same ratio).

2. If two triangles have their corresponding sides in the same ratio, then they are similar (and their corresponding angles are equal).

3. If two triangles have a pair of corresponding sides in proportion and the angle between these sides equal, then they are similar.

Congruent figures and similar figures

The idea of congruence and similarity can be extended to other plane figures, and to solids too. Figures or solids that are exactly the same shape are said to be *similar*.

Figures or solids that are exactly the same shape *and* the same size are said to be *congruent*. It may be necessary to rotate the figures or solids, or to reflect one of them in a mirror to determine whether they are similar or congruent.

Exercise

1. State which of the following shapes is not congruent to the other two.

   a) 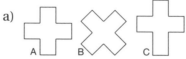   b)

2. a) (i) Draw two squares that are not congruent to one another.
      (ii) Is it possible to draw two squares that are not similar to one another? Give reasons for your answer.
   b) Is it possible to draw rectangles that are not similar to one another?
   c) A regular hexagon is divided into four triangles, as shown in the diagram. Which of the triangles are congruent to one another?
   d) Show how a regular hexagon can be divided into four triangles, three of which are congruent to one another.

Symmetry

Two-dimensional symmetry

When two sides of a shape or object are identical (each side is a mirror image of the other), the shape or object is symmetrical. There are two kinds of symmetry in flat shapes: line symmetry and rotational symmetry.

Line symmetry

Look at the drawing of the African mask on the left.

The drawing is said to have *line symmetry* and the dashed line is called the drawing's line of symmetry. The *line of symmetry* is sometimes called a 'mirror line'.

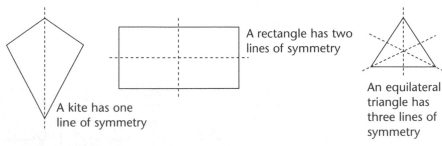

A kite has one line of symmetry

A rectangle has two lines of symmetry

An equilateral triangle has three lines of symmetry

Rotational symmetry

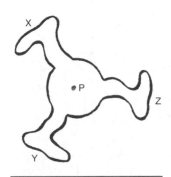

The shape on the left can be turned (or rotated), keeping its centre point P in a fixed position. The shape can be turned so that X is in position X, Y or Z, and the shape will still look the same. We say that the shape has *rotational symmetry*. In this case, it fits onto itself three times when rotated through 360° (one full revolution). We therefore say it has *rotational symmetry of order 3*.

Order of rotational symmetry = the number of times a shape looks the same when rotated through 360°.

Hint

There is another way of working out the order of rotational symmetry. If the smallest angle through which the shape can be rotated and still look the same is A°, then the order of rotational symmetry = $\frac{360°}{A°}$.

Exercise

State for each shape: How many lines of symmetry (if any) does it have? Does it have rotational symmetry? If so, of what order?

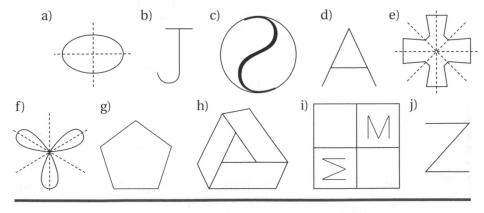

a) b) c) d) e)

f) g) h) i) j)

Three-dimensional symmetry

Solids can also have symmetry. There are two kinds of three-dimensional symmetry: plane symmetry and rotational symmetry.

Plane symmetry

A solid has plane symmetry if it can be cut into two halves and each half is the mirror image of the other half. The plane separating the two halves is called the *plane of symmetry*.

A cuboid (length, width and height all different) has 3 planes of symmetry.

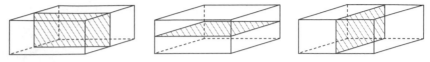

A cube has 9 planes of symmetry.

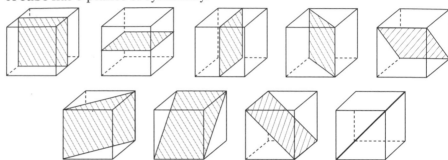

A sphere has an infinite number of planes of symmetry. It is symmetrical about any plane which passes through its centre.

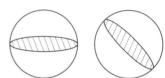

Rotational symmetry

A solid has rotational symmetry if it can be rotated about a line through an angle less than 360° and still look the same. The line about which the solid is rotated is called an *axis of symmetry*. A cuboid (length, width and height all different) has 3 axes of symmetry.

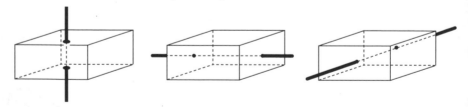

Exercise

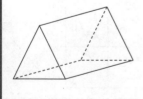

1. A prism with equilateral triangular ends has 4 planes of symmetry and 4 axes of rotational symmetry. Draw diagrams to show these planes and axes.
2. a) How many planes of symmetry does a right circular cone have?
    b) How many axes of symmetry does a right circular cone have?

Symmetry properties of a circle

A circle has line symmetry about any diameter and it has rotational symmetry about its centre. From these facts, we can deduce a number of results.

The perpendicular bisector of a chord passes through the centre

The perpendicular bisector of AB is the locus of points equidistant from A and B.

But centre O is equidistant from A and B (OA and AB radii of circle with centre O) ∴ O must be on the perpendicular bisector of AB.

This result can be expressed in other ways:

a) The perpendicular from the centre of a circle to a chord meets the chord at its mid-point.
b) The line joining the centre of a circle to the mid-point of a chord is perpendicular to the chord.

Equal chords are equidistant from the centre and chords equidistant from the centre are equal in length

If chords AB and CD are the same length, then OM = ON, and vice versa.

This follows from the fact that triangle OAM is congruent to triangle ODN. (Can you prove this?)

It is also a consequence of the fact that the circle has rotational symmetry about O.

The two tangents drawn to a circle from a point outside the circle are equal in length

A and B are the points of contact of the tangents drawn from P.

The result is PA = PB.

This is a consequence of the fact that the figure is symmetrical about the line OP. (It can also be obtained by proving that △OAP is congruent to △OBP. You need to use the 'tangent perpendicular to radius' property for this.)

Remember

The distance from a point to a line is always the perpendicular distance, which is the shortest distance from the point to the line.

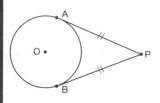

Note

Concentric circles are circles that have the same centre.

Exercise

1. P is a point inside a circle whose centre is O. Describe how to construct the chord that has P as its mid-point.
2. A straight line cuts two concentric circles at A, B, C and D (in that order). Prove that AB = CD.

More angles in circles

Circles have many useful angle properties that can be used to solve problems. You will now explore more of these properties. The following examples and theorems will help you to solve problems involving angles and circles.

The angle in a semi-circle is a right angle (90°)
The following example will help you to work out the size of an angle in a semi-circle.

■ Example
AB is the diameter of a circle. C is the centre. D is any point on the circumference. Remember that all radii of a circle are equal. Work out the size of AD̂B.

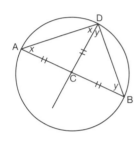

AC = CB = CD (radii of circle)
∴ △ACD and △BCD are isosceles.
∴ CÂD = AD̂C = x
∴ CD̂B = DB̂C = y
But $2x + 2y = 180°$ (sum ∠s △ABD)
∴ $x + y = 90°$
∴ AD̂B = 90°

The angle between the tangent and radius is 90°

■ Example
The diameter divides the circle evenly into two equal parts.
 So $a = b$
and $a + b = 180°$ (angles on a straight line)
 ∴ $a = b = 90°$

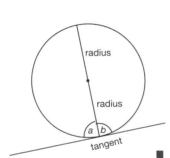

Exercise

In the diagram, BCF and BAE are the tangents to the circle at C and A respectively. AD is a diameter and AB̂C = 40°.
a) Explain why △ABC is isosceles.
b) Calculate the size of:
 (i) CÂB
 (ii) DÂC
 (iii) AD̂C.

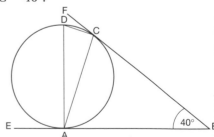

Note

If you are following the core syllabus, continue on page 169.

The angle subtended at the centre of a circle by an arc is twice the angle subtended at the circumference by the arc

 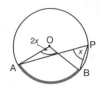

AB is an arc of a circle with centre O. P is a point on the circumference, but not on the arc AB. The angle at the centre theorem states that
   $A\hat{O}B = 2 \times A\hat{P}B$.

As you saw before, this is also true when AB is a semi-circular arc. The angle at the centre theorem states that the angle in a semi-circle is 90°. This is because, in this case, $A\hat{O}B$ is a straight line (180°).

Angles in the same segment, subtended by the same arc, are equal
In these two diagrams, $p = q$. Each of the angles p and q is half the angle subtended by the arc AB at the centre of the circle.

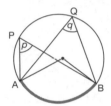 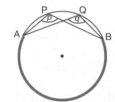

Note

A cyclic quadrilateral is one that has all four vertices touching the circumference of a circle.

Opposite angles of a cyclic quadrilateral add up to 180°

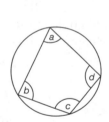 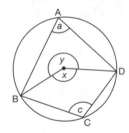

$a + c = 180°$ and $b + d = 180°$
$x = 2a$ ($\angle$ at centre theorem, minor arc BD)
$y = 2c$ ($\angle$ at centre theorem, major arc BD)
   $\therefore x + y = 2a + 2c$
but $x + y = 360°$ ($\angle$s around a point)
   $\therefore a + c = 180°$ ($\angle$ sum of a quadrilateral)

Each exterior angle of a cyclic quadrilateral is equal to the interior angle opposite to it

■ Example

Prove that $x = a$.

$x + B\hat{C}D = 180°$ ($\angle$s on a straight line)

$a + B\hat{C}D = 180°$ (opp. int. $\angle$s of a cyclic quadrilateral)

∴ $x = a$

Exercise

1. Find the size of each lettered angle in these sketches.
   When it is marked, O is the centre of the circle.

   a)   b)   c)   d)

   e) 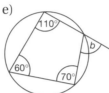   f) 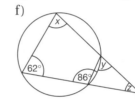   g)

2. In the diagram on the left, SAT is the tangent to the circle at point A. The points B and C lie on the circle and O is the centre of the circle. If $A\hat{C}B = x$, express in terms of x, the size of:

   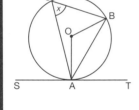

   a) $A\hat{O}B$ b) $O\hat{A}B$ c) $B\hat{A}T$.

3. Find the size of each lettered angle in these sketches.
   When it is marked, O is the centre of the circle.

   a)   b)   c)

4. In the diagram, TA and TB are the tangents from T to the circle whose centre is O. AC is a diameter of the circle and $A\hat{C}B = x$.

   a) Find $C\hat{A}B$ in terms of x.

   b) Find $A\hat{T}B$ in terms of x.

   c) The point P on the circumference of the circle is such that BP is parallel to CA. Express $P\hat{B}T$ in terms of x.

Scale drawing

Sometimes you have to draw a diagram to represent something that is much bigger or much smaller than you can draw it, such as a plan of a building, a map of a country, or the design of a microchip. In order to do an accurate diagram, you can do a scale drawing.

The lines in the drawing are all the same fraction of the lines they represent in reality. This fraction is called the *scale* of the drawing.

The scale of a diagram, or a map, may be given as a fraction or as a ratio, such as $\frac{1}{50\,000}$ or $1 : 50\,000$.

A scale of $\frac{1}{50\,000}$ means that every line in the diagram has a length which is $\frac{1}{50\,000}$ of the length of the line it represents in real life. Hence, 1 cm in the diagram represents 50 000 cm in real life. In other words, 1 cm represents 500 m or 2 cm represents 1 km.

■ Example

A rectangular field is 100 m long and 45 m wide. A scale drawing of the field is made with a scale of 1 cm to 10 m. What are the length and width of the field in the drawing?

    10 m is represented by 1 cm

∴ 100 m is represented by (100 ÷ 10) cm = 10 cm

    ∴ 45 m is represented by (45 ÷ 10) cm = 4.5 cm

        ∴ length = 10 cm   width = 4.5 cm

Exercise

1. On the plan of a house, the living room is 3.4 cm long and 2.6 cm wide. The scale of the plan is 1 cm to 2 m. Calculate the actual length and width of the room.

2. The actual distance between two villages is 12 km. Calculate the distance between the villages on a map whose scale is:
   a) 1 cm to 4 km   b) 1 cm to 5 km.

3. A car ramp is 28 m long and makes an angle of 15° with the horizontal. A scale drawing is to be made of the ramp using a scale of 1 cm to 5 m.
   a) How long will the ramp be on the drawing?
   b) What angle will the ramp make with the horizontal on the drawing?

Angle of elevation and angle of depression

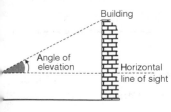

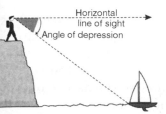

Scale drawing questions often involve the observation of objects that are higher than you or lower than you, for example, the top of a building, an aeroplane, or a ship in a harbour.

In these cases, the angle of elevation or depression is the angle between the horizontal line of sight and the object.

Angles of elevation and depression are *always* measured from the *horizontal.*

Drawing diagrams to scale

Here are some tips for drawing diagrams to scale:
- Draw a rough sketch, showing the lengths of lines and sizes of given angles.
- When you are told to use a particular scale, use it. Otherwise, choose a suitable scale to use, that is one that:
  a) is easy to use, for example one that has a number such as 1, 2, 5, 10, ... in it (these numbers are easy to divide by)
  b) makes the drawing as large as possible and still fits on the page.
- Make a clean, tidy and accurate scale drawing using appropriate geometrical instruments. Show on it the given lengths and angles. Write the scale next to the drawing.
- Measure lengths and angles in the drawing to find the answers to the problem. Remember to change the lengths to *full size* using the scale. Remember that the full-size angles are the same as the angles in the scale drawing.

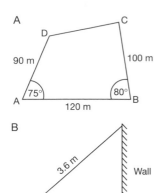

Exercise

1. Diagram A is a rough sketch of a field ABCD.
   a) Using a scale of 1 cm to 20 m, make an accurate scale drawing of the field.
   b) Find the sizes of Ĉ and D̂ at the corners of the field.
   c) Find the length of the side CD of the field.

2. A ladder of length 3.6 m stands on horizontal ground and leans against a vertical wall at an angle of 70° to the horizontal (diagram B).
   a) What is the size of the angle the ladder makes with the wall?
   b) Use a scale drawing with a scale of 1 cm to 50 cm to find how far the ladder reaches up the wall.

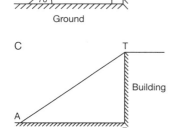

3. An accurate scale drawing (diagram C) represents the vertical wall TF of a building that stands on horizontal ground. It is drawn to a scale of 1 cm to 8 m.
   a) Find the height of the building.
   b) Find the distance of the point A from the foot (F) of the building.
   c) Find the angle of elevation of the top (T) of the building from the point A.

Bearings

To give directions on the Earth's surface, you can use the compass dirctions north, east, west, south. However, they are not accurate enough, for example, for surveying or for flying an aeroplane.

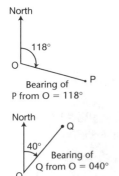

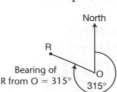

Directions nowadays are usually given as *three-figure bearings*. The three-figure bearing of a point P from an observer O is the angle, measured in degrees in a *clockwise* direction, from *north* to the line OP. The bearing must contain *three figures*, so you have to put zeros on the left-hand side if you need to.

■ Example

The bearing of Town B from City A is 048°.
What is the bearing of City A from Town B?

In the second diagram, the two north lines are parallel.
Hence, angle $\theta = 48°$
(corresponding angles are equal).

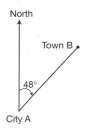 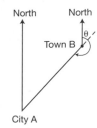

The bearing of City A from
Town B = 48° + 180° = 228°.

Notice that the difference between
the two bearings (048° and 228°)
is 180°.

Exercise

1. Give the three-figure bearing corresponding to:
    a) west   b) south-east   c) north-east.

2. Write down the three-figure bearings of A from B.

    a) 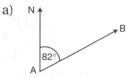   b)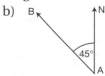

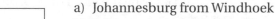

3. Use the map of southern Africa to find the three-figure bearing of:

a) Johannesburg from Windhoek
b) Johannesburg from Cape Town
c) Cape Town from Johannesburg
d) Lusaka from Cape Town
e) Kimberley from Durban.

4. Townsville is 140 km west and 45 km north of
    Beeton. Using a scale drawing with a scale of
    1 cm to 20 km, find:
    a) the bearing of Beeton from Townsville
    b) the bearing of Townsville from Beeton
    c) the direct distance from Beeton to Townsville.

5. Village Q is 7 km from village P on a bearing of
    060°. Village R is 5 km from village P on a bearing
    of 315°. Using a scale drawing with a scale of 1 cm
    to 1 km, find:
    a) the direct distance from village Q to village R
    b) the bearing of village Q from village R.

Loci

'Locus' comes from the Latin for 'position' or 'place'. Its plural is 'loci'.

A locus is a set of points that satisfy a given condition (or rule). Look at the map on page 171 again. You could identify the set of points in South Africa that are exactly 30 km away from Johannesburg, or the places that are further from Cape Town than they are from Kimberley.

■ Examples

1. The position of a point A is fixed and a point P is to be marked so that AP = 2 cm.

   Describe the locus of possible positions of P.

   Some of the possible positions of P are marked in the diagram on the left.

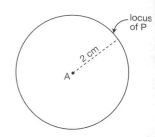

   The more positions are marked, the clearer it becomes that the locus of P is the circumference of the circle with centre at A and with radius 2 cm.

Hint

Revise linear inequalities in Module 2, page 82 if you find this difficult.

   Thus circle centre A is the locus of P.

2. The positions of two points, D and E, are fixed. A point P is to be marked so that it is nearer to D than it is to E.

   Describe the locus of possible positions of P.

   The rule given is equivalent to the inequality DP < EP.

   Because the rule is an *inequality*, the locus is a *region* of the plane.

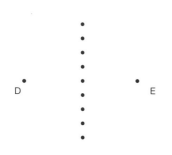

   The boundary of the region is the set of points that satisfy the equality DP = EP, that is, the set of points that are the same distance from D as they are from E (they are said to be *equidistant* from D and E).

   The first diagram on the left shows some of the points on the boundary. It is clear that the locus of these points includes the mid-point of DE and, by symmetry, it is perpendicular to DE. In other words, it is the perpendicular bisector of DE.

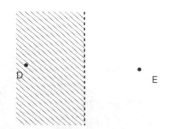

   The shaded region in the second diagram is the locus of points that are nearer to D than they are to E. (The boundary is shown as a dashed line because it is not included in the region.)

Locus of points at a given distance (d) from a given fixed point (C)

The locus is the circumference of a circle whose centre is the given fixed point (C) and whose radius is the given distance (d).

Locus of points at a given distance (d) from a given line (l)

The locus is two straight lines that are both parallel to the given line (l) and a distance d from it.

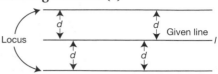

Locus of points at a given distance (d) from a given line segment AB

The locus is shown in this diagram.
It is in four parts:
- a line segment, distance d above AB,
- a line segment, distance d below AB,
- a semi-circle, radius d, centre A,
- a semi-circle, radius d, centre B.

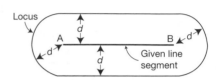

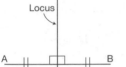

Locus of points equidistant from two given fixed points A and B

Equidistant from A and B means that every point P in the locus is the same distance from A as it is from B, that is PA = PB.

The locus is the perpendicular bisector of the line segment AB.

Locus of points equidistant from the arms of a given angle

In this diagram, AB and AC are the arms of the given angle. Every point P in the locus is the same perpendicular distance from AB as it is from AC.

The locus is the bisector of the given angle.

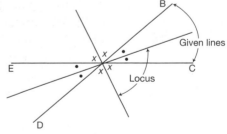

Locus of points equidistant from two given intersecting straight lines

This is similar to the situation above – the arms of the given angle BAC have been extended beyond A.

In the diagram, BD and CE are the given lines. Every point P in the locus is the same perpendicular distance from BD as it is from CE.

The locus consists of the two straight lines that bisect the angles between the given lines.

The two lines that make up the locus are at right angles to each other. (Can you see why?)

Check your progress

1. Look at these diagrams of six solids.

A   B   C   D   E   F

   a) Write down the mathematical name given to:
   (i) solid A   (ii) solid B   (iii) solid C.

   b) Make a table like the one below. Under 'solids', fill in A, B, C and so on. Then calculate and fill in the number of edges, faces and vertices of each solid shown here.

Solid	Number of faces	Number of vertices	Number of edges

   c) For these solids, the number of faces + the number of vertices is related to the number of edges. What is the relationship?

2. a) Draw a line to divide this L-shaped diagram into two congruent shapes.

   b) Draw lines to divide this L-shaped diagram into three congruent shapes.

   c) Draw lines to divide this L-shaped diagram into six congruent shapes.

3. If you take a strip of paper with parallel edges, tie a knot in it, and carefully press the knot flat, you will find that a regular pentagon is made. The diagram shows the knot. The hidden edges of the strip of paper are shown as dashed lines. ABCDE is the regular pentagon.

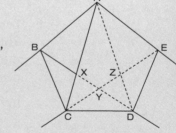

   a) Write down the triangles in the diagram that are congruent to △ABX.

   b) Write down the triangles in the diagram that are similar to △ABX but not congruent to it.

4. The diagram is the net of a solid.

   a) How many planes of symmetry does the solid have?

   b) How many axes of rotational symmetry does the solid have?

5. In this diagram, the lines AB and CDEF are parallel. Calculate:
   a) BD̂E
   b) AB̂D
   c) BÊF.

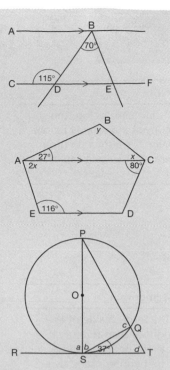

6. In this diagram, AC is parallel to ED.
   CÂE is twice the size of AĈB,
   BÂC = 27°, AÊD = 116°,
   AĈD = 80°.

   Work out the values of *x* and *y*.

7. RST is a tangent to the circle with centre O.
   PS is a diameter. Q is a point on the
   circumference and PQT is a straight line.
   QŜT = 37°.

   Write down the values of *a*, *b*, *c* and *d*.

8. The diagram on the right represents a garden ABCDE.
   AB = 2.5 m, AE = 7 m, ED = 5.2 m, DC = 6.9 m,
   EÂB = 120°, DÊA = 90° and ED̂C = 110°.
   a) (i)   Using a scale of 1 cm to represent 1 m,
            construct an accurate plan of the garden.
      (ii) Construct the locus of points equidistant
            from CD and CB.
      (iii) Construct the locus of points 6 m from A.
   b) A fountain is to be placed nearer to CD than to CB and no more than 6 m
      from A. Shade and label R, the region within which the fountain could be
      placed in the garden.
9. In the scale drawing below, the rectangle ABCD represents a sheet of glass
   standing vertically on horizontal ground on its edge AB. The sheet of glass is
   to be rotated in a vertical plane, without slipping, about the point B, until it is
   standing on the edge BC. Draw accurately:

   a) the new position of ABCD, labelling
      clearly the vertices A, C and D
   b) the locus of the point D for this
      rotation.

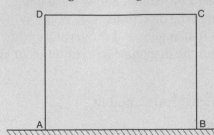

Trigonometry

Note

BCE means before the current era, that is before year 1. CE means current era, that is the years from 1 to the present

The word trigonometry means 'triangle measurement'. The development of this branch of mathematics can be traced back to Ancient Greece in the period 330 BCE to 150 CE. Today trigonometry is the basis of surveying, navigation, engineering and astronomy. Trigonometrical expressions and calculations are also used to solve problems related to electricity, magnetism and electronics.

Pythagoras' theorem

Pythagoras' theorem, which relates to the sides of right-angled triangles, is probably one of the best-known theorems in mathematics.

The theorem proves that in any right-angled triangle, the area of the square on the hypotenuse is equal to the sum of the areas of the squares on the other two sides. This can be represented in a diagram like this:

Remember

The longest side of any right-angled triangle is called the hypotenuse. The hypotenuse is always opposite the right angle.

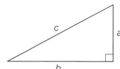

$$c^2 = a^2 + b^2$$
$$\therefore a^2 = c^2 - b^2$$
$$\therefore b^2 = c^2 - a^2$$

There are hundreds of proofs and demonstrations of Pythagoras' theorem. It is so well known that even stamps have been printed to celebrate it. Look carefully at the stamp on the left. If you count the number of black and white squares on the longest side (hypotenuse) you will see that this number is equal to the sum of the black and white squares on the other two sides. In effect, this is a proof of the theorem.

Pythagoras' theorem means that if you know the lengths of any two sides of a right-angled triangle, you can work out the length of the other side. This is extremely useful for solving real-life problems. In order to use the theorem, you have to recognise that right angles are found in various places (where you have a right angle you can construct a right-angled triangle). For example:

• the angles of squares and rectangles
• the angle between a horizontal and vertical line
• the angles between the directions N, S, W, E
• the angle between a tangent and radius at its point of contact
• the angle between the diagonals in a rhombus or kite.

■ Examples

1. Given right-angled $\triangle ABC$, find BC.
   Let BC $= x$
   $$\therefore x^2 = 12^2 + 35^2 \text{ (Pythagoras' theorem)}$$
   $$= 144 + 1\ 225$$
   $$= 1\ 369$$
   $$x = \sqrt{1\ 369} = 37 \qquad \therefore BC = 37 \text{ cm}$$

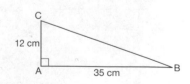

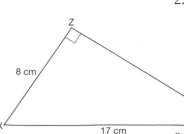

2. Work out the length of YZ.

Let YZ = a

$17^2 - 8^2 = a^2$ (Pythagoras' theorem)

$a^2 = 289 - 64 = 225$

$a = \sqrt{225} = 15$

∴ YZ = 15 cm

3. Kurt has a circular log of wood, of diameter 40 cm. He saws it in half along AD. He then cuts off another piece, whose cross-section is ABC, so that AC = 32 cm.

a) Give a reason why AĈD = 90°.

AD is a diameter, so AĈD = 90° (∠ in a semi-circle)

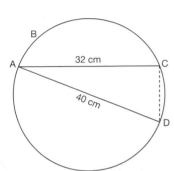

b) Calculate the straight line distance from C to D.

Let CD = x cm

$40^2 = 32^2 + x^2$ (Pythagoras' theorem)

$40^2 - 32^2 = 1\,600 - 1\,024 = x^2$

$x^2 = 576$

$x = \sqrt{576} = 24$

∴ CD is 24 cm

4. Mohamed takes a short cut from his home (H) to the bus stop (B) along a footpath HB.

How much further would it be for Mohamed to walk to the bus stop by going from H to the corner (C) and then from C to B?

Give your answer in metres.

If Mohamed takes the alternative route, the distance he walks is HC + CB.

HCB is a right-angled triangle with hypotenuse HB = 521 m.

Let HC = x m

$(521)^2 = (350)^2 + x^2$ (Pythagoras' theorem)

$x^2 = (521)^2 - (350)^2$

$= 271\,441 - 122\,500$

$= 148\,941$

$x = \sqrt{148\,941} = 385.9287$

Length of alternative route = (385.92878 + 350) m

= 735.9287 m

Extra distance walked = (735.9287 − 521) m

= 214.9287 m

= 215 m (to 3 significant figures)

Exercise

1. For each of the triangles shown below, work out the length of the lettered side.

   a)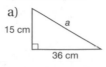

   b)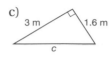

   c)

   d)

2. 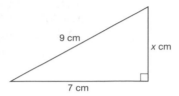   Work out the value of x.

3. 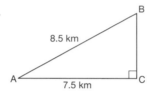   In the diagram, AB = 8.5 km, AC = 7.5 km and $A\hat{C}B = 90°$.

   Calculate the length of BC.

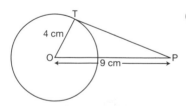

4. A ladder is standing on horizontal ground and rests against a vertical wall. The ladder is 4.5 m long and its foot is 1.6 m from the wall. Calculate how far up the wall the ladder will reach. Give your answer correct to 3 significant figures.

5. Grisley is 90 km due north of Ford. Highton is 64 km due east of Ford. Calculate the distance of Highton from Grisley, giving your answer correct to 3 significant figures.

6. A circle has centre O and radius 4 cm. A point P is 9 cm from O. A tangent is drawn from P to the circle and its point of contact is T.
   a) Give a reason why $O\hat{T}P = 90°$.
   b) Calculate the length of PT.

7. In the diagram, the rectangle ABCD represents the floor of a room. P represents a power point at floor level. Calculate the distance of the power point from:
   a) the corner A of the room
   b) the corner B of the room.

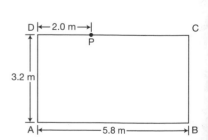

A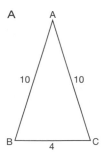

B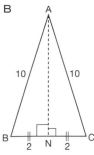

Applying Pythagoras' theorem

In order to use Pythagoras' theorem to solve problems, you may have to create a right-angled triangle. For example, in an isosceles triangle as in diagram A, you need the perpendicular height to calculate the area. You can work this out by creating two symmetrical right-angled triangles as in diagram B. Once you have done this, you can work out AN and find the area of the triangle by applying the formula $A = \frac{1}{2}bh$.

■ Examples

1. A rectangular box has a base with internal dimensions 21 cm by 28 cm, and an internal height of 12 cm. Calculate the length of the longest straight thin rod that will fit:
   a) on the base of the box
   b) in the box.

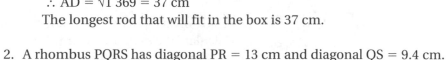

Rough sketch

   a) Longest rod on base is diagonal AC
      △ABC is right-angled ($\hat{B} = 90°$)
      ∴ $28^2 + 21^2 = AC^2$ (Pythagoras' theorem)
      $AC^2 = 784 + 441 = 1\ 225$
      $AC = \sqrt{1\ 225} = 35$ cm
      ∴ Longest rod on base is 35 cm.
   b) Longest rod in box is diagonal AD.
      $A\hat{C}D = 90°$; AD ⇒ hypotenuse
      ∴ $AD^2 = (35)^2 + 12^2$ (Pythagoras' theorem)
      $= 1\ 225 + 144$
      $= 1\ 369$
      ∴ $AD = \sqrt{1\ 369} = 37$ cm
      The longest rod that will fit in the box is 37 cm.

2. A rhombus PQRS has diagonal PR = 13 cm and diagonal QS = 9.4 cm. Calculate the lengths of the sides of the rhombus.
   △POS is right-angled (diagonals ⊥)
   ∴ $PS^2 = (6.5)^2 + (4.7)^2$ (Pythagoras' theorem)
   $= 42.25 + 22.09 = 64.34$
   ∴ $PS = \sqrt{64.34} = 8.02122...$
   PS = SR = RQ = QP = 8.02 (to 3 significant figures)
   (sides of rhombus are equal)

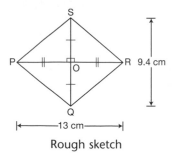

9.4 cm

13 cm

Rough sketch

The converse of Pythagoras' theorem

The converse of a theorem is the statement expressed in the opposite way. The converse of Pythagoras' theorem allows us to find out whether a triangle is right-angled. The converse is:

> If the square on the longest side of a triangle is equal to the sum of the squares on the other two sides, then the triangle has a right angle.

Exam tip

The converse of Pythagoras' theorem uses the term 'longest side' and not 'hypotenuse'. This is because it is not yet proven that the triangle is right-angled.

■ Examples

1. Is a triangle with sides 5.6 cm, 4.5 cm and 3.1 cm right-angled?

   Longest side = 5.6 cm
   $$(5.6)^2 = 31.36$$

   Sum of other two sides squared:
   $$(4.5)^2 + (3.1)^2 = 20.25 + 9.61$$
   $$= 29.86$$
   $$(5.6)^2 \neq (4.5)^2 + (3.1)^2$$
   $\therefore \triangle$ is not right-angled

2. Is a triangle with sides 7.2 cm, 9 cm and 5.4 cm right-angled?

   Longest side $= 9$
   $$9^2 = 81$$

   Sum of other sides squared:
   $$(7.2)^2 + (5.4)^2 = 51.84 + 29.16$$
   $$= 81$$
   $$9^2 = (7.2)^2 + (5.4)^2$$
   $\therefore \triangle$ is right-angled

If the square of the longest side is greater than the sum of squares of the other two sides, we can say the triangle is obtuse-angled.

If the square of the longest sides is less than the sum of the squares of the other two sides, we can say the triangle is acute-angled.

Exercise

1. Each side of an equilateral triangle is 10 cm long.
    a) Calculate the perpendicular height of the triangle.
    b) Calculate the area of the triangle.
2. Each edge of a cube is 20 cm long.
    a) Calculate the length of a diagonal of a face of the cube.
    b) Calculate the length of a diagonal of the cube (the distance between a vertex and the opposite vertex).
3. Find, by calculation, which of the following triangles are right-angled.

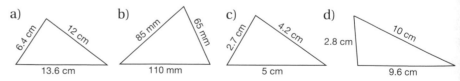

a) 6.4 cm, 12 cm, 13.6 cm
b) 85 mm, 65 mm, 110 mm
c) 2.7 cm, 4.2 cm, 5 cm
d) 10 cm, 2.8 cm, 9.6 cm

4. A triangle has sides of length 53 m, 82 m and 65 m. Investigate whether the triangle is acute-angled, right-angled or obtuse-angled.

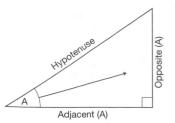

The tangent ratio

Before you work with the ratios of angles to sides in triangles, you need to remember how sides and angles are named in a right-angled triangle. The important terms are 'hypotenuse', 'adjacent' and 'opposite'. The diagram on the left demonstrates this using angle A. Make sure that you understand this naming system before you move on.

Naming the sides of right-angled triangles

■ Examples

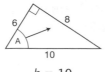

$h = 10$
opp(A) = 8
adj(A) = 6

$h = 29$
opp(B) = 20
adj(B) = 21

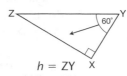

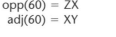

$h = ZY$
opp(60) = ZX
adj(60) = XY

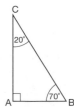

$h = BC$
opp(70) = AC
adj(70) = AB

opp(20) = AB
adj(20) = AC

Note

Opposite (A) and Adjacent (A) are normally shortened to opp(A) and adj(A).

Exercise

1. For each of these triangles, write down the length of the hypotenuse and the values of opp(A) and adj(A).

   a)

   b)

   c)

   d)

2.

   For this triangle, write down the length of the hypotenuse and the values of opp(60°), adj(60°), opp(30°) and adj(30°).

Calculating the tangent ratio

The ratio $\dfrac{\text{opp(angle)}}{\text{adj(angle)}}$ is constant in similar right-angled triangles. This ratio is called the tangent ratio. The name is normally shortened to tan and the ratio is written as:

$$\tan(\text{angle}) = \frac{\text{opp(angle)}}{\text{adj(angle)}}$$

It is possible to calculate the value of tan(angle) for any angle. The values for two similar triangles have been calculated in the table on page 182.

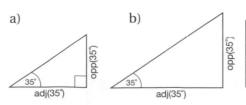

a)   b)

Triangle	opp(35°)	adj(35°)	$\frac{\text{opp}(35°)}{\text{adj}(35°)}$
a	1.4 cm	2 cm	0.70
b	2.1 cm	3 cm	0.70

Note that tan(angle) is equal for both triangles.

Using your calculator

You can find the value of tan(angle) on any scientific calculator. Before you do this, make sure your calculator is set to display degrees and that you know whether to press **tan** before or after the value of the angle.

■ Examples

1. Use your calculator to find:
   a) tan(35°)   b) tan(76°)   c) tan(57.3°)

   | 0.0700207538 | | 4.010780934 | | 1.557660082 |

2. Use your calculator to find, correct to 4 decimal places:
   a) tan(5.8°)   b) tan(25°)   c) tan(50°)

   | 0.101576296 | | 0.466307658 | | 1.19175393 |

   tan(5.8°) = 0.1016   tan(25°) = 0.4663   tan(50°) = 1.1918

3. The angle of approach of an aircraft to the runway should be exactly 3°.

   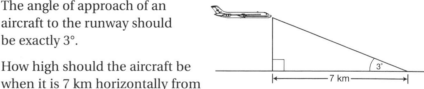

   How high should the aircraft be when it is 7 km horizontally from the runway?

   Give your answer in metres.

   Let the height of the aircraft be h km.

   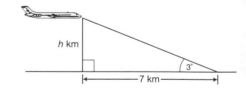

   Then $\tan(3°) = \dfrac{\text{opp}(3°)}{\text{adj}(3°)} = \dfrac{h}{7}$

   Hence $h = 7 \times \tan(3°)$
   $= 7 \times 0.052407779$
   $= 0.366854455$

   The height of the aircraft
   $= 0.366854455$ km
   $= 366.854455$ m
   $= 367$ m (to 3 significant figures)

Exercise

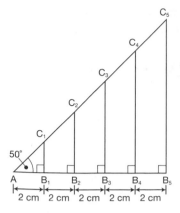

This diagram is a sketch of five overlapping right-angled triangles, AB_1C_1, AB_2C_2, AB_3C_3, AB_4C_4, AB_5C_5.

1. a) Draw the diagram accurately. (You will find it easier if you use graph paper.)
    b) Measure the lengths of B_1C_1, B_2C_2, B_3C_3, B_4C_4 and B_5C_5.
    c) For each of the five right-angled triangles, work out the value of $\dfrac{\text{opp}(50°)}{\text{adj}(50°)}$ to 2 significant figures. Use your answer to part c) to estimate the value of $\tan(50°)$.

2. Use your calculator to find, correct to 4 decimal places, the value of:
    a) $\tan(50°)$
    b) $\tan(37.5°)$
    c) $\tan(52.5°)$
    d) $\tan(79°)$.

3. Use your calculator to find, correct to 4 significant figures, the value of:
    a) $\tan(1°)$
    b) $\tan(10°)$
    c) $\tan(46.5°)$
    d) $\tan(84.3°)$.

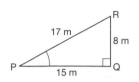

4. a) Look at $\triangle PQR$. Write down the value of $\tan(P)$ as a fraction.
    b) Find the value of $\tan(P)$ as a decimal correct to 4 places.

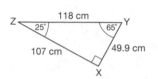

5. From $\triangle XYZ$ calculate, correct to 3 significant figures, the value of:
    a) $\tan(65°)$
    b) $\tan(25°)$.

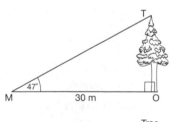

6. a) Use your calculator to find the value of $\tan(47°)$ correct to 4 decimal places.
    b) The diagram shows a vertical tree, OT, whose base, O, is 30 m horizontally from point M. The angle of elevation of T from M is 47°. Calculate the height of the tree.

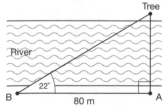

7. Devon wants to estimate the width of a river which has parallel banks. He starts at point A on one bank, directly opposite a tree on the other bank. He walks 80 m along the bank to point B and then looks back at the tree. He finds that the line between B and the tree makes an angle of 22° with the bank. Calculate the width of the river.

8. Given $\tan(45°) = 1$. What can you say about the value of $\tan(A)$ when A is an acute angle:
    a) $<45°$
    b) $>45°$?

9. The right-angled $\triangle ABC$ in which $B\hat{A}C = 30°$ can be regarded as half an equilateral $\triangle ABD$.
    a) Taking the length of BC to be 1 unit:
        (i) write down the length of AB
        (ii) use Pythagoras' theorem to obtain the length of AC.
    b) Write down the exact value of $\dfrac{BC}{AC}$.
    c) Work out the value of $\tan(30°)$ correct to 4 significant figures.

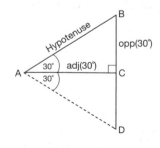

Calculating angles

If you know the size of an angle, you can find the tangent of the angle using the tan key of your calculator. You can also use your calculator to deal with the inverse, or reverse, of this – if you know the tangent of the angle, then you can find the size of the angle.

To do this, you need to use the key marked tan⁻¹ . This is normally a secondary function, so you may need to use the 2ndF or shift key as well.

Tan⁻¹ is short for 'inverse tan' and it means 'the angle whose tangent is'. Thus, tan⁻¹(2) means 'the angle whose tangent is 2'.

■ Examples

1. Find, correct to 1 decimal place, the acute angle with a tangent of:
   a) 0.1234   b) 5   c) 2.765

   | 7.034735756 | | 78.69006753 | | 70.11678432 |

   ∴ tan⁻¹(0.1234) = 7.0°   ∴ tan⁻¹(5) = 78.7°   ∴ tan⁻¹(2.765) = 70.1°
   (to 1 decimal place)   (to 1 decimal place)   (to 1 decimal place)

2. Calculate, correct to 1 decimal place, the lettered angles in the diagrams a) to c).

   a) 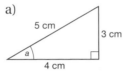   b)    c)

   a) $\tan(a) = \dfrac{\text{opp}(a)}{\text{adj}(a)} = \dfrac{3}{4} = 0.75$ b) $\tan(b) = \dfrac{\text{opp}(b)}{\text{adj}(b)} = \dfrac{12}{5} = 2.4$

   $\qquad a = \tan^{-1}(0.75)$ $\qquad\qquad\qquad\qquad b = \tan^{-1}(2.4)$
   $\qquad\quad = 36.86989765°$ $\qquad\qquad\qquad\qquad\quad = 67.38013505°$
   $\qquad a = 36.9°$ (to 1 decimal place) $\qquad b = 67.4°$ (to 1 decimal place)

   c) $\tan(c) = \dfrac{\text{opp}(c)}{\text{adj}(c)} = \dfrac{24}{7}$

   $\qquad c = \tan^{-1}\left(\dfrac{24}{7}\right)$

   $\qquad\quad = 73.73979529°$
   $\qquad c = 73.7°$ (to 1 decimal place)

   To find angle d, we could use the fact that the angle sum of a triangle is 180°. This gives $d = 180° - (90° + 73.7°) = 16.3°$.
   Alternatively, we could use trigonometry:

   $\qquad \tan(d) = \dfrac{\text{opp}(d)}{\text{adj}(d)} = \dfrac{7}{24}$

   $\qquad\quad d = \tan^{-1}\left(\dfrac{7}{24}\right)$

   $\qquad\qquad = 16.26020471°$ (calculator display)
   $\qquad\; d = 16.3°$ (to 1 decimal place)

Exercise

1. Find, correct to 1 decimal place, the acute angle that has a tangent of:
   a) 0.85
   b) 1.2345
   c) 3.56
   d) 10.

2. Find, correct to the nearest degree, the acute angle that has a tangent of:
   a) $\frac{2}{5}$
   b) $\frac{7}{9}$
   c) $\frac{25}{32}$
   d) $2\frac{3}{4}$.

3. Find, correct to 1 decimal place, the lettered angles in these diagrams.

   a)
   b)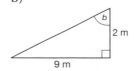
   c)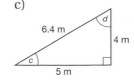

4. A ladder stands on horizontal ground and leans against a vertical wall. The foot of the ladder is 2.8 m from the base of the wall and the ladder reaches 8.5 m up the wall. Calculate the angle that the ladder makes with the ground.

   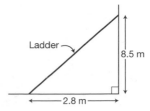

5. The top of a vertical cliff is 68 m above sea level. A ship is 175 m from the foot of the cliff. Calculate the angle of elevation of the top of the cliff from the ship.

6. Limpo is 48 km north and 64 km east of Onjo. Calculate the three-figure bearing of Limpo from Onjo.

   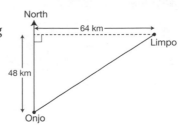

7. O is the centre of a circle with OM = 12 cm.
   a) Calculate AM.
   b) Calculate AB.

   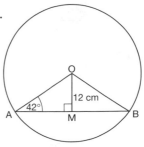

Hint

Sine is shortened to sin.
Cosine is shortened to cos.
Pronounce sin as 'sign', cos
as 'coz'.

Sine and cosine ratios

The sine and cosine ratios of a right-angled triangle allow us to solve problems that require the use of, or length of, the hypotenuse.

The definitions of the sine and cosine ratios are:

$$\sin(\text{angle}) = \frac{\text{opp(angle)}}{\text{hypotenuse}} \qquad \cos(\text{angle}) = \frac{\text{adj(angle)}}{\text{hypotenuse}}$$

For acute angles, the value of tan(angle) can be any size, but the values of sin(angle) and cos(angle) must be less than 1. This is because the hypotenuse is the longest side in the triangle.

Sine and cosine ratios are calculated in the same way as tangent ratios. Use the **sin** and **cos** keys on your calculator to find the values for acute angles.

If the sine or cosine is known, you can find the size of the angle using the inverse keys **sin⁻¹** or **cos⁻¹**.

■ Examples

1. For each of these triangles, write down the value of:
    (i) sin(A)   (ii) cos(A).

a)   b)   c)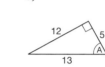

a) (i) $\sin(A) = \dfrac{\text{opp}(A)}{\text{hypotenuse}} = \dfrac{16}{20}$

    (ii) $\cos(A) = \dfrac{\text{adj}(A)}{\text{hypotenuse}} = \dfrac{12}{20}$

b) (i) $\sin(A) = \dfrac{\text{opp}(A)}{\text{hypotenuse}} = \dfrac{7}{25}$

    (ii) $\cos(A) = \dfrac{\text{adj}(A)}{\text{hypotenuse}} = \dfrac{24}{25}$

c) (i) $\sin(A) = \dfrac{\text{opp}(A)}{\text{hypotenuse}} = \dfrac{12}{13}$

    (ii) $\cos(A) = \dfrac{\text{adj}(A)}{\text{hypotenuse}} = \dfrac{5}{13}$

2. For each of these triangles, write down the ratio of sides corresponding to the trig ratio names.

a)   b) 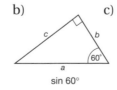   c) 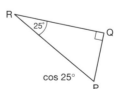   d)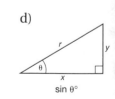

cos 42°   sin 60°   cos 25°   sin θ°

a) $\cos(42°) = \dfrac{\text{adj}(42°)}{\text{hypotenuse}} = \dfrac{g}{e}$ b) $\sin(60°) = \dfrac{\text{opp}(60°)}{\text{hypotenuse}} = \dfrac{c}{a}$

c) $\cos(25°) = \dfrac{\text{adj}(25°)}{\text{hypotenuse}} = \dfrac{QR}{PR}$ d) $\sin(\theta) = \dfrac{\text{opp}(\theta)}{\text{hypotenuse}} = \dfrac{y}{r}$

3. For each of these triangles, calculate, correct to 3 significant figures, the length of the lettered side.

a) The lettered side is *opposite* the angle of 25°, so we use

$$\sin(25°) = \frac{\text{opp}(25°)}{\text{hypotenuse}}$$

$$\sin(25°) = \frac{a}{20}$$

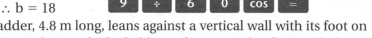

$$a = 20 \times \sin(25°)$$
$$= 8.452365235$$
$$a = 8.45 \text{ (to 3 significant figures)}$$

b) The lettered side is the hypotenuse and the adjacent side is 9, so we use

$$\frac{9}{b} = \cos(60°)$$

$$\therefore b = \frac{9}{\cos(60°)}$$

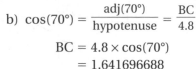

$$\therefore b = 18$$

4. A ladder, 4.8 m long, leans against a vertical wall with its foot on horizontal ground. The ladder makes an angle of 70° with the ground.
   a) How far up the wall does the ladder reach?
   b) How far is the foot of the ladder from the wall?

   In the diagram, AC is the hypotenuse of right-angled △ABC, AB = opp(70°) and BC = adj(70°).

   a) $\sin(70°) = \dfrac{\text{opp}(70°)}{\text{hypotenuse}} = \dfrac{AB}{4.8}$

   $$AB = 4.8 \times \sin(70°)$$
   $$= 4.51052458$$

   The ladder reaches 4.51 m up the wall.

   b) $\cos(70°) = \dfrac{\text{adj}(70°)}{\text{hypotenuse}} = \dfrac{BC}{4.8}$

   $$BC = 4.8 \times \cos(70°)$$
   $$= 1.641696688$$

   The foot of the ladder is 1.64 m from the wall (to 3 significant figures).

Exam tip

We could have used Pythagoras' theorem to calculate BC from the lengths of AC and AB. However, it is better to use trigonometry (in case we had made a mistake in calculating AB).

Exercise

1. For each of these triangles a) to d), write down the value of:
   (i) sin(A) (ii) cos(A).

   a) 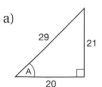 b)   c) 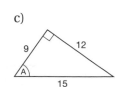 d)

2. Use your calculator to find, correct to 4 decimal places, the value of:
   a) sin(5°)   b) sin(30°)
   c) sin(60°)   d) sin(85°).

3. Use your calculator to find, correct to 4 decimal places, the value of:
   a) cos(5°)   b) cos(30°)
   c) cos(60°)   d) cos(85°).

4. For each of these triangles, write down the ratio of the side corresponding to the trig ratio named.

a)
cos 48°

b)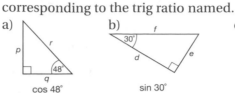
sin 30°

c)
cos 35°

d)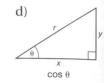
cos θ

5. For each of these triangles, calculate, correct to 3 significant figures, the length of the lettered side.

a)

b)

c)

d)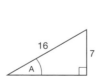

6. Use your calculator to find, correct to 1 decimal place:
   a) the acute angle whose sine is 0.99
   b) the acute angle whose cosine is 0.5432
   c) the acute angle whose sine is $\frac{3}{8}$
   d) the acute angle whose cosine is $\frac{10}{23}$.

7. Find, correct to the nearest degree, the size of the lettered angle in each of the following diagrams.

a)

b)

c)

d)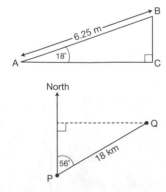

8. This diagram shows a ramp, AB, which makes an angle of 18° with the horizontal. The ramp is 6.25 m long. Calculate the difference in height between A and B. (This is the length of BC in the diagram.)

9. Village Q is 18 km from village P, on a bearing of 056°.
   a) Calculate the distance Q is north of P
   b) Calculate the distance Q is east of P.

10. A 15 m ladder is leaning against a wall. The base of the ladder is at 70° to the ground.
    a) At what height is the top of the ladder against the wall?
    b) How far is the base of the ladder from the wall?

Using trigonometry to solve problems

When you are asked to calculate lengths and/or angles, you should not use scale drawing. You may use Pythagoras' theorem or trigonometry. In order to do this, you must know the following:

- Pythagoras' theorem and its converse
- $\tan(\text{angle}) = \dfrac{\text{opp(angle)}}{\text{adj(angle)}}$ and if you know the tangent of an angle you can use $\tan^{-1}$ to find its size
- $\sin(\text{angle}) = \dfrac{\text{opp(angle)}}{\text{hypotenuse}}$ and $\sin^{-1}$ will give you the angle if you know its sine
- $\cos(\text{angle}) = \dfrac{\text{adj(angle)}}{\text{hypotenuse}}$ and $\cos^{-1}$ will give you the angle if you know its cosine.

Exam tip

One of the skills you need to demonstrate is the ability to select appropriate methods to solve problems. The steps given here will help you to be more systematic about doing this.

Some hints on solving trigonometry problems

- If no diagram is given, draw one yourself.
- In the diagram, mark the right angles.
- Show the sizes of the other angles that are known and the lengths of any lines that are known.
- Mark the angles or sides you have to calculate.
- Identify the right-angled triangle(s) that contain(s) the angles or sides you have to calculate.
- If it is a 3-dimensional problem, draw separate diagrams of the right-angled triangles you are going to use.
- Consider whether you need to create right-angled triangles by drawing extra lines in your diagram(s). For example, an isosceles triangles can be divided into two congruent right-angled triangles.
- Decide on the steps you will take to solve the problem – will you need Pythagoras' theorem, sine, cosine and/or tangent?
- Remember that there is often more than one way of solving a problem – if you can, choose the shortest method using the given information.
- If you have to use a length or angle you have already calculated, use the most accurate value you have – do not use a 'rounded' value.
- Show how you obtained your answer(s) – in examinations, marks are given for correct methods as well as for correct answers.
- Give sensible answer(s) to the problem – usually 3-figure accuracy for lengths and 1 decimal place accuracy for angles (unless the answer is exact). Remember that your calculator will give many more figures than is sensible in a practical problem.
- Check that your answer is reasonable – remember that the hypotenuse is the longest side in a right-angled triangle, and that the shortest side is opposite the smallest angle.

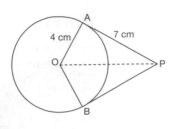

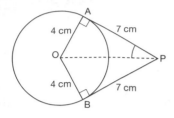

■ Examples

1. Tangents are drawn from P to touch the circle, centre O, at A and B. PA = 7 cm and OA = 4 cm.

   Calculate AP̂B.

   The angle between a tangent and the radius to the point of contact is 90°, so OÂP = 90° and OB̂P = 90°.

   Join O to P to form two congruent, right-angled triangles OAP and OBP.

   Let AP̂O = x and AP̂B = y.

   In △APO, $\tan x = \dfrac{\text{opp}(x)}{\text{adj}(x)} = \dfrac{4}{7}$

   $$\therefore x = \tan^{-1}\left(\frac{4}{7}\right)$$
   $$= 29.7448813°$$

   $$y = 2 \times x$$
   $$= 59.4897626°$$
   $$\therefore y = 59.5° \text{ (to 1 decimal place)}$$

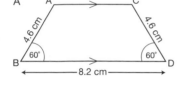

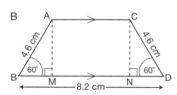

2. Diagram A represents an isosceles trapezium ABDC.

   Calculate the area of the trapezium.

   The area of a trapezium = (average of the parallel sides) × (perpendicular distance between them)

   In diagram B, we need to calculate the lengths of AM (or CN) and AC.

   AC = MN and we can find the length of MN if we calculate the lengths of BM and ND.

   In △ABM,
   $$\sin(60°) = \frac{\text{opp}(60°)}{\text{hypotenuse}} = \frac{AM}{4.6} \text{ and}$$
   $$\cos(60°) = \frac{\text{adj}(60°)}{\text{hypotenuse}} = \frac{BM}{4.6}$$

   Hence, AM = 4.6 × sin(60°) and
   BM = 4.6 × cos(60°)
   AM = 3.983716857 cm and
   BM = 2.3 cm

   By symmetry, ND = BM = 2.3 cm and
   ∴ MN = 8.2 − (2.3 + 2.3) = 3.6 cm

   Hence, AC = 3.6 cm and AM = CN = 3.983716857 cm

   The area of ABDC $= \left(\dfrac{AC + BD}{2}\right) \times AM$

   $$= \left(\frac{3.6 + 8.2}{2}\right) \times 3.983716857 \text{ cm}^2$$
   $$= 23.5039296 \text{ cm}^2$$

   Area of ABDC = 23.5 cm² (to 3 significant figures)

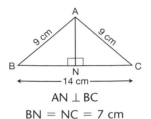

AN ⊥ BC

BN = NC = 7 cm

3. Calculate the angles of an isosceles triangle that has sides of length 9 cm, 9 cm and 14 cm.

In △ABN, $\cos(B) = \dfrac{\text{adj(B)}}{\text{hypotenuse}} = \dfrac{BN}{AB} = \dfrac{7}{9}$

$$\hat{B} = \cos^{-1}\left(\frac{7}{9}\right)$$
$$\hat{B} = 38.94244127°$$
$$\text{Similarly } \hat{C} = 38.94244127°$$
$$\hat{BAC} = 180° - (\hat{B} + \hat{C}) = 102.1151175°$$

Correct to 1 decimal place, the angles of the isosceles triangle ABC are 102.1°, 38.9° and 38.9°.

These values do not add up to 180° because each of them is approximate – that is, correct to 1 decimal place. If we give them correct to the nearest degree, they are 102°, 39° and 39°, which do add up to 180°!

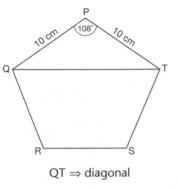

QT ⇒ diagonal

4. Find the length of a diagonal of a regular pentagon that has sides of length 10 cm.

Sum of ∠s = 540° ∴ QP̂T = 108°
By symmetry,
$$\hat{QPN} = \hat{TPN} = \frac{108°}{2} = 54°$$
and QN = NT

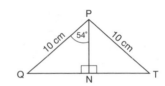

In △PQN,
$$\sin(54°) = \frac{\text{opp(54°)}}{\text{hypotenuse}} = \frac{QN}{10}$$

$$QN = 10 \times \sin(54°)$$
$$= 8.090169944$$
$$\therefore QT = 2 \times QN = 16.18033989$$

The length of each diagonal = 16.2 cm (to 3 significant figures)

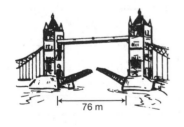

5. The span between the towers of Tower Bridge in London is 76 m.

Here is a simplified drawing of the bridge, showing the two halves raised to 35°.

How wide is the gap?

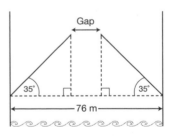

The gap = BD = MN
and MN = AC – (AM + NC).

The right-angled triangles ABM and CDN are congruent, so AM = NC.

When the two halves are lowered, they must meet in the middle.

$$\therefore AB = CD = \frac{76\text{ m}}{2} = 38\text{ m}$$

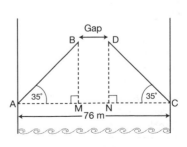

In $\triangle$ABM, $\cos(35°) = \dfrac{\text{adj}(35°)}{\text{hypotenuse}} = \dfrac{\text{AM}}{38}$

$$\begin{aligned} \text{AM} &= 38 \times \cos(35°) \\ &= 31.1277 \ldots \text{ m} \\ \therefore \text{MN} &= (76 - (31.1277 \ldots + 31.1277 \ldots)) \text{ m} \\ &= 13.744 \ldots \text{ m} \end{aligned}$$

The gap BD = 13.7 m (to 3 significant figures)

Exercise

1. The diagram represents a ramp AB for a lifeboat. AC is vertical and CB is horizontal.
   a) Calculate the size of AĈB correct to 1 decimal place.
   b) Calculate the length of BC correct to 3 significant figures.

2. AB is a chord of a circle, centre O, radius 8 cm. AÔB = 120°. Calculate the length of AB.

3. The diagram represents a tent in the shape of a triangular prism. The front of the tent, ABD, is an isosceles triangle with AB = AD. The width, BD, is 1.8 m and the supporting pole AC is perpendicular to BD and 1.5 m high. The tent is 3 m long. Calculate:
   a) the angle between AB and BD
   b) the length of AB
   c) the volume inside the tent.

4. The sketch represents a field PQRS on level ground.
   The sides PQ and SR run due east.
   a) Write down the bearing of S from P.
   b) Calculate the shortest distance between SR and PQ.
   c) Calculate, in square metres, the area of the field PQRS.

Note

If you are following the core syllabus, you have completed your work on trigonometry. Turn to page 208 and do the appropriate questions in Check your progress.

5. In the isosceles triangle DEF, Ê = F̂ = 35° and side EF = 10 cm.
   a) Calculate the perpendicular distance from D to EF.
   b) Calculate the length of the side DE.

In trig formulae, capital letters refer to angles and small letters refer to sides of a triangle.

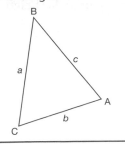

Trigonometry for any triangle

In the previous section, you learnt the trig formulae of acute angles found in right-angled triangles. In triangles without a right angle, one of the angles may be obtuse. In order to work with such triangles, you need to know the trig ratios of an obtuse angle.

Sine and cosine of an obtuse angle

Village Q is D kilometres from Village P on a bearing of A°.
How far east and how far north is Q from P?
If A is acute, then you work out sin(A) and cos (A).
From the diagrams on the right, you can see that:

$$\sin(A) = \frac{RQ}{d} \text{ and}$$

$$\cos(A) = \frac{PR}{d}$$

$\therefore$ RQ = $d\sin(A)$ and PR = $d\cos(A)$

Now, suppose A is obtuse.
$\hat{QPR} = 180° - A$
$\therefore \sin(180° - A) = \frac{RQ}{d}$ and

$$\cos(180° - A) = \frac{PR}{d}$$

$\therefore$ RQ = $d\sin(180° - A)$ and PR = $d\cos(180° - A)$

The sin and cosine rule for obtuse angles is:

If A is obtuse, sin A = + sin(180° − A) and cos A = −cos(180° − A).

These definitions are programmed into all scientific calculators.
Use you calculator to check that:

$$\sin(100°) = +0.984807753 = \sin(80°)$$
$$\cos(100°) = -0.173648177 = -\cos(80°)$$
$$\sin(150°) = +0.5 = \sin(30°)$$
$$\cos(150°) = -0.866025403 = -\cos(30°)$$

Your calculator will also give these results:

$$\sin(0°) = 0 \qquad \cos(0°) = 1$$
$$\sin(90°) = 0 \qquad \cos(90°) = 0$$
$$\sin(180°) = 0 \qquad \cos(180°) = -1$$

Area of a triangle

How can you find the area of a triangle when you know the lengths of two sides and the size of the angle between these sides?

Suppose the known sides are a and b, and C is the angle between them.

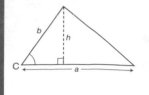

a) If C is acute and side a is taken as the base, the perpendicular height (h) of the triangle is given by $\sin(C) = \dfrac{h}{b}$.
Hence $h = b\sin(C)$.

The area of the triangle
$= \frac{1}{2}$ base $\times$ perpendicular height
$= \frac{1}{2}\, a \times b\sin(C)$
$= \frac{1}{2}\, ab\sin(C)$

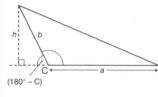

b) If C is obtuse, h is given by $\sin(180° - C) = \dfrac{h}{b}$.
Hence $h = b\sin(180° - C)$.

The area of the triangle
$= \frac{1}{2}$ base $\times$ perpendicular height
$= \frac{1}{2}\, a \times b\sin(180° - C)$
$= \frac{1}{2}\, ab\sin(180° - C)$

However, we have already defined the sine of an obtuse angle to be the same as the sine of (180° – the angle), so $\frac{1}{2}ab\sin(180° - C) = \frac{1}{2}ab\sin(C)$.

We can therefore use the following formula for all triangles:

> The area of a triangle $= \frac{1}{2}\, ab\sin(C)$.

Note the following:
1. Sin(C) is usually written as sin C, but the brackets in sin(180° – C) cannot be omitted.
2. Because sin 90° = 1, the formula applies to right-angled triangles.
3. The formula may be written in words:
   Area of a triangle
   = half (product of any two sides) $\times$ (sine of the angle between them).
   The area of $\triangle$ABC can be written as
   $\frac{1}{2}ab\sin C$ or $\frac{1}{2}bc\sin A$ or $\frac{1}{2}ca\sin B$.

Notice that, starting with the first version, the letters have moved around cyclically to obtain the other versions.

Similarly, area of a triangle PQR $= \frac{1}{2}pq\sin R = \frac{1}{2}qr\sin P = \frac{1}{2}rp\sin Q$.

a)

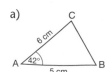

b)

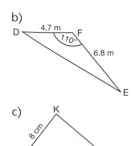

c)

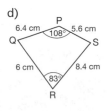

d)

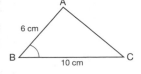

■ Examples

1. Find the area of shapes a) to d).

   a) A = 42°, b = 6 cm and c = 5 cm.

   Area of triangle = $\frac{1}{2}bc\sin A = \frac{1}{2} \times 6 \times 5 \times \sin 42°$

   $= 10.0369591$

   $= 10.0$ cm^2 (to 3 significant figures)

   b) F = 110°, d = 6.8 m and e = 4.7 m.

   Area of triangle = $\frac{1}{2}de\sin F = \frac{1}{2} \times 6.8 \times 4.7 \times \sin 110°$

   $= 15.01628808$

   $= 15.0$ m^2 (to 3 significant figures)

   c) H = 60°, j = 8 cm and k = 15 cm.

   Area of triangle = $\frac{1}{2} \times 8 \times 15 \times \sin 60°$

   $= 51.96152423$

   $= 52.0$ cm^2 (to 3 significant figures)

   d) Divide the quadrilateral into two triangles, PQS and RQS.

   Area of △PQS $= \frac{1}{2} \times 6.4 \times 5.6 \times \sin 108°$

   $= 17.04293277$ cm^2

   Area of △RQS $= \frac{1}{2} \times 6 \times 8.4 \times \sin 83°$

   $= 25.01216302$

   Area of quadrilateral PQRS $= 42.95509579$ cm^2

   $= 42.1$ cm^2 (to 3 significant figures)

2. In an acute-angled triangle DEF, the sine of D̂ is 0.45.
   Find D̂ correct to the nearest degree.

   On the calculator, press:

   | . | 4 | 5 | shift | sin^{-1} |

   The display is:

   | DEG | 2 | 6 | . | 7 | 4 | 3 | 6 | 8 | 3 | 9 | 5 |

   ∴ D̂ = 27° (to the nearest degree)

3. In △ABC, side AB = 6 cm and side BC = 10 cm.
   The area of the triangle is 15 cm^2.
   What are the possible sizes of B̂?

   Area of △ABC = $\frac{1}{2}ca\sin B$

   where a = 10 cm and c = 6 cm.

   ∴ $\frac{1}{2} \times 6 \times 10 \times \sin B = 15$

   $30 \times \sin B = 15$

   $\sin B = 0.5$

On the calculator, press the keys [.] [5] [shift] [sin⁻¹]

The display is [DEG] [3] [0] [.]

This means that B̂ could be 30°.
However, sin 150° = sin 30° because 180° − 30° = 150°.
∴ B̂ is either 30° or 150°.

Scientific calculators are programmed to give the smallest possible value of the angle in cases such as this. You will have to decide whether there is more than one possible answer. The problem does not arise when you are working with cosines. The *cosine* of an *obtuse* angle is *negative*, whereas the cosine of an acute angle is positive.

Exercise

1. Use your calculator to find, correct to 4 decimal places:
   a) sin 145°   b) cos 145°   c) sin 150°   d) cos 150°.
2. Find the area of the shapes below:

   a)   b) 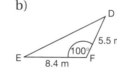   c)   d)

3. Find the area of parallelogram ABCD in which AB = 9 cm, AD = 12 cm and Â = 95°.
4. a) Given that sin A = 0.83, find the value of Â to the nearest degree:
      (i)   when Â is acute
      (ii) when Â is obtuse.
   b) Find, correct to 1 decimal place, the obtuse angle that has a cosine of −0.48.
5. The diagram shows △PQR, which has an area of 630 cm².
   a) Explain how you can tell that Q̂ is acute.
   b) Use the formula $A = \frac{1}{2} pr \sin Q$ to find Q̂ correct to 1 decimal place.
   c) Find P̂ correct to 1 decimal place.

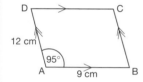

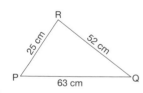

Solving a triangle

You can construct a triangle if you know:
- the lengths of three sides
- the lengths of two sides and the size of the angle between them
- the length of one side and the sizes of two angles.

If a triangle can be constructed, it is possible to calculate its remaining sides and angles. This is called solving the triangle. The formulae used to solve a triangle are called the sine rule and the cosine rule.

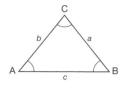

Note

Both versions of the sine rule indicate that the sides of a triangle are proportional to the sines of the angles opposite to them.

The sine rule

The area of $\triangle ABC$ can be written as $\frac{1}{2}bc\sin A$ or $\frac{1}{2}ca\sin B$ or $\frac{1}{2}ab\sin C$.

It follows that

$\frac{1}{2}bc\sin A = \frac{1}{2}ca\sin B = \frac{1}{2}ab\sin C$.

Dividing by $\frac{1}{2}abc$ gives us $\dfrac{\sin A}{a} = \dfrac{\sin B}{b} = \dfrac{\sin C}{c}$.

This is a version of the sine rule that is convenient for calculating angles. Inverting each fraction gives us the version that is more convenient for calculating sides:

$$\frac{a}{\sin A} = \frac{b}{\sin B} = \frac{c}{\sin C}$$

You need to remember the following:

> For any triangle ABC, $\dfrac{a}{\sin A} = \dfrac{b}{\sin B} = \dfrac{c}{\sin C}$ and $\dfrac{\sin A}{a} = \dfrac{\sin B}{b} = \dfrac{\sin C}{c}$.

Of course, you may have to use different letters in different triangles. For example, in $\triangle PQR$,

$$\frac{p}{\sin P} = \frac{q}{\sin Q} = \frac{r}{\sin R} \text{ and } \frac{\sin P}{p} = \frac{\sin Q}{q} = \frac{\sin R}{r}.$$

To make use of the sine rule, you need to know the numerator and denominator of one of the fractions, and the numerator or denominator of one of the other fractions.

■ Examples

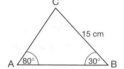

1. In $\triangle ABC$, $\hat{A} = 80°$, $\hat{B} = 30°$ and side BC = 15 cm.

   Calculate the size of $\hat{C}$ and the lengths of the sides AB and AC.

   $\hat{C} = 70°$ ($\angle$ sum of $\triangle$)

   Sine rule: $\dfrac{a}{\sin A} = \dfrac{b}{\sin B} = \dfrac{c}{\sin C}$

   $\therefore \dfrac{15}{\sin 80°} = \dfrac{AC}{\sin 30°} = \dfrac{AB}{\sin 70°}$

   $AC = \dfrac{15 \times \sin 30°}{\sin 80°} = 7.615699589$

   $AB = \dfrac{15 \times \sin 70°}{\sin 80°} = 14.31283341$

   $\therefore$ AB = 14.3 cm and AC = 7.62 cm (to 3 significant figures)

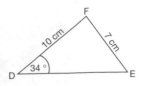

2. In △DEF, DF = 10 cm, EF = 7 cm and D̂ = 34°.
   Calculate, to the nearest degree, the possible size of
   a) Ê and
   b) F̂.

   a) $\dfrac{\sin D}{d} = \dfrac{\sin E}{e} = \dfrac{\sin F}{f}$

   Find Ê first because you know the opposite length.

   Remember that d = EF, e = DF and f = DE,

   $\therefore \dfrac{\sin 34°}{7} = \dfrac{\sin E}{10} = \dfrac{\sin F}{f}$.

   Using the first and second fractions, $\sin E = \dfrac{10 \times \sin 34°}{7}$

   $= 0.798847005$

   Ê could be the acute angle 53.0201406°, but it could also be the obtuse angle 126.9798599°. (There is no reason to reject either of these values.)

   b) F̂ is found by using the angle sum of a triangle = 180°.

   To the nearest degree, you can obtain Ê = 53° and F̂ = 93° (this gives △DFE₁ in the diagram) or Ê = 127° and F̂ = 19° (this gives △DFE₂ in the diagram).

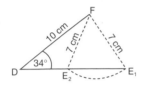

Exercise

1.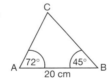

   In △ABC, Â = 72°, B̂ = 45° and side AB = 20 cm. Calculate the size of Ĉ and the lengths of the sides AC and BC.

2.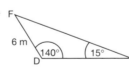

   In △DEF, D̂ = 140°, Ê = 15° and side DF = 6 m. Calculate the size of F̂ and the lengths of the sides DE and EF.

3.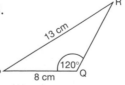

   In △PQR, Q̂ = 120°, side PQ = 8 cm and side PR = 13 cm. Calculate the size of R̂, the size of P̂, and the length of side QR.

4.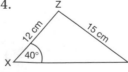

   In △XYZ, X̂ = 40°, side XZ = 12 cm and side YZ = 15 cm.
   a) Explain why Ŷ must be less than 40°.
   b) Calculate, correct to 1 decimal place, Ŷ and Ẑ.
   c) Calculate the length of the side XY.

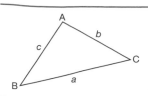

The cosine rule

When you know the lengths of three sides of a triangle or the lengths of two sides and the size of the angle between them, the sine rule is not useful because you do not have enough information to use it. In these cases, you use the cosine rule.

The cosine rule is a generalisation of Pythagoras' theorem and it can be proved using the theorem. For the IGCSE examination, you do not need to know the proof. However, you do need to learn the cosine rule and be able to apply it.

The cosine rule is:

> For any $\triangle ABC$, $c^2 = a^2 + b^2 - 2ab \cos C$.

Hints for using the cosine rule

- To avoid the most common source of error in using the cosine rule, you should write it as $c^2 = a^2 + b^2 - (2ab \cos C)$. The values of a^2, b^2 and $(2ab \cos C)$ should be worked out before any addition or subtraction is done.
- As with formulae for the area of a triangle, the letters can be moved around cyclically to obtain other versions of the cosine rule:
  $a^2 = b^2 + c^2 - (2bc \cos A)$ and $b^2 = c^2 + a^2 - (2ca \cos B)$.
- The cosine rule gives you the square of one side. You must work out the square root if you need the length of a side.
- When you know the lengths of all three sides of a triangle and you need the size of an angle, you can use the cosine rule in the form:
  $$\cos C = \frac{a^2 + b^2 - c^2}{2ab}.$$

■ Examples

1. In $\triangle ABC$, $\hat{B} = 50°$, side $AB = 9$ cm and side $BC = 18$ cm.

   Calculate the length of AC.

   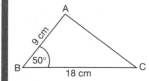

   Notice that $AC = b$ and we know that $\hat{B} = 50°$.
   We use the cosine rule in the form
   $b^2 = c^2 + a^2 - (2ca \cos B)$

   $b^2 = 9^2 + 18^2 - (2 \times 9 \times 18 \times \cos 50°)$
   $= 81 + 324 - (208.2631855)$
   $= 196.7368145$

   $b = \sqrt{196.7368145}$
   $= 14.02629012$

   Length of $AC = 14.0$ cm (to 3 significant figures)

2. In △DEF, F̂ = 120°, side EF = 25 m and side FD = 34 m.

   Calculate the length of side DE.

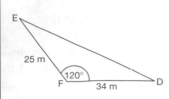

   DE = f, so we use the cosine rule in the form
   $$f^2 = d^2 + e^2 - (2de\cos F)$$
   $$f^2 = 25^2 + 34^2 - (2 \times 25 \times 34 \times \cos 120°)$$
   $$= 625 + 1\ 156 - (-850) \text{ (notice that cos 120° is negative)}$$
   $$= 625 + 1\ 156 + 850$$
   $$= 2\ 631$$
   $$\therefore f = \sqrt{2\ 631} = 51.29327441$$

   Length of DE = 51.3 m (to 3 significant figures)

3. In △PQR, R̂ = 100°, side PR = 8 cm and side RQ = 5 cm.
   a) Calculate the length of side PQ.
   b) Calculate, correct to the nearest degree, P̂ and Q̂.

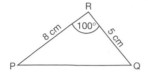

   a) PQ = r, so we use the cosine rule in the form
   $$r^2 = p^2 + q^2 - (2pq\cos R)$$
   $$= 5^2 + 8^2 - (2 \times 5 \times 8 \times \cos 100°)$$
   $$= 25 + 64 - (-13.89185421) \text{ (notice that cos 100° is negative)}$$
   $$= 102.8918542$$
   $$r = \sqrt{102.8918542} = 10.14356221$$

   Length of PQ = 10.1 cm (to 3 significant figures)

   b) Now we know the values of r and R̂, we can make use of the sine rule
   $$\frac{\sin P}{p} = \frac{\sin Q}{q} = \frac{\sin R}{r}$$
   $$\frac{\sin P}{5} = \frac{\sin Q}{8} = \frac{\sin 100°}{10.14356...}$$

   Using the first and third fractions, $\sin P = \frac{5 \times \sin 100°}{10.14356...} = 0.48534866$

   R̂ is obtuse so P̂ is acute, and P̂ = 29.04096759°
   P̂ = 29° (to the nearest degree)

   To find Q̂ we can use the angle sum of a triangle = 180°:
   $$Q̂ = 180° - (100° + 29°)$$
   $$\therefore Q̂ = 51° \text{ (to the nearest degree)}$$

   Alternatively, we could go back to the sine rule statement and obtain:
   $$\sin Q̂ = \frac{8 \times \sin 100°}{10.14356...} = 0.776695955$$

   This gives Q̂ = 50.95904775°
   $$\therefore Q̂ = 51° \text{ (to the nearest degree)}$$

4. a) Change the subject of the formula $c^2 = a^2 + b^2 - (2ab\cos C)$ to cos C.

b) Use your answer to part a) to find the smallest angle in the triangle which has sides of length 7 m, 8 m and 13 m.

a) $$c^2 = a^2 + b^2 - (2ab\cos C)$$
$$(2ab\cos C) = a^2 + b^2 - c^2$$
$$\cos C = \frac{a^2 + b^2 - c^2}{2ab}$$

b) The smallest angle in a triangle is opposite the shortest side.
In the given triangle, the smallest angle is opposite the 7 m side.
Let this angle be C. Then $c = 7$, and we can take $a = 8$ and $b = 13$.
Using the result of part a),
$$\cos C = \frac{8^2 + 13^2 - 7^2}{2 \times 8 \times 13}$$
$$\cos C = \frac{64 + 169 - 49}{208}$$
$$\cos C = \frac{184}{208}$$
$$\hat{C} = \cos^{-1}\left(\frac{184}{208}\right) = 27.7957725°$$

The smallest angle of the triangle = 27.8° (to 1 decimal place)

Exercise

1. In $\triangle ABC$, $\hat{B} = 45°$, side AB = 10 cm and side BC = 12 cm. Calculate the length of side AC.

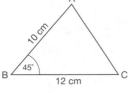

2. In $\triangle DEF$, $\hat{F} = 150°$, side EF = 9 m and side FD = 14 m. Calculate the length of side DE.

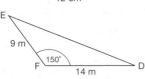

3. In $\triangle PQR$, side PQ = 11 cm, side QR = 9 cm and side RP = 8 cm. Calculate the size of $\hat{P}$ correct to 1 decimal place.

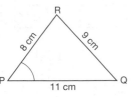

4. In $\triangle STU$, $\hat{S} = 95°$, side ST = 10 m and side SU = 15 m.
    a) Calculate the length of side TU.
    b) Calculate $\hat{U}$.
    c) Calculate $\hat{T}$.

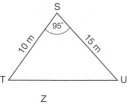

5. In $\triangle XYZ$, side XY = 15 cm, side YZ = 13 cm and side ZX = 8 cm. Calculate the size of:
    a) $\hat{X}$
    b) $\hat{Y}$
    c) $\hat{Z}$.

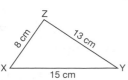

For reflex angles the relationships of sines and cosines are:

For $180° \leqslant A \leqslant 270°$
sin A = –sin (A – 180°)
cos A = –cos (A – 180°)

For $270° \leqslant A \leqslant 360°$
sin A = –sin (360° – A)
cos A = cos (360° – A)

Sine and cosine functions

We have worked with the sine and cosine of angles up to 180°. You can work out the sine and cosine of angles that are greater than 180°. Such angles occur in bearings and in rotating shafts of machinery. You do not need to learn rules for working these out, because your calculator is programmed for all sizes of angles. However, as you are going to graph sine and cosine functions, you need to round the values given by your calculator to 2 or 3 significant figures. Some values may be negative, so you also need to take note of the signs on the calculator display.

■ Examples

1. Draw the graph of $y = \sin x°$ for $0 \leq x \leq 360$.

   Use a calculator, and round values to 2 decimal places, to obtain this table.

x	0	30	60	90	120	150	180	210	240	270	300	330	360
sin x°	0	0.50	0.87	1	0.87	0.50	0	–0.50	–0.87	–1	–0.87	–0.50	0

Here is the graph.

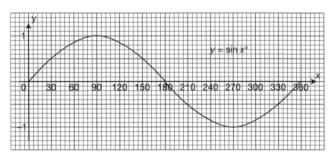

2. a) Draw the graph of $y = 1 + 2\cos x°$ for $0 \leq x \leq 360$.
   b) Use the graph to find two solutions of the equation $1 + 2\cos x° = 1.4$.

x	0	30	60	90	120	150	180	210	240	270	300	330	360
y	3	2.73	2	1	0	–0.73	–1	–0.73	0	1	2	2.73	3

a)

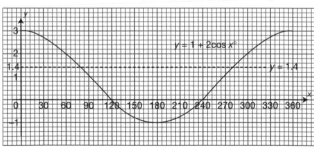

b) On the diagram, the line $y = 1.4$ has been drawn.
   It intersects the graph of $y = 1 + 2\cos x°$ at the points where $x = 78$ and $x = 282$. (Each small square on the x-axis is 6 units.)

Exercise

This is the graph of sin $x°$ for $0 \leq x \leq 360$. Note that the period of the graph is 360°. This means that the graph repeats every 360°.

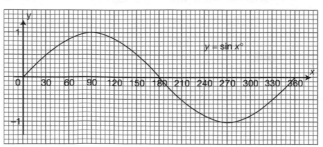

1. Find the value of:
   a) sin 70°
   b) sin 145°
   c) sin 275°
   d) sin 352°
   e) sin 400°
   f) sin 700°.

2. Find two values of z between 0° and 360° for:
   a) sin z = 0.906
   b) sin z = 0.5
   c) sin z = −0.087.

3. Draw the graph of $y = \cos x°$ for $0 \leq x \leq 360$.
   (Plot points for x = 0, 30, 60, 90, …, 360.)

4. a) Given that $y = \sin x° - \cos x°$, complete this table of values.
       (Give the values correct to 2 decimal places.)

x	0	15	30	45	60	75	90	105	120	135	150	165	180
y	−1	−0.71	−0.37	0		0.71	1		1.37	1.41		1.22	1

x	195	210	225	240	255	270	285	300	315	330	345	360
y		0.37	0		−0.71	−1		−1.37	−1.41		−1.22	−1

   b) Draw the graph of $y = \sin x° - \cos x°$ for $0 \leq x \leq 360$.

5. a) Given that $y = 1 + 2\sin x°$, complete this table of values.

x	0	30	60	90	120	150	180	210	240	270	300	330	360
y	1		2.73	3		2	1		−0.73	−1	−0.73		1

   b) On a set of axes, draw the graph of $y = 1 + 2 \sin x°$ for $0 \leq x \leq 360$.

Applied trigonometry

When you work with solids, you may need to calculate the angle between an edge, or a diagonal, and one of the faces. This is called the angle between a line and a plane.

Consider a line PQ, which meets a plane ABCD at point P. Through P draw lines PR_1, PR_2, PR_3 … in the plane and consider the angles QPR_1, QPR_2, QPR_3 …
- If PQ is perpendicular to the plane, all these angles will be right angles.
- If PQ is not perpendicular to the plane, these angles will vary in size.

It is the smallest of these angles which is called the angle between the line PQ and the plane ABCD.

To identify this angle, do the following:
- From Q draw a perpendicular to the plane. Call the foot of this perpendicular R.
- The angle between the line PQ and the plane is $Q\hat{P}R$.

PR is called the projection of PQ on the plane ABCD.

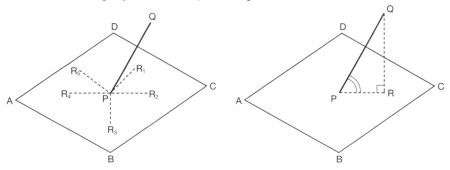

■ Example

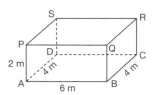

The diagram represents a room which has the shape of a cuboid. AB = 6 m, AD = 4 m and AP = 2 m. Calculate the angle between the diagonal BS and the floor ABCD.

First identify the angle required. B is the point where the diagonal BS meets the plane ABCD.

SD is the perpendicular from S to the plane ABCD and so DB is the projection of SB on the plane.

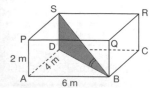

The angle required is $S\hat{B}D$.

We know that $\triangle SBD$ has a right angle at D and that SD = 2 m (equal to AP).

To find $S\hat{B}D$, we need to know the length of DB or the length of SB.

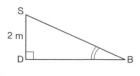

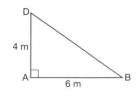

We can find the length of BD by using Pythagoras' theorem in $\triangle$ABD.

$$(BD)^2 = 6^2 + 4^2 = 36 + 16 = 52$$
$$BD = \sqrt{52}$$

∴ using right-angled $\triangle$SBD,

$$\tan S\hat{B}D = \frac{\text{opposite}}{\text{adjacent}} = \frac{SD}{BD} = \frac{2}{\sqrt{52}}$$

$$S\hat{B}D = \tan^{-1}\left(\frac{2}{\sqrt{52}}\right) = 15.50135957°$$

The angle between diagonal BS and the floor ABCD = 15.5° (to 1 decimal place)

Using formulae

To solve IGCSE problems, you will need to use a variety of trig formulae and techniques. Make sure you know the following:

The sine rule $\dfrac{a}{\sin A} = \dfrac{b}{\sin B} = \dfrac{c}{\sin C}$.

The cosine rule $c^2 = a^2 + b^2 - (2ab\cos C)$.

The area of a triangle $= \dfrac{1}{2}ab\sin C$.

These apply to any triangle, including right-angled triangles (provided you remember that sin 90° = 1 and cos 90° = 0).

However, for right-angled triangles it is usually better to use:

$$\text{sine} = \frac{\text{opposite}}{\text{hypotenuse}}, \quad \text{cosine} = \frac{\text{adjacent}}{\text{hypotenuse}}, \quad \text{tangent} = \frac{\text{opposite}}{\text{adjacent}}.$$

■ Examples

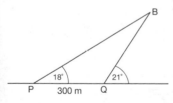

1. Two points, P and Q, are 300 m apart on horizontal ground. They are both due south of a stationary balloon, B. From P, the angle of elevation of B is 18° and, from Q, the angle of elevation is 21°. Calculate the height of the balloon.

Let N be the point on the ground that is vertically below the balloon.
$$P\hat{Q}B = 180° - 21° \; (\angle s \text{ on a straight line})$$
$$= 159°$$
$$P\hat{B}Q = 21° - 18° \; (\text{exterior } \angle \text{ of triangle})$$
$$= 3°$$

In $\triangle$PQB, $\dfrac{b}{\sin B} = \dfrac{p}{\sin P} = \dfrac{q}{\sin Q}$

$$\frac{300}{\sin 3°} = \frac{BQ}{\sin 18°} = \frac{PB}{\sin 159°}$$

$$\therefore BQ = \frac{300 \times \sin 18°}{\sin 3°} = 1\,771.346$$

$\triangle$BQN is right-angled at N, so

$$\sin 21° = \frac{\text{opposite}}{\text{hypotenuse}} = \frac{BN}{BQ}$$

$$BN = BQ \times \sin 21°$$
$$= 1771.346 \times \sin 21°$$
$$= 634.7937...$$

Height of the balloon = 635 m (to 3 significant figures)

2. The number of hours of daylight in the north of Iceland is given approximately by $12 - 12\cos(30t + 10)°$, where t is the time in months which has passed since 1 January.
   a) Calculate the number of hours of daylight on 1 August.
   b) Draw a graph of the function $h = 12 - 12\cos(30t + 10)°$ for $0 \le t \le 12$.
   c) When will there be 11 hours of daylight?

   a) 1 January to 1 August is 7 months.
      When $t = 7$, $12 - 12\cos(30t + 10)° = 12 - 12\cos 220°$
      $$= 12 - (-9.1925...)$$
      $$= 21.1925...$$

      Number of hours of daylight on 1 August = 21.2 (to 3 significant figures)

   b) This table gives values of h correct to 1 decimal place (obtained by using a calculator).

t	0	1	2	3	4	5	6	7	8	9	10	11	12
h	0.2	2.8	7.9	14.1	19.7	23.3	23.8	21.2	16.1	9.9	4.3	0.7	0.2

   Here is the graph.

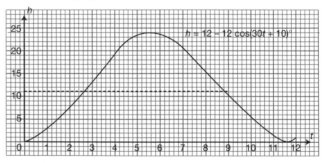

   c) From the graph, $h = 11$ when $t = 2.7$ and when $t = 8.9$. There will be (approximately) 11 hours of daylight on 21 March (2.7 months after 1 January) and on 27 September (8.9 months after 1 January).

Exercise

1. The diagram represents a triangular prism. The rectangular base, ABCD, is horizontal. AB = 20 cm and BC = 15 cm. The cross-section of the prism, BCE, is right-angled at C and EB̂C = 41°.

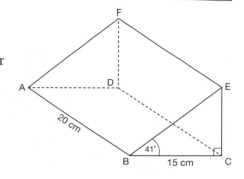

   a) Calculate the length of AC.
   b) Calculate the length of EC.
   c) Calculate the angle which the line AE makes with the horizontal.

2. To find the height, CD, of a building, Jim took measurements from two points, A and B, 50 m apart, as shown in the diagram. B, A and D are in a straight line on horizontal ground, and AD̂C = 90°. The angle

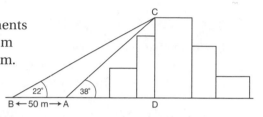

   of elevation of the top of the building from A is 38° and from B it is 22°.

   a) Calculate the distance BC.
   b) Calculate the height, CD, of the building.

3. The diagram shows the relative positions of Osaka (O), Tokyo (T) and Sapporo (S) in Japan. ST = 850 km, TO = 400 km and ST̂O = 110°.

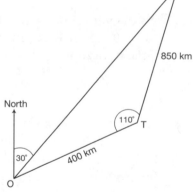

   a) Calculate OS, the distance from Osaka to Sapporo.
   b) Calculate SÔT, to the nearest degree.
   c) The bearing of Sapporo from Osaka is 030°. Find the bearing of Osaka from Tokyo.
   d) A plane flew from Sapporo to Tokyo at an average speed of 500 km/h. It left Sapporo at 0930. At what time did it arrive in Tokyo?

4. The depth, d metres, of water in a harbour on a certain day was given by $d = 4 - 3\sin(30t)°$, where t is the number of hours after midnight.

   a) Calculate the depth of water in the harbour at:
      (i) 2 a.m.
      (ii) 10 a.m.

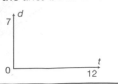
b) Complete this table, showing values of *d* correct to 1 decimal place.

t	0	1	2	3	4	5	6	7	8	9	10	11	12
d	4	2.5		1		2.5	4		6.6	7		5.5	4

c) On a sheet of graph paper, draw a graph of the function
$d = 4 - 3\sin(30t)°$.

d) Between which times was the depth of water more than 6 m?

Check your progress

1. Evaluate $15\cos(40°)$.
   a) Write down the answer as accurately as your calculator will allow.
   b) Round off your answer to 3 significant figures.

2. The diagram shows the cross-section of the roof of Mr Haziz's house.
   The house is 12 m wide, $C\hat{A}B = 35°$ and $A\hat{C}B = 90°$. Calculate the
   lengths of the two sides of the roof, AC and BC.

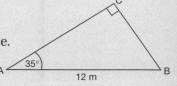

3. The diagram shows a trapezium ABCD in which $A\hat{B}C = B\hat{C}D = 90°$.
   $AB = 90$ mm, $BC = 72$ mm and $CD = 25$ mm. Calculate the perimeter
   of the trapezium.

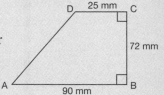

4. A girl, whose eyes are 1.5 m above the ground, stands 12 m away
   from a tall chimney. She has to raise her eyes 35° upwards from
   the horizontal to look directly at the top of the chimney. Calculate
   the height of the chimney.

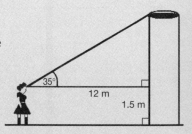

5. The diagram shows the cross-section PQRS of a cutting
   made for a road. PS and QR are horizontal. PQ makes
   an angle of 50° with the horizontal.
   a) Calculate the horizontal distance between P and Q
      (marked *x* in the diagram).
   b) Calculate the angle which RS makes with the horizontal (marked *y* in the diagram).

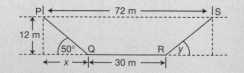

6. A game warden is standing at a point P alongside a road which runs north-south. There is a marker post at the point X, 60 m north of his position. The game warden sees a lion at Q on a bearing of 040° from him and due east of the marker post.

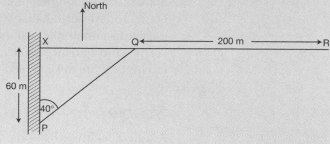

a) (i) Show by calculation that the distance, QX, of the lion from the road is 50.3 m, correct to 3 significant figures.

(ii) Calculate the distance, PQ, of the lion from the game warden.

b) Another lion appears at R, 200 m due east of the first one at Q.

(i) Write down the distance XR.

(ii) Calculate the distance, PR, of the second lion from the game warden.

(iii) Calculate the bearing of the second lion from the game warden, correct to the nearest degree.

7. In the △OAB, AÔB = 15°, OA = 3 m and OB = 8 m.
Calculate, correct to 2 decimal places:

a) the length of AB

b) the area of △OAB.

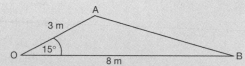

8. A pyramid, VPQRS, has a square base, PQRS, with sides of length 8 cm. Each sloping edge is 9 cm long.

a) Calculate the perpendicular height of the pyramid.

b) Calculate the angle the sloping edge VP makes with the base.

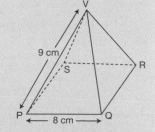

9. The diagram shows the graph of $y = \sin x°$ for $0 \le x \le 360$.

a) Write down the coordinates of A, the point on the graph where $x = 90°$.

b) Find the value of sin 270°.

c) Draw, on the diagram, the line $y = -\frac{1}{2}$ for $0 \le x \le 360°$.

d) How many solutions are there for the equation
$\sin x = -\frac{1}{2}$, for $0 \le x \le 360$?

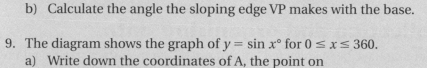

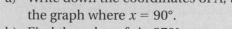

10. Two ships leave port P at the same time. One ship sails 60 km on a bearing of 030° to position A. The other ship sails 100 km on a bearing of 110° to position B.

a) Calculate:

(i) the distance AB

(ii) PÂB

(iii) the bearing of B from A.

b) Both ships took the same time, t hours, to reach their positions. The speed of the *faster* ship was 20 km/h. Write down:

(i) the value of t

(ii) the speed of the slower ship.

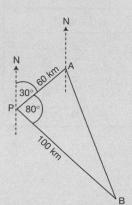

Statistics

Statistics is the branch of mathematics in which facts and information are collected, sorted, displayed and analysed. Statistics are used to make decisions and predict what may happen in the future.

The word statistics comes from the word 'state', largely because it was the job of the state to keep records and make decisions based on census results.

Stages in a statistical investigation

A statistical investigation normally involves the following steps:
- Identifying the problem that will be investigated. Problems can be simple (What is the most popular food for sale at a school tuckshop?) or more complicated (Does smoking cause cancer?).
- Collecting the data that you need for making a decision and solving the problem. Data can be in the forms of a list, a table or a graph. The data can be collected in various ways, but observation, interviewing people (surveys) and using questionnaires are the most popular methods. The number of things observed, or people surveyed or questioned is called the sample.
- Studying the data, drawing conclusions based on the data and making decisions.

Collecting and organising data

Data that have been collected but not organised in any way are called raw data. Table 1 contains raw data – the marks obtained by 25 students in a test in the order that they were gathered.

Table 1				
5	5	7	1	6
7	7	5	6	2
5	7	8	7	9
7	10	8	6	3
4	1	8	4	9

Raw data are difficult to interpret. If you look at Table 1, you might be able to find the minimum and maximum marks, you could also find an average mark by adding them together and dividing by 25. To get more information and to get it quickly, data need to be organised in some way. The simplest way to do this is by forming an array. In an array, data are arranged according to size. This can be in ascending or descending order. Table 2 is an array of the 25 marks from Table 1. It is arranged in ascending order.

Table 2				
1	1	2	3	4
4	5	5	5	5
6	6	7	7	7
7	7	7	7	8
8	8	9	9	10

An array has certain advantages over raw data. You can see quite quickly that the marks range from 1 to 10. You can also see a concentration of values near 7. The array also roughly reveals the distribution pattern of the marks.

However, the array is still a cumbersome and time-consuming method of organising data, especially if a very large sample is involved. Also, once a few types of information have been obtained, the array is not very useful. It is far more useful to compress data into a more manageable, shorter form.

Frequency tables

Raw data can be arranged in a frequency table. A frequency table shows the number of times (frequency) each value occurs. A frequency table of the raw data on page 210 would look like this:

Mark obtained	1	2	3	4	5	6	7	8	9	10
Frequency	2	1	1	2	4	3	6	3	2	1

Note the following:
- the first row contains the numerical data, in this case the marks students can obtain
- the row marked frequency tells you how many times each result appeared.

Tallying

Tallying is a system of recording and counting results using diagonal lines grouped in fives. Each time five is reached, a horizontal line is drawn through the tally marks to make a group of five. The next line starts a new group. For example:

1 → | 6 → ЖΗ|
4 → |||| 10 → ЖΗ ЖΗ
5 → ЖΗ   12 → ЖΗ ЖΗ ||

Tallying is convenient because it easy to count in fives; this reduces the chances of making mistakes.

Graphical representation

A diagram can help you see information at a glance. Most data are represented visually in the form of graphs. However, you need to understand the uses of different graphs and be able to choose the type that is most appropriate for the data you have collected. In this module, you will learn about pictograms, bar graphs, pie charts, histograms and line graphs.

You have already worked with graphs in Module 3, so you should know about axes and plotting points. When you work with statistical graphs, you should remember the following principles:
- The object of any graph is to present the data as simply and clearly as possible. Lines that criss-cross each other are confusing, so you should avoid using them.
- The axes of any graph must be graded and scales must be chosen carefully. If you exaggerate or compress the scale of a graph, you may distort the information shown on the graph.
- The axes must be labelled to show what is marked on them.
- Every graph should be given a title that tells what is being shown.

Pictograms

A pictogram is a simple way of representing data. Frequency is indicated by identical pictures (called symbols or motifs) arranged in rows or columns. Symbols may be divided into halves or other fractions to represent parts of a number.

Pictograms are mainly used in newspapers, magazines and reports in order to make a striking display. They are usually aimed at people who are unskilled in statistics or who have limited interest in the information shown, and are more suitable for comparisons than measurements.

Constructing pictograms

Follow these steps to construct a pictogram:

- Round off the figures if necessary (2 significant figures are normally acceptable).
- Decide on a convenient unit for the symbols; these are normally whole numbers such as hundreds, tens or millions.
- Select a symbol for each unit. Make sure it can be drawn easily.
- Provide a key to explain what the symbols represent.
- Give the pictogram a title.

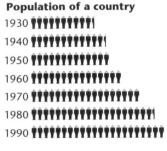

Population of a country

1930
1940
1950
1960
1970
1980
1990

represents 10 million people

Transport students use to get to school

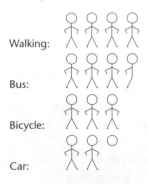

Walking:

Bus:

Bicycle:

Car:

represents 50 students (head 10, arms and legs 10 each)

■ Examples

1. Look at the example on the left. Notice that the figures are rounded to millions or half millions. Each figure represents 10 million and half a figure represents 5 million. The pictogram has a key and a title.

   You can see from this representation that the population of this country increased steadily from 1930 to 1990. You can also count the total population for each year.

2. Look at the bottom pictogram.
   a) How many students go to school by car?
      There are two full figures and one head, so 50 + 50 + 10 = 110 students.
   b) How many students travel by bus?
      There are three full figures plus one representing 30, so 3(50) + 30 = 180.

Bar graphs

In bar charts or bar graphs, data are represented in a series of bars that are equally wide. The width itself is not significant, but all the bars should be the same width. Sometimes the bars are just thick lines.

Bars can touch each other or be separated by gaps of equal width. The height of the bars represents the magnitude or frequency of the figures. Bars may be horizontal or vertical.

Bar charts are particularly useful for showing more than one set of facts. This makes them useful for comparing data.

Look at the two examples of bar graphs on page 213.

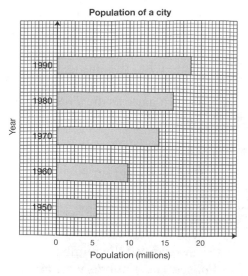

Population of a city

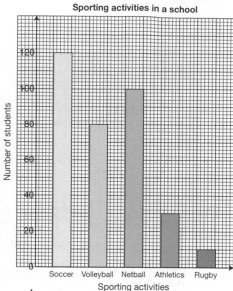

Sporting activities in a school

Constructing bar graphs

Follow these steps to construct a bar graph:

- Organise the data in a table.
- Round off the numbers if necessary (2 places are usually sufficient).
- Decide whether the bars will be vertical or horizontal.
- Draw the axes a convenient length.
- Choose an appropriate scale for the numerical axis (horizontal or vertical). Divide the axis equally according to the scale you have chosen and label the divisions.
- Decide how wide the bars will be and how much space, if any, you will leave between them.
- Construct the bars using the appropriate numerical values on the scale.
- Label the axes and give the graph a proper title.

Note

Also look at histograms on page 224, and make sure that you understand the difference between bar graphs and histograms.

■ Examples

1. The marks obtained by 50 students in a class test are given on the left. Make a frequency table for the given marks. Draw a bar graph to represent the data.

Raw Data				
10	3	6	4	7
7	4	5	6	9
4	8	6	7	5
5	6	7	5	4
6	5	6	9	1
8	2	3	4	1
7	5	4	6	7
6	4	5	6	8
7	5	6	1	6
5	4	6	7	7

Marks ($\overline{10}$)	Tally	Frequency
1					3
2			1
3				2
4									7
5												10
6														12
7											9
8					3
9				2
10			1

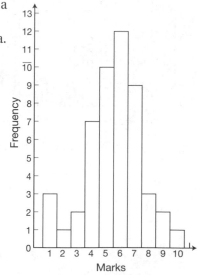

Test results

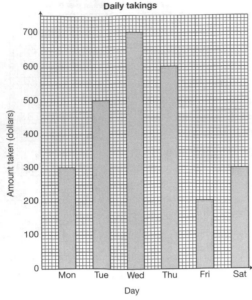

Daily takings

2. This bar graph shows the takings of a small business.
   a) On which day was the smallest amount of money collected?
      Friday – it has the shortest bar.

   b) On which day was the largest amount of money collected?
      Wednesday, it has the longest bar.

   c) Find the total money collected for the week.
      $300 + $500 + $700 + $600 + $200 + $300 = $2 600

Exercise

1. A survey recorded the number of people living in each of 40 houses. The numbers were as follows:

3	4	2	4	3	2	2	5	4	3
4	1	2	6	3	5	5	2	4	1
4	3	4	2	4	4	6	2	4	3
2	5	4	5	6	4	2	3	2	4

   a) Complete the following table.

Number of people	Tally marks	Number of houses
1		
2		
3		
4		
5		
6		

   b) Draw a bar graph to illustrate your results.
   c) What is the total number of people living in these 40 houses?

2. The bar graph on the left shows the shoe sizes of a group of 16-year-old boys.
   a) How many of these boys have a shoe size of 12?
   b) How many boys are there in the group altogether?
   c) Comment on the shape of the bar graph, saying whether or not this is the shape you would expect.

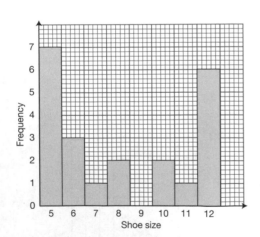

Pie charts (circle graphs)

A pie chart is a circle graph in which the angles of the sectors represent the frequency. When sectors are nearly the same size, it is difficult to compare them. In such cases, the measurements are usually given on the graph.

Constructing pie charts

Follow these steps to draw a pie chart:

- Add all the frequencies (items) and write each frequency as a fraction of all the frequencies.
- Change each fraction into a number of degrees by multiplying by 360°. (There are 360° in a circle.) The degrees tell you how big the angles of each sector in the pie chart will be.
- Tabulate the angles in ascending or descending order.
- Draw a circle of convenient size. Then draw a radius as a starting point.
- Use a protractor to construct the angles at the centre corresponding to each sector.
- Label each sector and give the graph a title.

■ Examples

1. The table below shows how a student spends her day.

Activity	School	Sleeping	Homework	Eating	Other
No. of hours	8	8	3	1	4

Show this on a pie chart.

Start by working out the fractions.
Total no. of hours = 24

School: $\frac{8}{24}$ Sleeping: $\frac{8}{24}$ Homework: $\frac{3}{24}$ Eating: $\frac{1}{24}$ Other: $\frac{4}{24}$

Change each fraction to degrees.

School: $\frac{8}{24} \times 360 = 120°$ Sleeping: $\frac{8}{24} \times 360° = 120°$

Homework: $\frac{3}{24} \times 360° = 45°$ Eating: $\frac{1}{24} \times 360° = 15°$

Other: $\frac{4}{24} \times 360° = 60°$

Tabulate the angles and use the step-by-step method below to draw the pie chart.

Activity	School	Sleeping	Other	Homework	Eating
Angle	120°	120°	60°	45°	15°

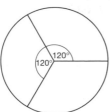

Step 1: Draw a radius as a starting point. Then draw the first 120° sector.

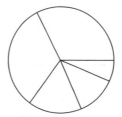

Step 2: Draw the next 120° sector.

Step 3: Draw the other sectors.

Time spent on activities in one day

Step 4: Label the graph and give it a title.

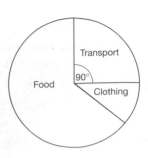

2. A person's expenditure each month is $1 200, split as shown in the pie graph on the left.
   a) If the angle at the centre in the transport sector is 90°, what amount of money is spent on transport?

   90° out of 360° represents a fraction of $1 200

   $$\frac{90°}{360°} \times \$1\ 200 = \$300$$

   b) If this person spends $700 on food, find the angle at the centre of this sector.

   $$\frac{\$700}{\$1\ 200} \times 360° = 210°$$

   c) What fraction of this person's expenditure is on clothing?

   $1 200 – $700 – $300 = $200

   $$\frac{\$200}{\$1\ 200} = \frac{1}{6}$$

Exercise

Faculty	Students
Science	495
Arts	375
Medicine	108
Engineering	54
Law	48

1. The table on the left shows the number of students in a university and their faculty. Draw a pie chart to illustrate this information.
2. The table below shows how an income of $400 was spent. Show these data on a bar graph and a pie chart.

	Food	Rent	Clothing	Transport	Savings
Amount	$120	$80	$40	$110	$50

3. The diagram on the right is a pie chart showing the expenses of a small manufacturing firm. The total expenses were $720 000. The angles at the centre of each sector are: wages 150°; raw materials 120°; fuel 40°; extras 50°. Work out how much was spent under each heading.

4. The pie chart on the right, which is not drawn to scale, shows the distribution of various types of land in a district. Calculate:
   a) the area of woodland as a fraction of the total area shown
   b) the angle of the urban sector
   c) the total area of the district.

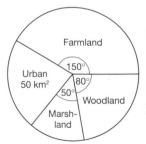

5. A number of students were asked to name their favourite sport. The results are shown in the pie chart on the left. $\frac{1}{4}$ of the students said tennis, $\frac{1}{8}$ said rugby, $\frac{1}{3}$ said football and the rest said swimming.
   a) What fraction said swimming?
   b) Calculate the value of x, the angle of the sector representing rugby in the pie chart.
   c) If 32 students chose football, how many said tennis?

6. A total of 48 football league matches were played on a Saturday. The pie chart on the left shows the proportion of home wins, away wins and draws.
   a) Measure and write down the angle of the sector representing draws.
   b) Calculate the number of home wins.

7. The pictogram shows the number of loaves of bread sold in a shop during each of three weeks. A total of 450 loaves were sold in week 1.

   a) Calculate the number of loaves sold in weeks 2 and 3.
   b) Complete the pictogram to show that 375 loaves were sold in the fourth week.

8. The pie chart on the left, which is accurately drawn, shows the nationalities of people staying in a holiday hotel.

   a) Which of these five nationalities had the smallest number of people in the hotel?
   b) (i) What fraction of the people in the hotel were French? Give the answer in its lowest terms.
      (ii) Write the answer to b)(i) as a percentage correct to the nearest whole number.
   c) Write the ratio $\dfrac{\text{number of Germans}}{\text{total number of people}}$ as a decimal.
   d) If there were 288 people in the hotel altogether, how many of them were Dutch?

Averages – mean, median and mode

Different kinds of averages are used in statistics, they are called the mean, median and mode. Each one has a specific purpose and is used differently to the others. For example, a student who was interested in her average performance on class tests would use the mean. The median is used to give the middle values in a set of data; it is useful when one piece of information in the data might cause the results to be distorted. A person who sells shoes would want to know the size that is most common; this is given by the mode.

The mean

The mean is the most common type of average. It is the number obtained by dividing the sum of all the items by the number of items.

$$\text{Mean} = \frac{\text{sum of all the items}}{\text{the number of items}}.$$

■ Example

Find the mean of 8, 9, 7, 6, 8 and 10.

$$\text{sum of items} = 8 + 9 + 7 + 6 + 8 + 10 = 48$$
$$\text{number of items} = 6$$
$$\text{mean} = \frac{48}{6} = 8$$

The mean can also be represented as a formula: $\bar{x} = \dfrac{X_1 + X_2 + X_3 \ldots X_n}{N}$

$\bar{x}$ is read 'x bar' → mean

$X_1, X_2, \ldots, X_n$ are the items

N represents the number of items

The mean of a frequency distribution

You can find the mean of a set of data using a frequency table.
Follow the worked examples to see how this is done.

■ Examples

1. Calculate the mean for the following frequency distribution.

Test marks	1	2	3	4	5	6	7	8	9	10
Frequency	2	3	4	10	5	3	2	1	0	1

Redraw the table with another row. Label the test marks x, the Frequency f and the last row fx. Add a column at the end for totals.

Calculate fx and fill in the figures in the last row.
Calculate the total for row (f and fx) and fill it in.

Test marks (x)	1	2	3	4	5	6	7	8	9	10	Total
Frequency (f)	2	3	4	10	5	3	2	1	0	1	31
fx		2	6	12	40	25	18	14	8	0	10	135

Can you see that you have actually found the total of all the marks by doing this? You have also found the number of marks by totalling the figures in row *f.*

You can now work out the mean:

$\frac{135}{31} = 4.35$

2. Tickets for a circus have been sold at the following prices: 180 @ $6.50, 215 @ $8, 124 @ $10.
   a) What is the total amount of money received for the tickets?
   b) What is the mean price of tickets sold (to 3 significant figures)?

Price of tickets (*x*)	No of tickets (*f*)	*fx*
$6.50	180	$1 170
$8	215	$1 720
$10	124	$1 240
Total	519	$4 130

a) The total amount (*fx*) = $4 130
b) The mean is $\frac{\$4\,130}{519} = \$7.95761$

$= \$7.96$ (to 3 significant figures)

The median

In order to find the median of a set of data, these must be arranged in ascending or descending order. The median is the central or middle figure.

For an odd number of items, the median is the value of the item that is in the middle.
For an even number of items, the median is the mean value of the two middle items.

■ Examples

1. Find the median of the following scores:

   20   70   50   30   35   45   75   15   90

   Arrange the data in ascending or descending order. Ascending order:

   15   20   30   35   45   50   70   75   90

   Find the middle value. There are 9 scores, so the 5th is the middle one. Thus the median is 45.

2. Find the median of:

   6   5   3   8   4   2

   Arrange in order:

   2   3   4   5   6   8

   There is an even number, so the median is the mean value of the 3rd and 4th values.

   $\frac{4 + 5}{2} = 4.5$

Remember

$\frac{n+1}{2}$ can be used to calculate the median position.

The median of a frequency distribution

This example shows you how the median can be found from a frequency table.

■ Example

The distribution of marks obtained by the students in a class is shown in the table below.

Mark obtained	0	1	2	3	4	5	6	7	8	9	10
Number of students	1	0	3	2	2	4	3	4	6	3	2

Find the median of this distribution.

The total number of students $= 1 + 0 + 3 + 2 + 2 + 4 + 3 + 4 + 6 + 3 + 2 = 30$

$$\frac{n+1}{2} = \frac{30+1}{2} = 15.5$$

The median is the mean of the 15th and 16th marks when the marks are placed in ascending (or descending) order.

   We could list all 30 marks thus: 0 2 2 2 3 3 4 4 … but this is tedious, and it would be very time-consuming if the class contained a very large number of students.

   An easier method is to add up the frequencies (numbers of students) in the table, starting at the left-hand end, until you reach the 15th and 16th students .

$1 + 0 + 3 + 2 = 6$ so the 6th student obtained 3 marks

$6 + 2 = 8$ so the 8th student obtained 4 marks.

$8 + 4 = 12$ so the 12th student obtained 5 marks.

$12 + 3 = 15$ so the 15th student obtained 6 marks.

The 16th student is in the next group (of 4) and so obtained 7 marks.

The median $= \frac{6+7}{2} = 6.5$.

The mode

The mode of a set of data is the value with the highest frequency. A distribution that has two modes is called bimodal. If there are more than two values that appear most frequently, then there is no mode – such a distribution is non-modal.

   The mode requires no calculation, only counting.

■ Examples

1. Find the mode for the following distribution:

   70   80   50   95   80   73   90   85

   80 appears twice, so it is the mode.

2. Find the mode of the following data:

3   6   3   5   3   8   1   8   5   4   2   8   10

Arrange the data in an array:

1   2   3   3   3   4   5   5   6   8   8   8   10

3 and 8 appear 3 times each. This is a bimodal distribution.

3. Find the mode of the given distribution:

Marks	0	1	2	3	4	5	6	7	8	9	10
Students	2	1	1	2	6	10	7	6	3	1	1

The highest frequency is 10. However, remember that the mode refers to the actual data, so the modal value is 5.

Exercise

1. Construct a frequency table for the following data and calculate the mean.

3   4   5   1   2   8   9   6   5   3   2   1   6   4   7   8   1
1   5   5   2   3   4   5   7   8   3   4   2   5   1   9   4   5
6   7   8   9   2   1   5   4   3   4   5   6   1   4   4   8

2. Find the median value of:
   a) 8, 1, 6, 7, 5, 2, 3
   b) 100, 75, 85, 95, 43, 99, 70, 60
   c) 2, 3, 1, 5, 6, 4
   d) 31, 28, 25, 21, 22, 20
   e) 41, 47, 42, 41, 47, 43, 45, 41.
3. Find the mode of each of the following sets of numbers:
   a) 4, 5, 5, 1, 2, 9, 5, 6, 4, 5, 7, 5, 5
   b) 1, 8, 19, 12, 3, 4, 6, 9
   c) 2, 2, 3, 5, 8, 2, 5, 6, 6, 5
   d) 41, 47, 43, 41, 42, 45, 42.
4. A man kept count of the number of letters he received each day over a period of 60 days. The results are shown in the table below.

Number of letters per day	0	1	2	3	4	5
Frequency	28	21	6	3	1	1

For this distribution, find:
a) the mode
b) the median
c) the mean.

Grouped and continuous data

Classifying data

Data that can take any value in a given range are called continuous data. Examples of continuous data are heights, age, temperatures and mass. Such data are normally rounded off in a distribution.

Discrete data can only take particular values (normally whole or half numbers) such as the number of children in a family. There cannot be values such as 1.5 or 2.7 children.

Grouping continuous data

Data can be grouped into classes to make them easier to work with. The table below shows the heights of 50 students in ascending order. The heights have been rounded up to the nearest whole centimetre. This means that a height (h) of 141 cm is in the range $140 < h \leq 141$.

131	132	133	134	134	135	135	136	136	136
136	137	137	137	138	138	139	139	140	140
141	141	142	142	142	142	142	143	143	144
144	144	145	145	145	146	147	147	147	148
148	149	149	149	150	150	151	151	152	153

The heights can also be grouped as follows:
$130 < h \leq 135, 135 < h \leq 140, 140 < h \leq 145, 145 < h \leq 150, 150 < h \leq 155$.

Each group has a width of 5 cm.

A height that is a little more than 140 cm belongs to the group $140 < h \leq 145$, whereas a height of 140 cm belongs to the group $135 < h \leq 140$.

Heights, (h cm)	Frequency
$130 < h \leq 135$	7
$135 < h \leq 140$	13
$140 < h \leq 145$	15
$145 < h \leq 150$	11
$150 < h \leq 155$	4
Total	50

Now look at these frequency distributions of grouped data.

The table below shows the lengths of 50 pieces of wire used in a physics laboratory. Lengths have been measured to the nearest centimetre.

Length	26–30	31–35	36–40	41–45	46–50
Frequency	4	10	12	18	6

The interval '26–30' means 25.5 cm $\leq$ length < 30.5 cm.

The table below shows the masses of 50 small packets brought to a post office counter in a day.

Mass in grams	–100	–250	–500	–800
Frequency	8	12	20	10

The interval '–250' means 100 g $<$ mass $\leq$ 250 g. This includes any mass greater than 100 g and less than or equal to 250 g.

The class boundaries are:
0	100	250	500	800

Finding the mean of grouped data

In order to calculate the mean of grouped data, you need to:
- find the mid-point of each interval (x)
- multiply the frequency of each interval by its mid-point (fx).
- find the sum of all the products fx
- find the sum of all the frequencies
- divide the sum of the products fx by the sum of the frequencies.

■ Examples

1. The table below shows the lengths of 50 pieces of wire used in a physics laboratory. Lengths have been measured to the nearest centimetre. Find the mean.

Length	26–30	31–35	36–40	41–45	46–50
Frequency (f)	4	10	12	18	6

Find the mid-points of each interval:

Length	26–30	31–35	36–40	41–45	46–50
Frequency (f)	4	10	12	18	6
Interval	25.5–30.5	30.5–35.5	35.5–40.5	40.5–45.5	45.5–50.5
Mid-point (x)	28	33	38	43	48

Multiply f by x:

Length	26–30	31–35	36–40	41–45	46–50
Frequency (f)	4	10	12	18	6
Mid-point (x)	28	33	38	43	48
fx	28×4 = 112	33×10 = 330	38×12 = 456	43×18 = 774	48×6 = 288

Find the totals:
Sum of fx = 1 960
Sum of f = 50
The mean = (1 960 ÷ 50) cm = 39.2 cm

2. The table below shows the ages of the teachers in a secondary school to the nearest year.

Age in years	21–30	31–35	36–40	41–45	46–50	51–65
Frequency	3	6	12	15	6	7

Calculate the mean age of the teachers.

Age in years	21–30	31–35	36–40	41–45	46–50	51–65
Frequency (f)	3	6	12	15	6	7
Mid-point (x)	26	33.5	38.5	43.5	48.5	58.5
fx	78	201	462	652.5	291	409.5

Sum of fx = 2 094;

Sum of f = 49

Mean = $\frac{2094}{49}$ = 42.73469

Mean age of the teachers = 42.7 years (to 3 significant figures).

Histograms

Histograms are like vertical bar graphs of grouped data. There are no gaps between the bars. The area of the bars gives the number of items in the class interval. If all the class intervals are the same, then the bars will be the same width, and frequencies can be represented by the heights of the bars.

■ Examples

1. Draw a histogram to illustrate the following frequency table. The measurements were made to the nearest unit.

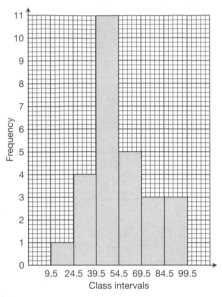

Class interval	10–24	25–39	40–54	55–69	70–84	85–100
Frequency	1	4	11	5	3	3

The class boundaries are:

9.5   24.5   39.5   54.5   69.5   84.5   99.5

The class widths are:

15   15   15   15   15   15

Area of rectangle = class width × height of rectangle.
As the class width is 15 for each interval, area of rectangle = 15 × height of rectangle. So, area ∝ height of rectangle.

If you make the height of each rectangle the same as the frequency, area ∝ frequency.

So you can draw the frequencies along the vertical axis and the class interval along the horizontal axis.

2. The frequency distribution gives the masses of 35 objects measured to the nearest kilogram. Draw a histogram to illustrate the data.

Mass in kg	6–8	9–11	12–17	18–20	21–29
Frequency	4	6	10	3	12

The class boundaries are: 5.5   8.5   11.5   17.5   20.5   29.5
The class widths are: 3   3   6   3   9

The class widths are not equal. Therefore we cannot make the height of each rectangle equal to the frequency. So we choose a convenient width as standard and adjust the heights of the rectangles accordingly.

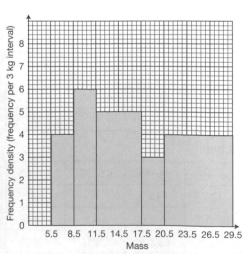

The first two class widths are 3 each and the heights of the first two rectangles can be taken as 4 units and 6 units respectively.

The third width is 6. So the height of this rectangle is half the frequency. The last interval is 9. This is 3×3. So the height of the rectangle is $\frac{1}{3}$ of the frequency.

In general, choose a width as 'standard' width.

If class width = $n \times$ standard width, then
the height of rectangle = $\frac{1}{n} \times$ corresponding frequency.

The mode of a grouped frequency distribution

You know that the mode of an ungrouped frequency distribution is the item that has the highest frequency. You cannot find the mode of grouped data. You can only find the modal class (or classes). The modal class is the class interval that has the largest frequency. However, the way in which the class intervals are chosen affects which class interval forms the modal class.

■ Example

Find the modal class of the frequency distribution on the left.
• The highest frequency is 9.
• The interval against this is 26–35. So the modal class is 26–35.

Marks	Frequency
6–15	2
16–25	7
26–35	9
36–45	3
46–55	4
56–65	2
66–75	1

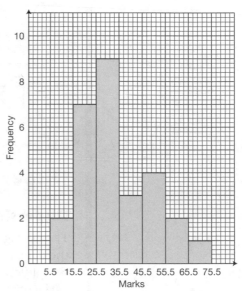

Exercise

1. The frequency table shows the distribution of the masses of some objects. Draw a histogram to illustrate this distribution.

Mass (in grams)	$60 \leq m < 63$	$63 \leq m < 64$	$64 \leq m < 65$
Frequency	9	12	15
Mass (in grams)	$65 \leq m < 66$	$66 \leq m < 68$	$68 \leq m < 72$
Frequency	17	10	8

2. The heights of 25 plants were measured to the nearest centimetre. The results are summarised in the table on the left.

Height in cm	Number of plants
5–14	4
15–19	8
20–24	7
25–39	6

a) The middle two classes are already shown in the histogram below. Copy and complete the histogram by adding the other two classes to it.

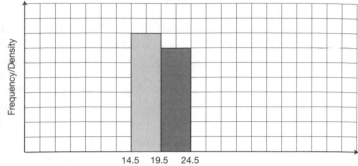

b) Copy and complete the table below. Hence calculate the mean height of the plants.

Height in cm	Mid-point (x)	Frequency (f)	fx
5–14		4	
15–19	17	8	
20–24	22	7	
25–39		6	

Dispersion and cumulative frequency

Dispersion

The average (mean, median or mode) gives a general idea of the size of the data, but two sets of numbers can have the same mean while being very different in other ways.

The other main statistic we need to find is a measure of *dispersion* or *spread*. There are several ways of measuring dispersion.

Range

The *range* is the easiest measure of dispersion to calculate. It is defined as the difference between the highest value and the lowest value. The range is a crude measure of dispersion since it makes no use of the intermediate values and it can be distorted by one or two extreme values.

Inter-quartile range

This is a measure of the middle half of the data, so it is more representative. A distribution is divided into four subgroups by three quartiles.

- The *first* or *lower quartile* (Q_1) is the point below which 25% of the items lie and above which 75% of the items lie.
- The *second quartile* (Q_2) is the point below which 50% of the items lie and above which 50% of the items lie. You will realise that the second quartile is the same as the median.
- The *third* or *upper quartile* (Q_3) is the point below which 75% of the items lie and above which 25% of the items lie.
- If there are n values, in ascending order, then the lower quartile Q_1 is the $\dfrac{(n+1)}{4}$th value, and the upper quartile Q_3 is the $\dfrac{3(n+1)}{4}$th value.

  The *inter-quartile range* $= Q_3 - Q_1$.
- In practice, quartiles are only required from cumulative frequency graphs or when dealing with the middle value of equal values in a table.

Percentiles

The median and quartiles divide a distribution into four parts. For a large mass of data, these may not give sufficient information. In such cases, two other sets of measures are useful. They are *deciles* and *percentiles*.

Deciles divide the distribution into 10 equal parts.
Percentiles divide the data into 100 equal parts.

For example, the 10th percentile $P_{10} = \frac{10}{100}(n+1)$th value.

The 90th percentile $P_{90} = \frac{90}{100}(n+1)$th value.

The 10 to 90 percentile range $= P_{90} - P_{10}$.

■ Example

A student's marks in ten subjects in two sets of tests are given below.

Test 1: 20   22   28   19   20   24   23   20   24   20
Test 2: 13   15   36   11   18   30   23   8   32   34

Find, for each set of tests:
a) the range
b) the median
c) the mean.

a) Range
   Test 1: $28 - 19 = 9$
   Test 2: $36 - 8 = 28$

b) Median

Test 1: $\frac{20 + 22}{2} = 21$

Test 2: $\frac{18 + 23}{2} = 20.5$

c) Mean

Test 1: $220 \div 10 = 22$

Test 2: $220 \div 10 = 22$

Cumulative frequency table

In statistics, we may have to answer questions such as 'How many students scored more than 75%?' or 'How many students scored less than 25%?'

To help answer such questions, it is convenient to add another column to the frequency table to show the total frequency up to and including the one corresponding to the value of the item we are interested in, that is, to show the frequency cumulatively.

The word 'cumulative' is related to the word 'accumulate', which means to 'pile up'. In other words, the *cumulative frequency* column shows the running total of the class frequencies. A cumulative frequency table gives the total up to each class boundary. This will be clear from the table below, which gives the marks obtained by 100 candidates in an examination.

The last number of the cumulative frequency column gives the total number of items or the sum of the frequencies. The cumulative frequency column shows us that 2 candidates scored 9 or less, 6 candidates scored 19 or less, and so on. The successive entries are obtained by adding the next figure in the frequency column to the previous total figure. This is what is implied by the word 'cumulative'.

Score	Frequency	Cumulative frequency
0–9	2	2
10–19	4	6
20–29	8	14
30–39	10	24
40–49	12	36
50–59	25	61
60–69	22	83
70–79	8	91
80–89	6	97
90–99	3	100

Let us see how we obtained the cumulative frequency column in the example.

The entries in the cumulative frequency were obtained as follows:

1st entry: Enter the 1st class frequency (2).
2nd entry: Add the 1st class frequency (2) to the 2nd class frequency (4).
$2 + 4 = 6$
3rd entry: Add the 1st, 2nd and 3rd class frequencies.
$2 + 4 + 8 = 14$
Or, add the 2nd cumulative frequency entry to the 3rd class frequency.
$6 + 8 = 14$
4th entry: Add the 1st, 2nd, 3rd and 4th class frequencies. Or, add the 3rd cumulative frequency entry to the 4th class frequency.
$14 + 10 = 24$

Continue this procedure to obtain the remaining entries.

Frequency	Cumulative frequency		
2	2	or	2
4	$2 + 4 = 6$		$4 + 2 = 6$
8	$2 + 4 + 8 = 14$		$8 + 6 = 14$
10	$2 + 4 + 8 + 10 = 24$		$10 + 14 = 24$
12	$2 + 4 + 8 + 10 + 12 = 36$		$12 + 24 = 36$
25	$2 + 4 + 8 + 10 + 12 + 25 = 61$		$25 + 36 = 61$
22	$2 + 4 + 8 + 10 + 12 + 25 + 22 = 83$		$22 + 61 = 83$
8	$2 + 4 + 8 + 10 + 12 + 25 + 22 + 8 = 91$		$8 + 83 = 91$
6	$2 + 4 + 8 + 10 + 12 + 25 + 22 + 8 + 6 = 97$		$6 + 91 = 97$
3	$2 + 4 + 8 + 10 + 12 + 25 + 22 + 8 + 6 + 3 = 100$		$3 + 97 = 100$

If you complete the cumulative frequency column correctly, the last entry will be equal to the sum of the class entries.

■ Example

The heights of flowers were measured during an experiment. The results are summarised in the table.

Height (h cm)	$0 < h \le 5$	$5 < h \le 10$	$10 < h \le 15$
Frequency	20	40	60
Height (h cm)	$15 < h \le 25$	$25 < h \le 50$	
Frequency	80	50	

a) Draw up a cumulative frequency table for this distribution.
b) Which class interval contains the median height?

a) The entries in the cumulative frequency table will be
20, 20 + 40, 20 + 40 + 60, 20 + 40 + 60 + 80, 20 + 40 + 60 + 80 + 50.

The table is as follows:

Height (h cm)	$h \leq 5$	$h \leq 10$	$h \leq 15$	$h \leq 25$	$h \leq 50$
Cumulative frequency	20	60	120	200	250

b) Altogether, there are 250 flowers, so the median is the mean of the heights of the 125th flower and the 126th flower. There are 120 flowers with heights less than or equal to 15 cm, and 200 flowers with heights less than or equal to 25 cm. Hence, the heights of the 125th and 126th flowers are each greater than 15 cm and less than or equal to 25 cm. The median height of the 250 flowers is in the interval $15 < h \leq 25$.

Cumulative frequency curve

The curve obtained by plotting cumulative frequencies against the upper boundaries of the classes is called a *cumulative frequency curve*. (The curve sometimes called an 'ogive' from a curve used in architecture.)

If the points are joined by straight lines, the graph is called a *cumulative frequency polygon*.

■ **Examples**

1. The table shows the examination marks of 300 students. Draw up a cumulative frequency table and draw the cumulative frequency graph. Then find the median mark.

Mark	Frequency
1–10	3
11–20	7
21–30	13
31–40	29
41–50	44
51–60	65
61–70	70
71–80	49
81–90	14
91–100	6

Draw up the cumulative frequency table as explained earlier.

Mark	Cumulative frequency
0	0
≤ 10	3
≤ 20	10
≤ 30	23
≤ 40	52
≤ 50	96
≤ 60	161
≤ 70	231
≤ 80	280
≤ 90	294
≤ 100	300

Draw a graph with cumulative frequency on the vertical axis against the upper boundary of each class on the horizontal axis.

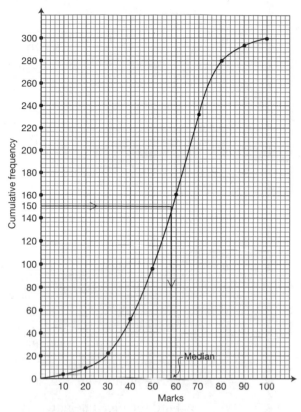

Total frequency is 300. The mark corresponding to the 150th candidate is the median mark.

Draw a line parallel to the mark axis from the 150 mark on the cumulative frequency axis so as to cut the graph.

Drop a perpendicular from the intersection to the mark axis. This gives the median mark: 58.

2. The cumulative frequency curve on the next page shows the journey times to school of some students.

   Use the curve to find:
   a) the total number of students
   b) the median journey time
   c) the number of students who took less than 10 minutes to get to school
   d) the number of students who had journey times greater than 30 minutes
   e) the number of students who took between 40 minutes and one hour to get to school.

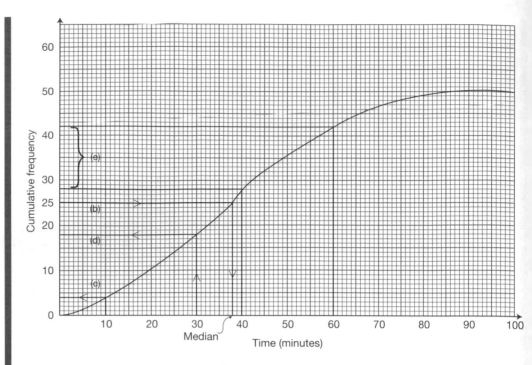

a) 50   b) 38 minutes   c) 4
d) 50 – 18 = 32   e) 42 – 28 = 14

3. Twenty bean seeds were planted for a biology experiment. The heights of the plants were measured after 3 weeks and recorded as below.

Find the mean height.

Draw a cumulative frequency curve and find the median height.

Heights (h cm)	$0 \leq h < 3$	$3 \leq h < 6$	$6 \leq h < 9$	$9 \leq h < 12$
Frequency	2	5	10	3

Heights (h cm)	Mid-point (x)	Frequency (f)	fx	Cumulative frequency
$0 \leq h < 3$	1.5	2	$2 \times 1.5 = 3$	2
$3 \leq h < 6$	4.5	5	$5 \times 4.5 = 22.5$	7
$6 \leq h < 9$	7.5	10	$10 \times 7.5 = 75$	17
$9 \leq h < 12$	10.5	3	$3 \times 10.5 = 31.5$	20
Total		20	132	

Mean height $= \frac{132}{20} = 6.6$ cm.

The median height = 7.0 cm (see graph).

Exam tip

Questions on cumulative
frequency may take up a
lot of time in the exam.
Make sure you do not
allocate them more time
than they are worth.

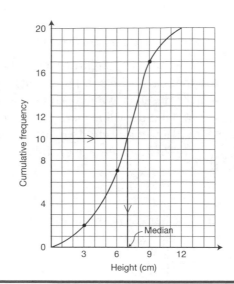

Exercise

1. The heights of 25 plants were measured to the nearest centimetre.
    The results are summarised in the table.

Height in cm	6–15	16–20	21–25	26–40
Number of plants	3	7	10	5

a) Draw up a cumulative frequency table for this distribution.
b) In which interval does the median plant height lie?
c) Estimate, to the nearest centimetre, the median plant height.

2. The table shows the amount of money, x, spent on books by a group
    of students.

Amount spent	$0 < x \leq 10$	$10 < x \leq 20$	$20 < x \leq 30$
No. of students	0	4	8
Amount spent	$30 < x \leq 40$	$40 < x \leq 50$	$50 < x \leq 60$
No. of students	12	11	5

a) Calculate an estimate of the mean amount of money per student
    spent on books.
b) Use the information in the table above to find the values of p, q
    and r in the following cumulative frequency table.

Amount spent	$x \leq 10$	$x \leq 20$	$x \leq 30$	$x \leq 40$	$x \leq 50$	$x \leq 60$
Cumulative frequency	0	4	p	q	r	40

c) Using a scale of 2 cm to represent 10 units on each axis, draw a
    cumulative frequency diagram.
d) Use your diagram:
    (i)   to estimate the median amount spent
    (ii) to find the upper and lower quartiles, and the inter-quartile range.

Check your progress

1. Two hundred children were asked to choose their favourite pet. The results are represented in the pie chart (which is not to scale).
   a) Find the value of *x*.
   b) How many children chose rabbits?
   c) What percentage of children chose dogs?

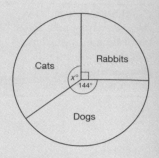

2. The bar graph shows the number of soccer teams scoring 0, 1, 2, 3, 4 or 5 goals on a particular day.
   a) How many teams scored 5 goals?
   b) How many teams were there altogether?
   c) How many goals were scored altogether?

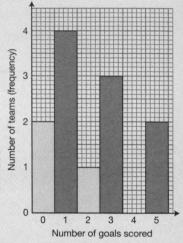

3. a) A bird-watcher recorded the number of eggs in each of 100 birds' nests.

Number of eggs per nest	1	2	3	4	5	6
Number of nests	5	18	28	34	9	6

   Draw a bar graph to display these results.
   b) Another bird-watcher looked into only 9 nests.
   He wrote down the number of eggs he saw in each as follows:
   5,   1,   3,   6,   5,   5,   2,   3,   1.
   For this set of figures:
   (i)   find the mode
   (ii)   find the median
   (iii)   calculate the mean, giving your answer correct to 1 decimal place.

4. Write down three positive integers with median 5 and mean 6.

5. After a morning's fishing, Arnold measured the mass, in grams, of each of the fish he had caught. The bar graph on the right represents his results.

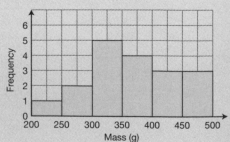

a) Use the bar graph to complete this table.

Mass (M) in grams	Number of fish	Classification
M < 300		Small
300 ≤ M < 400		Medium
M ≥ 400		Large

b) Represent the information in the table in a pie chart.
   Show clearly how you calculate the angles.

6. A survey of the number of children in 100 families gave the following distribution:

Number of children in family	0	1	2	3	4	5	6	7
Number of families	4	36	27	21	5	4	2	1

For this distribution, find:
a) the mode
b) the median
c) the mean.

7. The table shows the length of time of 100 telephone calls.

Time (t minutes)	$0 < t \le 1$	$1 < t \le 2$	$2 < t \le 4$
Number of calls	12	14	20
Time (t minutes)	$4 < t \le 6$	$6 < t \le 8$	$8 < t \le 10$
Number of calls	14	12	18
Time (t minutes)	$10 < t \le 15$		
Number of calls	10		

a) (i) Calculate an estimate of the mean time, in minutes, of a telephone call.
   (ii) Write your answer in minutes and seconds, to the nearest second.
b) Make a cumulative frequency table for the 100 calls. Start as shown on the right.
c) Draw a cumulative frequency diagram on a sheet of graph paper. Use a scale of 1 cm to represent 1 unit on the horizontal t-axis and 2 cm to represent 10 units on the vertical axis.

Time (t minutes)	Cumulative frequency
0	0
≤ 1	12
≤ 2	26
≤ 4	

d) Use your graph to find, correct to the nearest 0.1 minute:
   (i)   the median time
   (ii) the upper quartile
   (iii) the inter-quartile range.

Probability

The odds are that the USA will beat Russia in the race to get to Mars.

There is no chance that Mohammed Ali will ever box in a championship again.

It is very likely that the Black Tigers will win the cup.

You may have heard people say things similar to the statements above. In all of these cases, people are predicting the likelihood of something happening or not happening.

In mathematics, working out how likely or unlikely an event is to happen is called probability. An event in probability studies is simply something that happens, such as throwing a die or winning a race. An outcome is the result of an event such as throwing a 1 or a 6 when you roll a die.

Probability scale

Probability is measured on a scale from 0 to 1. A probability of 0 means that there is no chance of an event happening. A probability of 1 means that it is certain the event will happen.

Look at the spinner on the left. What is the probability that the pointer will stop on black?

Since half the spinner is black, the probability that the pointer will stop on black is $\frac{1}{2}$. Because a quarter of the spinner is red, the probability that it will stop on red is $\frac{1}{4}$.

A probability of $\frac{1}{2}$ is also called a 50–50 chance.

This does not mean that the spinner would land on black one out of every two times if you actually spun it! What we have worked out is the expected probability.

Experimental or relative probability

Note

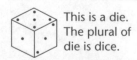

This is a die. The plural of die is dice.

Look at these results of two experiments in which a die was rolled 72 times and then the same die was rolled 240 times.

Experiment 1

Number on face	1	2	3	4	5	6
Number of times	14	12	8	13	9	16

Experiment 2

Number on face	1	2	3	4	5	6
Number of times	40	39	44	38	37	42

In the second experiment, each of the numbers on the die came up approximately the same number of times. That is, 40 out of 240 times. We can thus say that the experimental probability of rolling 1 is $\frac{40}{240}$. Generally, the more times the die is rolled, the closer the experimental probability will be to the expected probability.

> Experimental probability of an outcome $= \frac{\text{the number of outcomes}}{\text{the total number of events in the trial}}$.

Expected or theoretical probability

In an experiment such as the above, different people may get different results. However, in many cases, there is no reason why one outcome should happen more often than another. For example, if you toss a coin, there is no reason why heads should face up more often than tails. The outcome 'heads' or 'tails' is thus equally likely.

In the same way, if an unbiased die is rolled, each of the six numbers is equally likely to turn up. In this case, the expected or theoretical probability that any particular number on the die will turn up is $\frac{1}{6}$. Experimental probability normally approximates to expected probability.

Expected probability can be defined as:

> The probability of a favourable outcome $= \frac{\text{the number of favourable outcomes}}{\text{the number of possible outcomes}}$.

■ Examples

1. When an unbiased die is rolled, what is the probability of an even number turning up?

   Possible outcomes = 6 (all numbers are equally likely to turn up)
   Favourable outcomes = 3 (2, 4, 6 are all favourable in this case)
   ∴ the probability of an even number turning up is $\frac{3}{6} = \frac{1}{2}$.

   We can also express this as a percentage: There is a 50% chance of an even number turning up.

2. If a card is drawn at random from a pack of 52 cards, what is the probability that it will be an ace?

   Possible outcomes = 52
   Favourable outcomes = 4 (there are four aces in a pack of cards)
   ∴ the probability of obtaining an ace is $\frac{4}{52} = \frac{1}{13}$.

3. A bag contains identical beads. 10 are red and 6 are blue. What is the probability of randomly drawing a red bead from the bag?

Possible outcomes = 16 (there are 16 beads)

Favourable outcomes = 10 (there are 10 red beads)

∴ the probability of drawing a red bead is $\frac{10}{16} = \frac{5}{8}$.

4. The foot sizes of 11 women in a hockey team are shown in the table.

Foot size	38	39	40	41
Frequency	2	5	3	1

One woman is chosen at random.

What is the probability that her foot size is 40?

Possible outcomes = 11

Favourable outcomes = 3

∴ the probability that the woman chosen has a size 40 is $\frac{3}{11}$.

5. Margaret chooses a card at random from those shown on the left.

What is the probability that the card is:

a) an A   b) an A or a U   c) an E?

a) There are six cards and two of them are As, so the probability is $\frac{2}{6} = \frac{1}{3}$.

b) There are six cards and three of them are A or U, so the probability is $\frac{3}{6} = \frac{1}{2}$.

c) There are no Es, so the probability of one being chosen is $\frac{0}{6}$ or 0.

Probability of an event not happening

In the last example above, the probability of drawing an A was $\frac{1}{3}$. This means that you have a $\frac{2}{3}$ chance of not drawing an A.

> If the probability of an outcome is P, then the probability that the outcome will not happen is 1 – P.

The probability of an outcome is often denoted by P(outcome).

■ Examples

1. The probability of drawing a diamond card from a pack is $\frac{13}{52} = \frac{1}{4}$. What is the probability of not drawing a diamond?

P(card is a diamond) = $\frac{1}{4}$; ∴ P(card is not a diamond) = $1 - \frac{1}{4} = \frac{3}{4}$.

2. A bag contains blue, red and black marbles. The probability of drawing a black marble is $\frac{3}{7}$. What is the probability of drawing a marble that is not black?

P(marble is not black) = $1 - \frac{3}{7} = \frac{4}{7}$.

Exercise

1. What is the probability that a card drawn from a shuffled pack of playing cards will be a king?
2. What is the probability that, when an unbiased die is rolled, it will turn up a prime number?
3. A bag contains 2 red balls, 3 white balls and 5 black balls. A ball is chosen at random. What is the probability that it is:
   a) red   b) red or white   c) neither red nor black?
4. A bag contains 20 balls and 5 of the balls are blue.
   a) What is the probability that a ball chosen at random is blue?
   b) The probability of choosing a red ball is $\frac{1}{5}$. How many red balls are there in the bag?
5. The number of matches in each of 20 boxes of matches were counted. The results are shown in the table on the left.
   a) Complete this frequency table.

Number of matches	39	40	41	42	43	44	45
Frequency	5	3					

   b) If one of these boxes is selected at random, what is the probability that it contains more than 40 matches?
6. In any 50 s period, traffic lights are red for 25 s, green for 20 s and amber for 5 s. No two lights are on at the same time. A cyclist arrives at the traffic lights. What is the probability that the lights are:
   a) green   b) red?

Results				
39	43	42	40	41
41	42	40	45	42
43	40	39	41	39
39	43	41	39	43

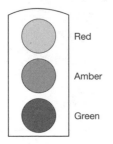

Red

Amber

Green

Mutually exclusive events

Events are mutually exclusive if they cannot happen at the same time. For example, if we toss a coin, either heads or tails might turn up, but they cannot happen at the same time. Similarly, in a single throw of a die, the numbers on the face are mutually exclusive events.

Let's say A and B are mutually exclusive events and the probabilities of them happening are P(A) and P(B). This means that the probability of A or B happening is P(A) + P(B).

P(A or B) = P(A) + P(B).

■ Examples

1. What is the probability of a die showing a 1 or a 6?

$P(1) = \frac{1}{6}$ $P(6) = \frac{1}{6}$

$P(1 \text{ or } 6) = P(1) + P(6) = \frac{1}{6} + \frac{1}{6} = \frac{2}{6} = \frac{1}{3}$

The probability of a die showing 1 or 6 is $\frac{1}{3}$.

Remember

Mutually exclusive events follow the 'OR' rule and we ADD to find P.

2. A bag contains 5 red, 3 white and 2 black balls that are all identical except for colour. A ball is drawn at random from the bag. What is the probability that it is either red or white?

$$P(\text{drawing a red ball}) = \frac{5}{10} = \frac{1}{2}$$

$$P(\text{drawing a white ball}) = \frac{3}{10}$$

$$P(\text{drawing a red or white ball}) = \frac{1}{2} + \frac{3}{10}$$

$$= \frac{8}{10}$$

$$= \frac{4}{5}$$

3. The probabilities of three teams, P, Q and R, winning a soccer competition are $\frac{1}{4}$, $\frac{1}{8}$ and $\frac{1}{10}$ respectively. Calculate the probability that
   a) either P or Q will win
   b) either P or Q or R will win
   c) none of these teams will win.

a) $P(\text{P or Q will win}) = \frac{1}{4} + \frac{1}{8}$

$$= \frac{3}{8}$$

b) $P(\text{P or Q or R will win}) = \frac{1}{4} + \frac{1}{8} + \frac{1}{10}$

$$= \frac{19}{40}$$

c) $P(\text{none will win}) = 1 - \frac{19}{40}$

$$= \frac{21}{40}$$

Exercise

1. If an unbiased die is thrown, what is the probability that it will show a 6 or an odd number?
2. What is the probability of a card drawn from a pack being a 4 or a king?
3. A box contains 30 red, 20 blue and 10 green straws. What is the probability of a straw drawn at random being red or green?
4. There are a number of red, white and blue beads in a bag. One bead is drawn at random. The probability of picking a red bead is $\frac{1}{3}$ and the probability of picking a blue bead is $\frac{1}{5}$.
   a) What is the probability of picking a bead that is red or blue?
   b) What is the probability of picking a white bead?
5. When three unbiased coins are tossed, the probability that they will show three heads is 0.125. What is the probability that they will show at least one tails?

Independent events

Events are independent if the outcome of one event does not affect the outcome of another. For example, if you throw a die and a coin, the score on the die does not affect whether the coin lands on heads or tails.

If A and B are two independent events, then the probability of A and B occurring is P(A occurring) × P(B occurring).

$$P(A + B) = P(A) \times P(B).$$

■ Examples

1. If two dice are thrown, find the probability of getting two 4s.

$$P(\text{getting a 4 on first throw}) = \tfrac{1}{6}$$

$$P(\text{getting a 4 on second throw}) = \tfrac{1}{6}$$

$$\therefore P(\text{two 4s}) = \tfrac{1}{6} \times \tfrac{1}{6} = \tfrac{1}{36}$$

2. Find the probability of scoring 18 when three dice are rolled.

$$P(\text{scoring 6 on each throw}) = \tfrac{1}{6}$$

$$P(\text{scoring 18}) = \tfrac{1}{6} \times \tfrac{1}{6} \times \tfrac{1}{6} = \tfrac{1}{216}$$

3. The probability that a student is left-handed is 0.15.
   a) What is the probability that the student is right-handed?
      P(right-handed) = 1 − 0.15 = 0.85
   b) Two students are chosen at random. What is the probability that they are both left-handed?

$$P(\text{first is left-handed}) = 0.15$$
$$P(\text{second is left-handed}) = 0.15$$
$$P(\text{both left-handed}) = 0.15 \times 0.15 = 0.0225$$

Remember

Independent events follow the 'AND' rule and we MULTIPLY to find P.

Hint

To score 18, you have to throw three 6s.

Representing probability using diagrams

In the die-rolling experiments earlier in this module, the results were shown in tables. Another way of showing results is to use a tree diagram. A probability tree shows all the possible events. The first event is represented by a dot. From this, branches are drawn to represent all possible outcomes of the event. The probability of each outcome is written on its branch.

Suppose an unbiased die is thrown and an unbiased coin is tossed.

We can represent the outcomes of the first event as shown on the left. Each branch represents a score on the die. The score is written at the end.

The probability of each outcome is $\tfrac{1}{6}$. This is written on the branch.

Throw of the die

$\tfrac{1}{6}$ 1
$\tfrac{1}{6}$ 2
$\tfrac{1}{6}$ 3
$\tfrac{1}{6}$ 4
$\tfrac{1}{6}$ 5
$\tfrac{1}{6}$ 6

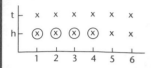
The second event is tossing the coin. This event is represented by a dot at the end of each branch because it is independent – in other words, it can follow any outcome of the first event.

The probability of tossing heads is $\frac{1}{2}$ and the probability of tossing tails is $\frac{1}{2}$. This is written on each branch. Your tree diagram will now look like this:

Suppose you want to know the probability of throwing a number less than 5 and heads.

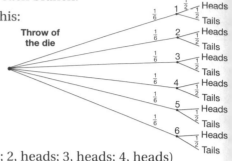

Favourable outcomes are: (1, heads; 2, heads; 3, heads; 4, heads)
So, we can calculate:

$(\frac{1}{6} \times \frac{1}{2}) + (\frac{1}{6} \times \frac{1}{2}) + (\frac{1}{6} \times \frac{1}{2}) + (\frac{1}{6} \times \frac{1}{2}) = \frac{4}{12} = \frac{1}{3}$.

When you are interested in specific probabilities, you can draw a tree diagram that only shows these. For example:

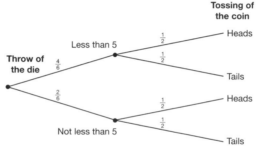

To find the probability of getting a number < 5 and heads, first locate the < 5 branch and then follow the heads branch. These are independent events, so you write down the probability on each branch and multiply them.

$P(< 5 \text{ and heads}) = \frac{4}{6} \times \frac{1}{2} = \frac{1}{3}$

■ Examples

1. Two unbiased coins are tossed together. Draw a tree diagram to find the probability of getting:
   a) two tails
   b) one heads and one tails.

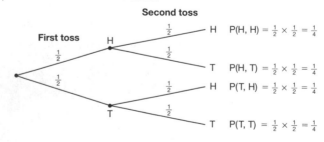

2. There are 21 students in a class. 12 are boys and 9 are girls.
   The teacher chooses two students at random.
   a) If the first student chosen is a boy, explain why the probability that
      the second student chosen is also a boy is $\frac{11}{20}$.

      If the first student chosen is a boy, there are 20 students left to choose
      from. 11 of these are boys (12 – 1) so P(2nd choice is a boy) $= \frac{11}{20}$.

   b) Draw a tree diagram to represent the situation.
   c) What is the probability that:
      (i)   both students are boys (B, B)
      (ii)   both students are girls (G, G)
      (iii)   one is a boy and one is a girl (B, G)?

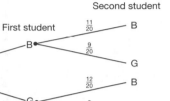

First student
Second student

   (i)   P(B, B) $= \frac{12}{21} \times \frac{11}{20} = \frac{11}{35}$

   (ii)   P(G, G) $= \frac{9}{21} \times \frac{8}{20} = \frac{6}{35}$

   (iii)   P(B, G) $= \frac{12}{21} \times \frac{9}{20} + \frac{9}{21} \times \frac{12}{20} = \frac{216}{420} = \frac{18}{35}$

   d) The teacher chooses a third student at random. What is the
      probability that:
      (i)   all three students are boys (B, B, B)?
      (ii)   at least one of the three students is a girl?

      (i)   P(B, B, B) $= \frac{12}{21} \times \frac{11}{20} \times \frac{10}{19} = \frac{22}{133}$

      (ii)   P(at least one G) $= 1 - $ P(all students are boys) $= 1 - \frac{22}{133} = \frac{111}{133}$

3. A family with three children could have a girl first, followed by a boy,
   followed by another boy. This can be written as GBB.
   a) Use the letters B and G to make a list of all the possible combina-
      tions in families with three children.
         GGG, GGB, GBG, GBB, BGG, BGB, BBG, BBB
   b) Assume that each of the combinations you have listed is equally
      likely. A family with three children is chosen at random. Find the
      probability that:
      (i)   it contains at least one girl
      (ii)   it contains two girls
      (iii)   the oldest and the youngest are the same gender.

      (i)   All combinations except BBB contain one girl, so P(at least
              one girl) $= \frac{7}{8}$.
      (ii)   There are 3 combinations with two girls, so P(two are girls) $= \frac{3}{8}$.
      (iii)   There are 4 combinations where the oldest (first) and
              youngest (last) are the same gender (GGG, GBG, BGB, BBB).
              So P(oldest and youngest same gender) $= \frac{4}{8} = \frac{1}{2}$.

Remember

Note that there are two
possibilities for each child
and the events are
independent, so the total
number of possibilities is
$2 \times 2 \times 2 = 8$.

Exercise

1. An unbiased coin is tossed twice. What is the probability of the two tosses giving the same results?

2. A bag contains 8 blue marbles and 2 red marbles. Two marbles are drawn at random. What is the probability of getting
   a) two red marbles
   b) one red marble and one blue marble
   c) two blue marbles?

3. A bag contains 12 beads. Five are red and the rest are white.
   Two beads are drawn at random. Find the probability that
   a) both beads are red
   b) both beads are white.

4. Mahmoud enjoys flying his kite. On any given day, the probability that there is a good wind is $\frac{3}{4}$. If there is a good wind, the probability that the kite will fly is $\frac{5}{8}$. If there is not a good wind, the probability that the kite will fly is $\frac{1}{16}$.

   a) (i)

   Copy the given tree diagram.
   Write on your diagram the probability for each branch.
   (ii) What is the probability of a good wind *and* the kite flying?
   (iii) Find the probability that, whatever the wind, the kite does *not* fly.

   b) If the kite flies, the probability that it gets stuck in a tree is $\frac{1}{2}$.

   Calculate the probability that, whatever the wind, the kite gets stuck in a tree.

Check your progress

1. There are 240 students at Denton School. The pie chart (which is not to scale) shows how they travel to school.
   a) Calculate the value of x.
   b) How many students walk to school?
   c) What is the probability that a student, chosen at random, travels to school by bus?

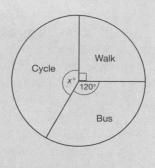

2. Sian has three cards, two of them black and one red. She places them side by side, in random order, on a table. One possible arrangement is red, black, black.
   a) Write down all the possible arrangements.
   b) Find the probability that the two black cards are next to one another. Give your answer as a fraction.

3. A die has the shape of a tetrahedron. The four faces are numbered 1, 2, 3 and 4. The die is thrown on the table. The probabilities of each of the four faces finishing flat on the table are as shown.

Face	1	2	3	4
Probability	$\frac{2}{9}$	$\frac{1}{3}$	$\frac{5}{18}$	$\frac{1}{6}$

   a) Fill in the four empty boxes with the probabilities changed to fractions with a common denominator.
   b) Which face is most likely to finish flat on the table?
   c) Find the sum of the four probabilities.
   d) What is the probability that face 3 does not finish flat on the table?

4. The diagram on the right shows all the possible outcomes when two fair dice are thrown.
   a) Explain clearly what outcome is represented by the cross P.
   b) Copy the diagram and ring the crosses representing all those outcomes with a total score on the two dice of 8.
   c) Find the probability, as a fraction in its lowest terms, that:
      (i) the two dice will show a total score of 8
      (ii) the two dice will show the same score as each other.

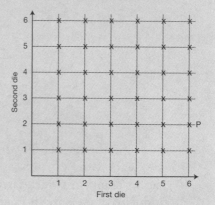

5. The probability that Jane will pass her English examination is $\frac{3}{4}$, and the probability that she will pass her mathematics examination is $\frac{4}{5}$. What is the probability that:
   a) she will not pass English?
   b) she will pass both subjects?

6. The Venn diagram on the right shows the number of students in each subset, in a school of 150 students.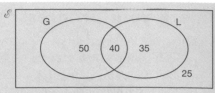

 a) (i) How many girls speak only one language?

  (ii) How many boys are there in the school?

 b) Give your answers to the following questions as fractions in their lowest terms.

  (i) A student is selected at random. What is the probability that this student speaks more than one language?

  (ii) A girl is selected at random. What is the probability that she speaks more than one language?

  (iii) A student who speaks more than one language is selected at random. What is the probability that this student is a girl?

  (iv) Two students are selected at random. What is the probability that they are both boys?

$\mathcal{E}$ = Students in the school
G = girls
L = Students who speak more than one language

7. In this question, give all your probabilities as fractions in their lowest terms.

 a) Six chairs are placed in a row. Alain is equally likely to sit on any one of the chairs.

  (i) What is the probability that Alain sits on one of the end chairs?

  (ii) After Alain has sat down, Bernard chooses any chair at random. What is the probability that Bernard sits next to Alain:
   • if Alain is sitting at an end
   • if Alain is not sitting at an end?

  (iii) Copy the tree diagram and write the probabilities on each branch.

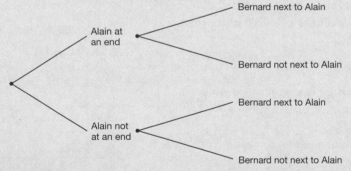

  (iv) Find the probability that Bernard sits next to Alain, wherever Alain sits.

 b) The six chairs are placed in a circle and Alain and Bernard sit down. What is the probability that Bernard sits next to Alain?

 c) There are n chairs in a circle and Alain and Bernard sit down. The probability that Bernard sits next to Alain is $\frac{1}{4}$. Find the value of n.

Transformations

If a point or a set of points (an object) moves from one place to another, it is said to have undergone a transformation. After the transformation, we talk of the set of points as the image. The pictures on the left show you four different types of transformation found in geometry. These are:
• translation (slide)
• reflection (mirror image)
• rotation (turn)
• enlargement.

The first three transformations shown in the pictures are isometric. This means that the image is identical (congruent) to the object that was transformed. Objects and images are directly congruent when the shape, size and orientation (the way the figure appears or faces) remain unchanged. If the orientation changes, the object and the image are said to be indirectly congruent. Some transformations, such as enlargement, change the shape and size of an object.

Points (such as the corner of the door) that remain in the same position before and after transformation are called invariant or fixed points.

Translation

Hint

See page 255 for more detail about vectors.

A translation, or slide, is the movement of an object over a specified distance along a line. The object is not twisted or turned. The movement is indicated by positive or negative signs according to the direction of movement along the axes of a plane. For example, movements to the left or down are negative and movements to the right or upwards are positive. A translation should be described by a column vector: $\begin{pmatrix} x \\ y \end{pmatrix}$.

■ Example

In this example, the triangle T is translated to five positions. Each translation is described below.

• Position 1 $\begin{pmatrix} 7 \\ 0 \end{pmatrix}$

• Position 2 $\begin{pmatrix} 0 \\ 4 \end{pmatrix}$

• Position 3 $\begin{pmatrix} -7 \\ 0 \end{pmatrix}$

• Position 4 $\begin{pmatrix} 0 \\ -8 \end{pmatrix}$

• Position 5 $\begin{pmatrix} 7 \\ -8 \end{pmatrix}$

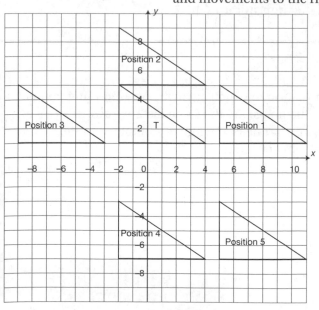

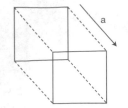

A translation of *a* units in the direction of the arrow

Properties of translation

- A translation moves the entire object the same distance in the same direction.
- Every point moves through the same distance in the same direction.
- To specify the translation, both the distance and direction of translation must be given by a column vector $\begin{pmatrix} x \\ y \end{pmatrix}$.
- The translation of an entire object can be named by specifying the translation undergone by any one point (see diagram).
- No part of the figure is invariant.
- The object and the image are directly congruent.

Exercise

1. Draw sketches to illustrate the following translations:
   a) a square is translated 6 cm to the left
   b) a triangle is translated 5 cm to the right.
2. Translate the triangle ABC given in the diagram:
   a) 3 units to the right
   b) 3 units to the left
   c) 3 units upwards
   d) 3 units downwards.

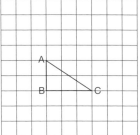

3. On squared paper, draw *x*- and *y*-axes and mark the points A(3, 5), B(2, 1) and C(–1, 4).
   a) Mark and label the triangle $\triangle A_1B_1C_1$, which is the image of $\triangle ABC$ under the translation $\begin{pmatrix} 2 \\ -3 \end{pmatrix}$.
   b) Mark and label the triangle $\triangle A_2B_2C_2$, which is the image of $\triangle ABC$ under the translation $\begin{pmatrix} 4 \\ 1 \end{pmatrix}$.

Reflection

A reflection is a transformation in which any two corresponding points (P, P′) in the object and the image are the same distance away from a line. The line *m* is called the mirror line or the line of symmetry. In the diagram below, P′ is the image of P by reflection.

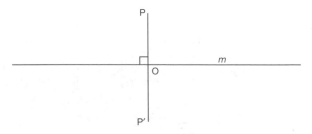

A

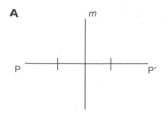

B

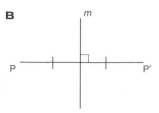

C

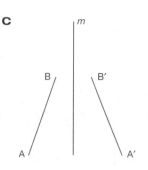

Properties of reflection

- A point (P) and its image (P′) are equidistant from the line m after reflection in line m (see figure A).
- The mirror line bisects the line joining a point and its image at right angles (see figure B).
- A line segment and its image are equal in length: AB = A′B′ (see figure C).
- A line and its image are equally inclined to the mirror line: AÔM = A′ÔM (see figure D).
- Points on the mirror line are their own images and are invariant (see figure E).
- A figure and its image are indirectly congruent (see figure F).

D 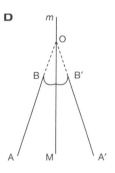   **E**    **F**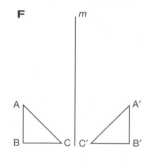

■ Examples

1. Reflect △ABC in the mirror line (indicated by the dashed line).

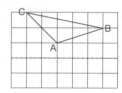

The mirror line is one of the grid lines. This makes it easy to reflect any point. You simply count the squares from the point to the mirror line, and the reflection is the same distance the other side of the mirror line.

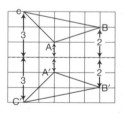

In the diagram, A is 1 unit from the mirror line, so its image A′ is also 1 unit from the mirror line. Point B is 2 units from the mirror line, so its image B′ is also 2 units from the mirror line. This principle also holds for C and its image C′.

The reflection of a straight line is a straight line. So, to obtain the reflection of △ABC, join A′ to B′, B′ to C′ and C′ to A′.

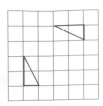

Note

You can draw the perpendicular bisector of the line joining any point and its image, e.g. P and P′ – the results will be the same.

2. A triangle and its reflection in a mirror line are shown. Draw the mirror line.

The mirror line is the perpendicular bisector of the line joining any point and its image.

In the diagram, P′ is the image of P. The mirror line is the perpendicular bisector of PP′. It is shown as a dashed line in the diagram.

Exercise

1. Copy the shapes and the mirror lines (indicated by dashed lines) onto squared paper and draw the image of each object.

a) 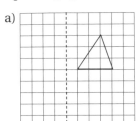   b) 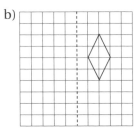   c)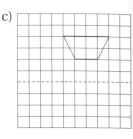

2. Copy the shapes given in parts a) to d). In each diagram, the shapes are reflections of one another. Draw the mirror line in each diagram.

a) 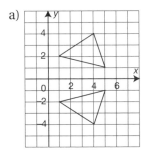   b)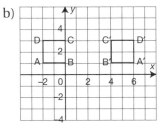

c) 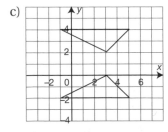   d)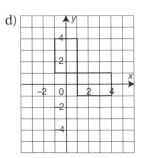

Rotation

Rotation occurs when an object is turned around a given point. You learnt about this in Module 4 – rotational symmetry. Rotation can be clockwise or anti-clockwise. The fixed point is called the centre of rotation and the angle through which the shape is rotated is called the angle of rotation.

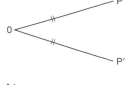

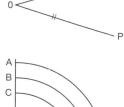

Properties of rotation

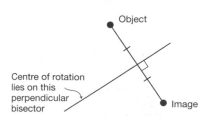

- Anti-clockwise rotation is positive and clockwise rotation is negative.
- A rotation through 180° is a half turn, a rotation through 90° is a quarter turn.
- A point and its image are equidistant from the centre of rotation.
- Each point of an object moves along the arc of a circle whose centre is the centre of rotation. So all the circles are concentric.

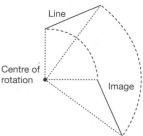

- Only the centre of rotation is invariant.
- The object and the image are directly congruent.
- The perpendicular bisector of a line joining a point and its image passes through the centre of rotation (see top right).
- A line segment and its image are equal in length (see above).

Finding the angle of rotation

To find the angle of rotation, choose any point on the given figure. Join the chosen point to the centre of rotation. Find the image of the chosen point after rotation. Join it to the centre of rotation and measure the angle between the two lines. Fix the sign of the angle according to the direction of rotation.

■ Example

In the figure (left), P′ is the image of P after a clockwise rotation. O is the centre of rotation. Find the angle of rotation.

Join OP and OP′. Let PÔP′ be $x°$. Remember that x is negative because the rotation was clockwise.

Finding the centre of rotation

A′B′ is the image of a line AB after rotation about the centre of rotation O. Find the position of the centre of rotation O. A and A′ are equidistant from the centre of rotation. The centre of rotation lies on the perpendicular bisector of AA′. Join AA′. Draw the perpendicular bisector of AA′. B and B′ are also equidistant from the centre of rotation. The centre of rotation lies on the perpendicular bisector of BB′ as well. Join BB′. Draw the perpendicular bisector of BB′. The point of intersection of the two perpendicular bisectors is O. O is the centre of the rotation.

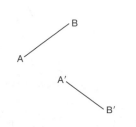

Exercise

1. Copy the diagrams in parts a) to c). Find the images of the given triangle under the rotations described.
   a) Centre of rotation (0, 0); angle of rotation 90° anti-clockwise.
   b) Centre of rotation (3, 1); angle of rotation 180°.
   c) Centre of rotation (−1, 0); angle of rotation 180°.

a)   b) 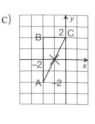 c)

2. For each diagram below, give the angle of the rotation that maps △ABC onto △A′B′C′.

a) 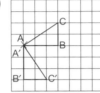 b) 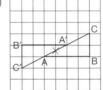 c)

Enlargement

Enlargement is a transformation in which each point of an object is mapped along a straight line drawn from a fixed point. The fixed point is called the centre of enlargement. The distance each point moves is a certain number multiplied by its distance from the centre of enlargement.

$$SF = \frac{\text{image length}}{\text{original length}}$$

Properties of enlargement

- The centre of enlargement can be anywhere (including inside the object or on the boundary).
- A scale factor greater than 1 enlarges the object whilst a scale factor smaller than 1 reduces the object, although this is still described as an enlargement (see the figures below).

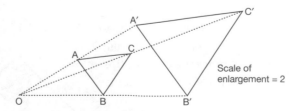

Scale of enlargement = 2

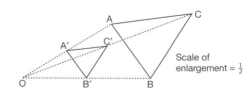

Scale of enlargement = $\frac{1}{2}$

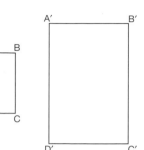

- If a point and its image are on opposite sides of the centre of enlargement, then the scale factor is negative.
- An object and its image are similar (not congruent) with sides in the ratio 1:k where k is the scale factor.
- The areas of the object and its image are in the ratio 1:k^2.
- Angles and orientation of the object are invariant.

Finding the centre of enlargement and scale factor

The figure on the left shows an object ABCD and its image A′B′C′D′ under an enlargement. To find the centre of enlargement and the scale factor, we work as follows: Join the point A and its image A′.

Extend AA′ in both directions. Similarly, draw and extend BB′, CC′ and DD′.

The point of intersection of these lines is the centre of enlargement, O.

Measure OA′ and OA.

The ratio OA′: OA gives the scale factor.

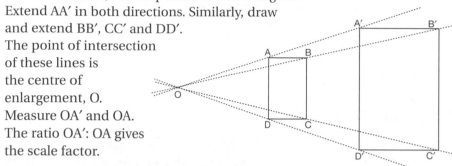

Drawing an image under enlargement

Draw the image of rectangle ABCD. O is the centre of enlargement and the scale factor is 2. Work as follows:

Join OA.
Measure OA.
Multiply the length of OA by 2.
Produce OA to A′, so that OA′ = 2OA.

Repeat for B and the other vertices of ABCD.
Join A′B′C′D′.

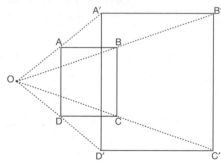

A′B′C′D′ is the image of ABCD under enlargement.

Exercise

1. Give the angle of rotation that maps △ABC onto △A'B'C'.

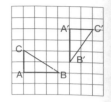

2. Draw the image of △ABC under an enlargement with scale factor 2 and centre of enlargement P(2, 1).

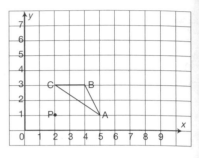

3. Draw the image of △DEF under an enlargement with scale factor –3 and centre of enlargement P(2, 0).

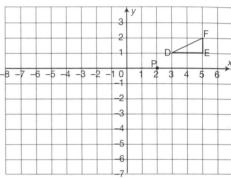

4. In the diagram, △G'H'I' is the image of △GHI under an enlargement. Find the scale factor of the enlargement and the coordinates of the centre of enlargement.

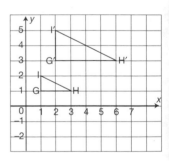

5. In the diagram, square P'Q'R'S' is the image of square PQRS under an enlargement. Find the scale factor of the enlargement and the coordinates of the centre of enlargement.

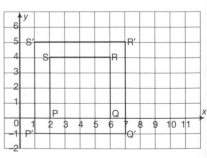

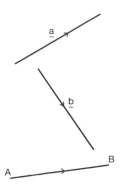

Vectors

A vector is an ordered pair of numbers that can be used to describe a translation. The ordered pair gives both magnitude and direction.

Vector notation

Vectors can be represented by a directed line segment as on the left. Note that the notation is a small letter with a wavy line beneath it. In textbooks, these vectors are denoted by bold, italic type: **a**, **b**, **u**, **v**. Vectors can also be represented by a named line such as AB (left).

In such cases, the vector is denoted by **AB** or $\overrightarrow{AB}$. The order of letters is important because they give the direction of the line. $\overrightarrow{AB}$ is not the same as $\overrightarrow{BA}$.

Writing vectors as number pairs

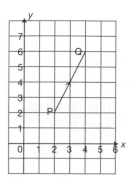

Vectors can be written using number pair notation. Look at line PQ on the diagram. This line represents the translation of P to Q. The translation is 2 units in the positive x-direction and 4 units in the positive y-direction.

This can be written as the ordered pair $\begin{pmatrix} 2 \\ 4 \end{pmatrix}$.

The top number shows the horizontal movement (parallel to the x-axis) and the bottom number shows the vertical movement (parallel to the y-axis). A negative sign indicates movements down or to the left.

We can thus write PQ = $\overrightarrow{PQ}$ = $\begin{pmatrix} 2 \\ 4 \end{pmatrix}$.

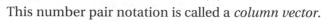

This number pair notation is called a *column vector*.

■ Examples

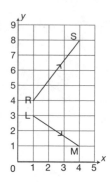

1. Express $\overrightarrow{RS}$ and $\overrightarrow{LM}$ in number pair form.

   The translation of R to S is 3 right and 4 up.
   $$\therefore \overrightarrow{RS} = \begin{pmatrix} 3 \\ 4 \end{pmatrix}$$
   The translation of L to M is 3 right and 2 down.
   $$\therefore \overrightarrow{LM} = \begin{pmatrix} 3 \\ -2 \end{pmatrix}$$

2. Draw the column vectors $\begin{pmatrix} 1 \\ 3 \end{pmatrix}$ and $\begin{pmatrix} -2 \\ -4 \end{pmatrix}$.

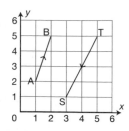

   Use squared paper and draw axes on it.
   Start at any point, for example A, and move 1 right and 3 up to B.
   $$\therefore \overrightarrow{AB} = \begin{pmatrix} 1 \\ 3 \end{pmatrix}$$
   Start at any point, for example T, and move 2 left then 4 down to S.
   $$\therefore \overrightarrow{TS} = \begin{pmatrix} -2 \\ -4 \end{pmatrix}$$

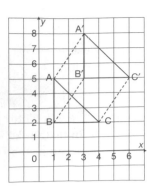

Translation described by a vector

Column vectors can be used to describe translations. In the example on the left, △ABC is translated to △A'B'C'. All points on the object have moved 2 units to the right and 3 units upwards. Therefore the column vector for this translation is $\begin{pmatrix} 2 \\ 3 \end{pmatrix}$.

■ Example

Square R is translated to square S.
Find the column vector for the translation.

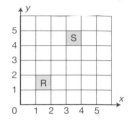

Each point on the square moves 2 units to the right and 3 units upwards. To make it easier, just consider one vertex of the square and its image.

Thus, the column vector is $\begin{pmatrix} 2 \\ 3 \end{pmatrix}$.

Exercise

1. Write column vectors for each of the vectors shown on the diagram given below.

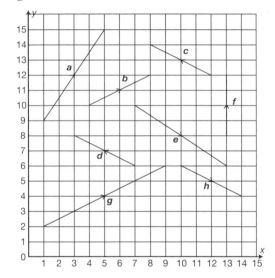

2. Represent these vectors on squared paper:

a) $\overrightarrow{AB} = \begin{pmatrix} 5 \\ 2 \end{pmatrix}$

b) $\overrightarrow{CD} = \begin{pmatrix} 2 \\ 2 \end{pmatrix}$

c) $\overrightarrow{PQ} = \begin{pmatrix} -1 \\ 3 \end{pmatrix}$

d) $\overrightarrow{RS} = \begin{pmatrix} 0 \\ 3 \end{pmatrix}$

e) $\overrightarrow{TU} = \begin{pmatrix} -2 \\ 0 \end{pmatrix}$

f) $\overrightarrow{MN} = \begin{pmatrix} -2 \\ -4 \end{pmatrix}$

g) $\overrightarrow{KL} = \begin{pmatrix} 0 \\ -5 \end{pmatrix}$ h) $\overrightarrow{VW} = \begin{pmatrix} -3 \\ -3 \end{pmatrix}$

i) $\overrightarrow{EF} = \begin{pmatrix} 4 \\ 0 \end{pmatrix}$ j) $\overrightarrow{JI} = \begin{pmatrix} -3 \\ -2 \end{pmatrix}$.

3. In the diagram on the left, ABCD is a parallelogram.
   Write column vectors for the following:
   a) $\overrightarrow{AB}$ and $\overrightarrow{DC}$
   b) $\overrightarrow{BC}$ and $\overrightarrow{AD}$.
   What can you say about the two pairs of vectors?

4. In the diagrams shown below, shapes A, B, C, D, E and F are mapped onto images A′, B′, C′, D′, E′ and F′ respectively by a translation. Find the column vector for the translation in each case.

a)
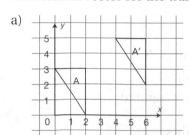

b)
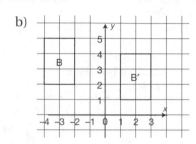

c)
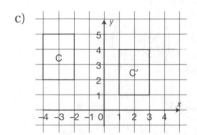

d)
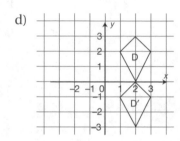

e)
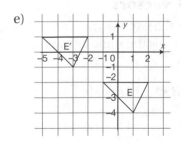

f)
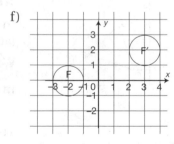

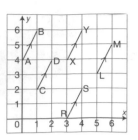

Equal vectors

Equal vectors have the same size (magnitude) and direction. As vectors are usually independent of position, they can start at any point. The same vector can be at many places in a diagram. In the figure on the left, $\overrightarrow{AB}$, $\overrightarrow{CD}$, $\overrightarrow{XY}$, $\overrightarrow{LM}$ and $\overrightarrow{RS}$ are equal vectors.

$$\overrightarrow{AB} = \overrightarrow{CD} = \overrightarrow{XY} = \overrightarrow{LM} = \overrightarrow{RS} = \begin{pmatrix} 1 \\ 2 \end{pmatrix}.$$

Multiplying a vector by a scalar

Vectors cannot be multiplied by each other, but they can be multiplied by a constant factor or scalar.

Vector a multiplied by 2 is the vector $2a$. Vector $2a$ is twice as long as vector a, but they have the same direction. In other words, they are either parallel or in a straight line.

If $a = \begin{pmatrix} 3 \\ 2 \end{pmatrix}$, then $2a = 2\begin{pmatrix} 3 \\ 2 \end{pmatrix} = \begin{pmatrix} 6 \\ 4 \end{pmatrix}$.

Vector a multiplied by -1 is the vector $-a$, opposite in direction to a but with the same magnitude as a.

■ Examples

If $u = \begin{pmatrix} 8 \\ -4 \end{pmatrix}$, find $\frac{1}{4}u$.

$$\frac{1}{4}u = \frac{1}{4}\begin{pmatrix} 8 \\ -4 \end{pmatrix} = \begin{pmatrix} \frac{1}{4} \times 8 \\ \frac{1}{4} \times (-4) \end{pmatrix}$$

$$= \begin{pmatrix} 2 \\ -1 \end{pmatrix}$$

If $v = \begin{pmatrix} -3 \\ 2 \end{pmatrix}$, find $-5v$.

$$-5v = -5\begin{pmatrix} -3 \\ 2 \end{pmatrix}$$

$$= \begin{pmatrix} (-5) \times (-3) \\ (-5) \times 2 \end{pmatrix}$$

$$= \begin{pmatrix} 15 \\ -10 \end{pmatrix}$$

Addition of vectors

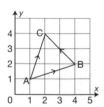

In the figure on the left, the translation from A to B followed by the translation from B to C is the same as the translation from A to C.
We write $\overrightarrow{AB} + \overrightarrow{BC} = \overrightarrow{AC}$.
In the figure, $\overrightarrow{AB} = \begin{pmatrix} 3 \\ 1 \end{pmatrix}$, $\overrightarrow{BC} = \begin{pmatrix} -2 \\ 2 \end{pmatrix}$ and $\overrightarrow{AC} = \begin{pmatrix} 1 \\ 3 \end{pmatrix}$.

That is $\begin{pmatrix} 3 \\ 1 \end{pmatrix} + \begin{pmatrix} -2 \\ 2 \end{pmatrix} = \begin{pmatrix} 1 \\ 3 \end{pmatrix}$.

In the figure, $\overrightarrow{AB}$ and $\overrightarrow{BC}$ are in the 'nose to tail' position, that is, the nose B of $\overrightarrow{AB}$ is connected to the tail B of $\overrightarrow{BC}$, and $\overrightarrow{AC}$ represents the sum of the vectors represented by $\overrightarrow{AB}$ and $\overrightarrow{BC}$. The rule is thus called the *nose to tail* rule of addition. Because AB, BC and AC form $\triangle$ABC, the rule $\overrightarrow{AB} + \overrightarrow{BC} = \overrightarrow{AC}$ is also called the *triangle law*.

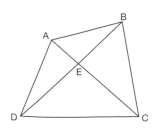

■ Examples

1. In the given figure, the various line segments are taken to represent vectors. Find, in the figure, directed line segments equal to the following:

   a) $\overrightarrow{AE} + \overrightarrow{EC}$ b) $\overrightarrow{DB} + \overrightarrow{BE}$

   c) $\overrightarrow{AD} + \overrightarrow{DB} + \overrightarrow{BC}$ d) $\overrightarrow{CB} + \overrightarrow{BE} + \overrightarrow{EA} + \overrightarrow{AD}$.

   a) $\overrightarrow{AE} + \overrightarrow{EC} = \overrightarrow{AC}$ b) $\overrightarrow{DB} + \overrightarrow{BE} = \overrightarrow{DE}$

   c) $\overrightarrow{AD} + \overrightarrow{DB} + \overrightarrow{BC} = \overrightarrow{AB} + \overrightarrow{BC} = \overrightarrow{AC}$ d) $\overrightarrow{CB} + \overrightarrow{BE} + \overrightarrow{EA} + \overrightarrow{AD}$
   $$= \overrightarrow{CE} + \overrightarrow{EA} + \overrightarrow{AD} = \overrightarrow{CA} + \overrightarrow{AD} = \overrightarrow{CD}$$

2. If $a = \begin{pmatrix} 3 \\ 4 \end{pmatrix}$ and $b = \begin{pmatrix} 2 \\ -1 \end{pmatrix}$, find the column vectors equal to $a + b$, $a - b$, $3a$, $a + 4b$, $2a - 3b$.

   $$a + b = \begin{pmatrix} 3 \\ 4 \end{pmatrix} + \begin{pmatrix} 2 \\ -1 \end{pmatrix} = \begin{pmatrix} 3 + 2 \\ 4 + (-1) \end{pmatrix} = \begin{pmatrix} 5 \\ 3 \end{pmatrix}$$

   $$a - b = \begin{pmatrix} 3 \\ 4 \end{pmatrix} - \begin{pmatrix} 2 \\ -1 \end{pmatrix} = \begin{pmatrix} 3 - 2 \\ 4 - (-1) \end{pmatrix} = \begin{pmatrix} 1 \\ 5 \end{pmatrix}$$

   $$3a = 3 \times \begin{pmatrix} 3 \\ 4 \end{pmatrix} = \begin{pmatrix} 3 \times 3 \\ 3 \times 4 \end{pmatrix} = \begin{pmatrix} 9 \\ 12 \end{pmatrix}$$

   $$a + 4b = \begin{pmatrix} 3 \\ 4 \end{pmatrix} + 4\begin{pmatrix} 2 \\ -1 \end{pmatrix} = \begin{pmatrix} 3 + 8 \\ 4 + (-4) \end{pmatrix} = \begin{pmatrix} 11 \\ 0 \end{pmatrix}$$

   $$2a - 3b = 2\begin{pmatrix} 3 \\ 4 \end{pmatrix} - 3\begin{pmatrix} 2 \\ -1 \end{pmatrix} = \begin{pmatrix} 0 \\ 11 \end{pmatrix}$$

3. OACB is a parallelogram in which $\overrightarrow{OA} = a$ and $\overrightarrow{OB} = b$. M is the mid-point of BC, and N is the mid-point of AC.

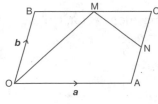

   a) Find, in terms of a and b: (i) $\overrightarrow{OM}$; (ii) $\overrightarrow{MN}$.

   b) Show that $\overrightarrow{OM} + \overrightarrow{MN} = \overrightarrow{OA} + \overrightarrow{AN}$.

   a) (i) $\overrightarrow{OM} = \overrightarrow{OB} + \overrightarrow{BM}$

          $\overrightarrow{OB} = b$

          M is mid-point BC, so $\overrightarrow{BM} = \frac{1}{2}\overrightarrow{BC}$

          $\therefore \overrightarrow{OM} = b + \frac{1}{2}a$

       (ii) $\overrightarrow{MN} = \overrightarrow{MC} + \overrightarrow{CN} = \frac{1}{2}\overrightarrow{BC} + \frac{1}{2}\overrightarrow{CA}$

              $= \frac{1}{2}a + (-\frac{1}{2}b)$ $(\overrightarrow{CA} = -\overrightarrow{AC} = -\overrightarrow{OB})$

              $= \frac{1}{2}a - \frac{1}{2}b = \frac{1}{2}(a - b)$

   b) $\overrightarrow{OM} + \overrightarrow{MN} = (b + \frac{1}{2}a) + (\frac{1}{2}a - \frac{1}{2}b)$

              $= a + \frac{1}{2}b$

       $OA + \overrightarrow{AN} = a + \frac{1}{2}b$

       $\overrightarrow{OM} + \overrightarrow{MN} = \overrightarrow{OA} + \overrightarrow{AN}$

Subtraction of vectors

Subtracting a vector is the same as adding its negative. So $a - b = a + (-b)$.

Consider $\overrightarrow{AC} - \overrightarrow{AB}$:

Adding the negative of $\overrightarrow{AB}$ is the same as adding $\overrightarrow{BA}$.

$$\therefore \overrightarrow{AC} - \overrightarrow{AB} = \overrightarrow{AC} + \overrightarrow{BA}$$

If we rearrange the vectors, then we can apply the nose to tail rule:

$$\overrightarrow{AC} - \overrightarrow{AB} = \overrightarrow{BA} + \overrightarrow{AC} = \overrightarrow{BC}.$$

Exercise

1. $p = \begin{pmatrix} 4 \\ -2 \end{pmatrix}$ and $q = \begin{pmatrix} -1 \\ -3 \end{pmatrix}$.

   Express in column vector form:

   a) $3p$　　　b) $p + q$.

2. Given that $a = \begin{pmatrix} 4 \\ -2 \end{pmatrix}$ and $b = \begin{pmatrix} -4 \\ 3 \end{pmatrix}$, express $2a - b$ as a column vector.

3. In the diagram, BCE and ACD are straight lines. $\overrightarrow{AB} = 2a$ and $\overrightarrow{BC} = 3b$.
   The point C divides AD in the ratio 2:1 and divides BE in the ratio 3:1.
   Express, in terms of a and b, the vectors:

   a) $\overrightarrow{AC}$

   b) $\overrightarrow{CD}$

   c) $\overrightarrow{CE}$

   d) $\overrightarrow{ED}$.

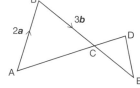

Note

If you are following the core syllabus, you have completed this module. Turn to page 274 and do the appropriate questions in Check your progress.

Position vectors

A vector that starts from the origin (O) is called a position vector. On the diagram on the left, point A has the position vector a.

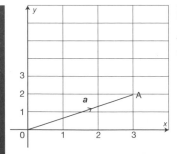

If $a = \begin{pmatrix} 3 \\ 2 \end{pmatrix}$, then the coordinates of A will be (3, 2).

■ Example

The position vector of A is $\begin{pmatrix} 3 \\ 2 \end{pmatrix}$ and the position vector of B is $\begin{pmatrix} -2 \\ 4 \end{pmatrix}$.
Find the vector $2\overrightarrow{AB}$.

$$\overrightarrow{OA} = \begin{pmatrix} 3 \\ 2 \end{pmatrix} \quad \overrightarrow{OB} = \begin{pmatrix} -2 \\ 4 \end{pmatrix}$$

$$\overrightarrow{AB} = -\overrightarrow{OA} + \overrightarrow{OB}$$
$$= \overrightarrow{OB} - \overrightarrow{OA}$$

That is, $\overrightarrow{AB} = \begin{pmatrix} -2 \\ 4 \end{pmatrix} - \begin{pmatrix} 3 \\ 2 \end{pmatrix} = \begin{pmatrix} -5 \\ 2 \end{pmatrix}$

So $2\overrightarrow{AB} = 2\begin{pmatrix} -5 \\ 2 \end{pmatrix} = \begin{pmatrix} -10 \\ 4 \end{pmatrix}$

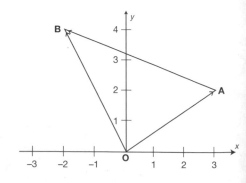

You could also find the column vector for $\overrightarrow{AB}$ by counting the movements parallel to the x-axis followed by those parallel to the y-axis.

Movement parallel to x-axis = –5 units
Movement parallel to y-axis = 2 units

$$\therefore \overrightarrow{AB} = \begin{pmatrix} -5 \\ 2 \end{pmatrix}$$

$$\therefore 2\overrightarrow{AB} = \begin{pmatrix} -10 \\ 4 \end{pmatrix}$$

The magnitude of a vector

The length of a vector is called the magnitude or modulus of the vector.
The magnitude of vector $\overrightarrow{AB}$ is written as $|AB|$.
The magnitude of $\overrightarrow{AB}$ is $|AB|$ and the magnitude of $\underset{\sim}{a}$ is $|\underset{\sim}{a}|$.

If $\overrightarrow{AB} = \begin{pmatrix} x \\ y \end{pmatrix}$, then $|AB| = \sqrt{(x^2 + y^2)}$ (Can you see how Pythagoras' theorem is used?)

■ Examples

If $\underset{\sim}{a} = \begin{pmatrix} -5 \\ 12 \end{pmatrix}$, find $|\underset{\sim}{a}|$

$$|\underset{\sim}{a}| = \sqrt{(-5)^2 + (12^2)}$$
$$= \sqrt{169}$$
$$= 13$$

If A is point (–1, –2) and B is (5, 6), find $|AB|$.

$$\overrightarrow{OA} = \begin{pmatrix} -1 \\ -2 \end{pmatrix} \quad \overrightarrow{OB} = \begin{pmatrix} 5 \\ 6 \end{pmatrix}$$
$$\overrightarrow{AB} = \overrightarrow{AO} + \overrightarrow{OB} = -\overrightarrow{OA} + \overrightarrow{OB}$$
$$= \begin{pmatrix} -(-1) + 5 \\ -(-2) + 6 \end{pmatrix} = \begin{pmatrix} 6 \\ 8 \end{pmatrix}$$
$$|AB| = \sqrt{6^2 + 8^2} = 10$$

Exercise

1. O is the point (0, 0), P is (3, 4), Q is (–5, 12) and R is (–8, –15).
   Find the values of $|OP|, |OQ|$ and $|OR|$.

2. $\overrightarrow{OA} = \begin{pmatrix} 4 \\ 2 \end{pmatrix}$, $\overrightarrow{OB} = \begin{pmatrix} -1 \\ 3 \end{pmatrix}$ and $\overrightarrow{OC} = \begin{pmatrix} 6 \\ -2 \end{pmatrix}$.

   a) Write down the coordinates of A, B and C.
   b) Write down the vectors $\overrightarrow{AB}$, $\overrightarrow{CB}$ and $\overrightarrow{AC}$ in column vector form.

3. OACB is a parallelogram in which $\overrightarrow{OA} = 2p$ and $\overrightarrow{OB} = 2q$.

   M is the mid-point of BC, and N is the mid-point of AC.
   Find, in terms of p and q:

   a) $\overrightarrow{AB}$
   b) $\overrightarrow{ON}$
   c) $\overrightarrow{NM}$.

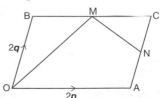

4. OATB is a parallelogram. M, N and P are mid-points of BT, AT and MN respectively. O is the origin, and the position vectors of A and B are *a* and *b* respectively. Find, in terms of *a* and/or *b*:

a) $\overrightarrow{MT}$

b) $\overrightarrow{TN}$

c) $\overrightarrow{MN}$

d) The position vector of P, giving your answer in its simplest form.

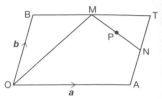

Further transformations

Shear

The transformation that keeps one line fixed, moves all other points parallel to this line and maps every straight line onto a straight line is called a *shear*. The fixed line is called the *invariant line*. The invariant line can be anywhere, in the figure or outside the figure. A shear is not an isometry, because the shape of the figure is distorted.

■ Examples

1. In the diagram on the left, the rectangle ABCD maps onto ABC′D′; under a shear.

   ABCD → ABC′D′

   AB → AB

   AB is the invariant line, that is, the points on the line AB are fixed.

   CD → C′D′

   length CD = length C′D′

   All the points (except those on AB) move parallel to the invariant line AB.

   Area of ABCD = area of ABC′D′.

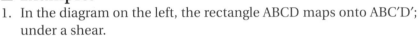

2. In this figure, the invariant line is PQ.

   ABCD → A′B′C′D′

   AB → A′B′; length AB = length A′B′

   CD → C′D′; length CD = length C′D′

   The points on one side of the invariant line move to the left while the points on the other side move to the right.

   Area of ABCD = area of A′B′C′D′.

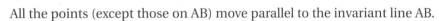

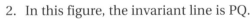

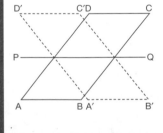

Properties of shear

- The points on the invariant line are fixed. Points on the line are unaffected by the shear.
- All points not on the invariant line move parallel to the invariant line (see figure on the left).

- The perpendicular distance of points from the invariant line remains constant.
- The image of a straight line is a straight line.
- A straight line and its image meet at a point on the invariant line unless the line is parallel to the invariant line. See how DA and D'A' produced meet at P in the diagram on the previous page.
- Points on opposite sides of the invariant line are displaced in opposite directions.
- The area of a figure remains constant.
- The distance a point moves is proportional to its distance from the invariant line. This constant is called the shear factor.

$$\text{Shear factor} = \frac{\text{distance a point moves}}{\text{distance of the point from the invariant line}}.$$

■ Examples

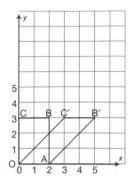

1. In the figure on the left, rectangle OABC is mapped onto parallelogram OAB'C' under a shear. Find the invariant line and the shear factor.

   The points on OA do not move, so OA is the invariant line.
   C has moved onto C'. The distance of C from OA = 3 units. C has also moved 3 units.

   $$\frac{\text{distance moved by C}}{\text{distance of C from OA}} = \frac{3}{3} = 1 \therefore \text{ the shear factor is 1.}$$

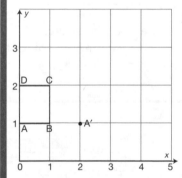

2. Find the image of ABCD under a shear which keeps the x-axis fixed and maps A to A'. A moves 2 units to A'.
   Distance of A from the invariant line (the x-axis) = 1 unit.

   $$\frac{\text{distance moved by A}}{\text{distance of A from } x\text{-axis (invariant line)}} = \frac{2}{1} = 2, \text{ so the shear factor} = 2.$$

   $$\frac{\text{distance moved by B}}{\text{distance of B from } x\text{-axis (invariant line)}} = \text{shear factor} = 2.$$

   Distance moved by B = $2 \times$ distance of B from the x-axis.
   Distance of B from the x-axis = 1 unit.
   Distance moved by B = $2 \times 1 = 2$ units.
   Coordinates of B' (the image of B) = (3, 1).

   $$\frac{\text{distance moved by C}}{\text{distance of C from } x\text{-axis (invariant line)}} = \text{shear factor} = 2.$$

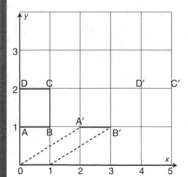

   Distance moved by C = $2 \times$ distance of C from the x-axis.
   Distance of C from x-axis = 2 units.
   Distance moved by C = $2 \times 2 = 4$ units.
   Coordinates of C' (the image of C) = (5, 2).

   $$\frac{\text{distance moved by D}}{\text{distance of D from } x\text{-axis (invariant line)}} = \text{shear factor} = 2.$$

   Distance of D from x-axis = 2 units.
   Distance moved by D = 4 units.
   Coordinates of D' (the image of D) = (4, 2).

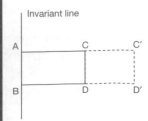

Invariant line

Stretch

A one-way stretch is an enlargement in one direction from a given line. The given line is the invariant line or axis.

$$\text{scale factor} = \frac{\text{distance of the image of a point from the invariant line}}{\text{distance of the point from the invariant line}}$$

$$\text{scale factor} = \frac{AC'}{AC} = \frac{BD'}{BD}$$

Properties of a stretch
- A one-way stretch with a scale factor (–1) is equivalent to a reflection in the invariant line.
- A two-way stretch is a combination of two one-way stretches with perpendicular invariant lines. The only invariant point is the intersection of the two lines. A two-way stretch affects both coordinates of each point by the respective scale factors.
- A two-way stretch with equal scale factors is an enlargement with the centre of enlargement at the point of intersection of the invariant lines, and the same scale factor.

A

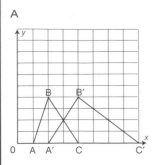

B

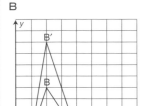

■ Examples

1. Figure A shows △ABC and its image A′B′C′ after a one-way stretch of scale factor 2 with the y-axis as invariant line. The x-coordinate of each point is doubled while the y-coordinate is unchanged.

2. Figure B shows △ABC and its image AB′C after a one-way stretch of scale factor 2 with the x-axis as invariant line. The y-coordinate of each point is multiplied by 2 while the x-coordinate remains unchanged.

Exercise

1. Copy this diagram onto graph paper. Draw the image A′B′C′ of △ABC under the shear that has the x-axis as invariant line and the scale factor 2.

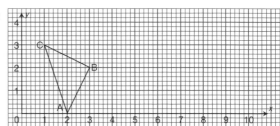

2. On the diagram, draw the image P′Q′R′ of △PQR under the shear that has the line l as invariant line and maps P onto the point P′ shown.

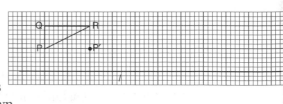

3. a) Describe the single transfor-
      mation that maps the
      rectangle OABC onto
      parallelogram OAPQ.
   b) Describe the single transfor-
      mation that maps the
      rectangle OABC onto the
      rectangle OAKL.

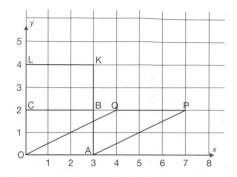

Matrices and matrix transformations

Describing and working with matrices

A set of numbers arranged in rows and columns enclosed in round or square brackets is called a matrix. You will use round brackets in this course.

The matrix $\begin{pmatrix} 2 & 2 & 4 \\ 10 & 8 & 12 \\ 3 & 2 & 3 \\ 1 & 0 & 1 \end{pmatrix}$ has 4 rows and 3 columns.

Column 1 ⟶ Column 2 ↓ Column 3 ⟶

Row 1 ⟶ $\begin{pmatrix} 2 & 2 & 4 \\ 10 & 8 & 12 \\ 3 & 2 & 3 \\ 1 & 0 & 1 \end{pmatrix}$
Row 2 ⟶
Row 3 ⟶
Row 4 ⟶

Each number in the array is called an *entry* or an *element* of the matrix.

A matrix is often denoted by a capital letter. For example, we can call the above matrix **A**.

$$\mathbf{A} = \begin{pmatrix} 2 & 2 & 4 \\ 10 & 8 & 12 \\ 3 & 2 & 3 \\ 1 & 0 & 1 \end{pmatrix}$$

The *order of a matrix* gives the number of rows followed by the number of columns in a matrix. A matrix with 3 rows and 2 columns is a 3 by 2 matrix or a 3×2 matrix. The order of such a matrix is 3×2.

A matrix of order 4×5 has 4 rows and 5 columns.

The matrix $\begin{pmatrix} 2 & 0 & 5 \\ 1 & 3 & 4 \end{pmatrix}$ is of the order 2×3 (you will say 'of the order 2 by 3').

$$P = \begin{pmatrix} 3 & 1 \\ 2 & 4 \end{pmatrix}$$

$$Q = \begin{pmatrix} 4 & 1 & 0 \\ 5 & 3 & 1 \\ 1 & 2 & 5 \end{pmatrix}$$

$$C = \begin{pmatrix} 4 & 0 & 0 \\ 0 & 3 & 0 \\ 0 & 0 & 5 \end{pmatrix}$$

$$I = \begin{pmatrix} 1 & 0 \\ 0 & 1 \end{pmatrix}$$

$$O = \begin{pmatrix} 0 & 0 \\ 0 & 0 \end{pmatrix}$$

Types of matrix

A matrix with an equal number of rows and columns is called a *square matrix*.

The matrix **P** has 2 rows and 2 columns. It is a square matrix of order 2×2.
The matrix **Q** is a square matrix of order 3×3.

A *diagonal matrix* has all its elements zero except for those in the leading diagonal (from top left to bottom right). The leading diagonal elements in matrix **C** are 4, 3 and 5. All other elements are zeros.

If the elements of a diagonal matrix are all equal to 1, it is called a *unit matrix*. It is denoted by the letter **I**. **I** is a 2×2 unit matrix.

If all the elements in a matrix are zero, then it is called a *zero matrix* or a *null matrix*. It is denoted by the letter **O**.

Equal matrices

Two matrices are equal if, and only if, they are identical. This means they must be of the same order and the respective elements must be identical.

Matrices **A** and **B** are equal matrices.

$$A = \begin{pmatrix} 4 & 1 \\ 3 & 5 \end{pmatrix} \quad B = \begin{pmatrix} 4 & 1 \\ 3 & 5 \end{pmatrix} \quad C = \begin{pmatrix} 5 & 3 \\ 2 & 4 \end{pmatrix} \quad D = \begin{pmatrix} 5 & 2 \\ 3 & 4 \end{pmatrix}$$

Matrices **C** and **D** are not equal.

Exercise

1. Write down the order of the following matrices.

   a) $\begin{pmatrix} 6 & 3 \\ 1 & 4 \end{pmatrix}$
   b) $\begin{pmatrix} 6 & 3 & 0 \\ 1 & 4 & 1 \end{pmatrix}$
   c) $\begin{pmatrix} 6 & 3 \\ 1 & 4 \\ 2 & 5 \end{pmatrix}$

   d) $\begin{pmatrix} 2 & 7 \end{pmatrix}$
   e) $\begin{pmatrix} 1 & 3 & 4 \end{pmatrix}$
   f) $\begin{pmatrix} 2 \end{pmatrix}$

   g) $\begin{pmatrix} 5 \\ 1 \end{pmatrix}$
   h) $\begin{pmatrix} 2 \\ 1 \\ 4 \end{pmatrix}$
   i) $\begin{pmatrix} 2 & 1 & -3 \\ 3 & 4 & 0 \\ 5 & 1 & 1 \end{pmatrix}$

2. Write down the equal matrices in this list.

   $$A = \begin{pmatrix} 4 & 5 \end{pmatrix} \quad B = \begin{pmatrix} 5 & 4 \end{pmatrix} \quad C = \begin{pmatrix} 4 & 5 & 6 \end{pmatrix}$$

   $$D = \begin{pmatrix} 4 \\ 5 \\ 6 \end{pmatrix} \quad E = \begin{pmatrix} 4 & 5 & 6 \end{pmatrix} \quad F = \begin{pmatrix} 6 & 5 & 4 \end{pmatrix}$$

   $$G = \begin{pmatrix} 6 & 3 \\ 1 & 4 \end{pmatrix} \quad H = \begin{pmatrix} 6 & 3 \\ 1 & 4 \end{pmatrix}$$

Addition and subtraction of matrices

You can only add or subtract matrices of the same order. To add, you simply add the corresponding elements in each matrix. To subtract, you subtract the corresponding elements in each matrix.

■ Examples

$$\begin{pmatrix} 3 & 2 & 1 \\ 0 & -4 & 5 \end{pmatrix} + \begin{pmatrix} -1 & 1 & 3 \\ 2 & 4 & 0 \end{pmatrix}$$

$$= \begin{pmatrix} 3+(-1) & 2+1 & 1+3 \\ 0+2 & -4+4 & 5+0 \end{pmatrix}$$

$$= \begin{pmatrix} 3 & 3 & 4 \\ 2 & 0 & 5 \end{pmatrix}$$

$$\begin{pmatrix} 3 & 2 & 4 \\ 0 & -2 & 5 \end{pmatrix} - \begin{pmatrix} 5 & 2 & 1 \\ 1 & -4 & 5 \end{pmatrix}$$

$$= \begin{pmatrix} 3-5 & 2-2 & 4-1 \\ 0-1 & -2-(-4) & 5-5 \end{pmatrix}$$

$$= \begin{pmatrix} -2 & 0 & 3 \\ -1 & 2 & 0 \end{pmatrix}$$

Scalar multiplication

You can multiply a matrix by a number. Each element of the matrix must be multiplied by the number.

■ Examples

$$3\begin{pmatrix} 6 & -1 \\ 3 & 2 \end{pmatrix}$$

$$= \begin{pmatrix} 3 \times 6 & 3 \times (-1) \\ 3 \times 3 & 3 \times 2 \end{pmatrix}$$

$$= \begin{pmatrix} 18 & -3 \\ 9 & 6 \end{pmatrix}$$

$$4\begin{pmatrix} 3x & 2y \\ -5x & y \end{pmatrix}$$

$$= \begin{pmatrix} 4 \times 3x & 4 \times 2y \\ 4 \times (-5x) & 4 \times y \end{pmatrix}$$

$$= \begin{pmatrix} 12x & 8y \\ -20x & 4y \end{pmatrix}$$

Exercise

Find:

1. $3\begin{pmatrix} 3 & -1 \end{pmatrix}$

2. $3\begin{pmatrix} 5 \\ -2 \end{pmatrix}$

3. $2\begin{pmatrix} 5 & -3 \\ 0 & 1 \end{pmatrix}$

4. $-4\begin{pmatrix} 1 & -2 \\ 3 & 0 \end{pmatrix}$

5. $\frac{1}{2}\begin{pmatrix} 4 & 0 \\ 2 & 1 \end{pmatrix}$

6. $3\begin{pmatrix} 2a & -2b \\ 3a & 4b \end{pmatrix}$

7. $-2\begin{pmatrix} 0 & 0 \\ 0 & -1 \end{pmatrix}$

8. $2\begin{pmatrix} 4 \\ -1 \end{pmatrix} - 3\begin{pmatrix} 1 \\ -2 \end{pmatrix}.$

Multiplication of matrices

It is possible to work out the product of two matrices according to the following rules:

- the number of columns in the first matrix must equal the number of rows in the second matrix
- the order of the product of the matrices is the number of rows in the first matrix multiplied by the number of columns in the second
- when multiplying, multiply the elements of a row of the first matrix by the elements in a column of the second matrix and add the products. For example, the sum of the products of the elements of the third row of the first matrix and the second column of the second matrix gives the elements in the third row and second column of the product matrix.

Follow these examples carefully to see how this works.

■ Examples

1. Find the product of $(2 \quad 5) \begin{pmatrix} 4 \\ 3 \end{pmatrix}$:

$$\left. \begin{array}{l} 2 \times 4 = 8 \\ 5 \times 3 = \underline{15} \\ \hphantom{5 \times 3 = } 23 \end{array} \right\} + \quad \therefore \text{ product matrix is } (23)$$

2. Given that $\begin{pmatrix} x & 2 \\ 3 & y \end{pmatrix} \begin{pmatrix} 4 \\ -1 \end{pmatrix} = \begin{pmatrix} 8 \\ 5 \end{pmatrix}$, find the value of x and y.

$$\begin{pmatrix} 4x & -2 \\ 12 & -y \end{pmatrix} = \begin{pmatrix} 8 \\ 5 \end{pmatrix} \text{ (multiply LHS)}$$

So $4x - 2 = 8$

$\hphantom{So } 12 - y = 5$

$\hphantom{So } \therefore 4x = 10 \quad$ and $\quad 12 - 5 = y$

$\hphantom{So \therefore 4} x = 2.5 \hphantom{aaaaaaaaaa} y = 7$

3. Find the product of $\begin{pmatrix} 2 & 3 \\ 1 & 0 \end{pmatrix} \begin{pmatrix} 3 & 1 \\ 6 & 4 \end{pmatrix}$.

$$(2 \quad 3) \begin{pmatrix} 3 \\ 6 \end{pmatrix} = 2 \times 3 + 3 \times 6 = 6 + 18 = 24$$

$$(2 \quad 3) \begin{pmatrix} 1 \\ 4 \end{pmatrix} = 2 \times 1 + 3 \times 4 = 2 + 12 = 14$$

$$(1 \quad 0) \begin{pmatrix} 3 \\ 6 \end{pmatrix} = 1 \times 3 + 0 \times 6 = 3 + 0 = 3$$

$$(1 \quad 0) \begin{pmatrix} 1 \\ 4 \end{pmatrix} = 1 \times 1 + 0 \times 4 = 1 + 0 = 1$$

$$\therefore \text{ product is } \begin{pmatrix} 24 & 14 \\ 3 & 1 \end{pmatrix}$$

Multiplication by a unit matrix

Because multiplication by 1 leaves the numbers unchanged, a square matrix is unchanged when multiplied by a unit matrix.

Order of matrices in multiplication

If **A** and **B** are two matrices, then **AB** is not generally equal to **BA**. In other words, multiplication of matrices is not commutative.

Exercise

1. Find the following matrix products:

   a) $(4 \quad 5) \begin{pmatrix} 3 \\ 2 \end{pmatrix}$

   b) $(-2 \quad 1) \begin{pmatrix} -3 \\ 0 \end{pmatrix}$

   c) $(-3 \quad -2) \begin{pmatrix} 3 \\ 4 \end{pmatrix}$

   d) $(3 \quad -2) \begin{pmatrix} -2 \\ 3 \end{pmatrix}$

   e) $(a \quad 3) \begin{pmatrix} 1 \\ 3 \end{pmatrix}$.

2. Carry out the following matrix multiplications:

   a) $\begin{pmatrix} 5 & 1 \\ 2 & 0 \end{pmatrix} \begin{pmatrix} 2 \\ 3 \end{pmatrix}$

   b) $\begin{pmatrix} -1 & -2 \\ -3 & 0 \end{pmatrix} \begin{pmatrix} -2 \\ -3 \end{pmatrix}$

   c) $\begin{pmatrix} 2 & 6 \\ 0 & 3 \end{pmatrix} \begin{pmatrix} 1 & 3 \\ 1 & 4 \end{pmatrix}$

   d) $\begin{pmatrix} 1 & -2 \\ 3 & 5 \end{pmatrix} \begin{pmatrix} 3 & -1 \\ -1 & 1 \end{pmatrix}$.

3. Find the value of x in each of the following matrix equations:

   a) $(x \quad 2) \begin{pmatrix} 1 \\ 3 \end{pmatrix} = (10)$

   b) $(3 \quad x) \begin{pmatrix} 5 \\ 2 \end{pmatrix} = (17)$

   c) $(x \quad -5) \begin{pmatrix} 2 \\ 3 \end{pmatrix} = (-1)$.

4. If $A = \begin{pmatrix} 0 & 5 \\ 2 & -2 \end{pmatrix}$, $B = \begin{pmatrix} 1 & 3 \\ -1 & 2 \end{pmatrix}$ and $I = \begin{pmatrix} 1 & 0 \\ 0 & 1 \end{pmatrix}$, find:

   a) **AB**   b) **BA**   c) **AI**
   d) **IA**   e) **B³**.

5. What is the image of (5 4) under a translation of $\begin{pmatrix} 1 \\ 2 \end{pmatrix}$?

6. a) Find the products of:

   (i) $\begin{pmatrix} -1 & 2 \\ 2 & 1 \end{pmatrix} \begin{pmatrix} 3 & 0 \\ -1 & 2 \end{pmatrix}$

   (ii) $\begin{pmatrix} 3 & 0 \\ -1 & 2 \end{pmatrix} \begin{pmatrix} -1 & 2 \\ 2 & 1 \end{pmatrix}$.

   b) Is the multiplication of matrices commutative, according to your answers?

   c) Can you explain why or why not?

Determinant of a matrix

If $A = \begin{pmatrix} 2 & 1 \\ 3 & 5 \end{pmatrix}$, then the number $(2 \times 5) - (1 \times 3)$ is the determinant of the matrix. In this case, the determinant is 7. The notation $|A|$ is used to denote the determinant.

If $A = \begin{pmatrix} a & b \\ c & d \end{pmatrix}$, $|A| = ad - bc$.

$|A|$ is the product of elements in the leading diagonal minus the product of the elements in the other diagonal.

■ Examples

If $M = \begin{pmatrix} 5 & -2 \\ 1 & 3 \end{pmatrix}$, find $|M|$.

$5 \times 3 = 15$
$1 \times (-2) = -2$

$\begin{aligned} |M| &= 15 - (-2) \\ &= 15 + 2 \\ &= 17 \end{aligned}$

Find x if the determinant of
$\begin{pmatrix} x & 3 \\ 4 & 1 \end{pmatrix} = 2$.

Determinant:
$(x \times 1) - (3 \times 4) = x - 12$
But this must $= 2$
$\therefore x - 12 = 2$
$x = 14$

Exercise

1. Find the determinants of the following matrices:

   a) $\begin{pmatrix} 4 & 1 \\ 3 & 5 \end{pmatrix}$ b) $\begin{pmatrix} 6 & 8 \\ 3 & 4 \end{pmatrix}$ c) $\begin{pmatrix} -3 & -2 \\ -1 & 1 \end{pmatrix}$

   d) $\begin{pmatrix} 2 & -1 \\ 3 & 9 \end{pmatrix}$ e) $\begin{pmatrix} -1 & 0 \\ 0 & 1 \end{pmatrix}$.

2. If $B = \begin{pmatrix} x & 2 \\ 3 & 1 \end{pmatrix}$ and $|B| = 3$, find the value of x.

3. Given that $C = \begin{pmatrix} 1 & 4 \\ -2 & x \end{pmatrix}$ and $|C| = 5$, find the value of x.

4. Find y if the determinant of the matrix $\begin{pmatrix} y & 2 \\ 3 & 2 \end{pmatrix} = 2$.

5. Given that $A = \begin{pmatrix} -2 & p \\ -2 & 3 \end{pmatrix}$ and $|A| = 12$, find the value of p.

Remember

A number multiplied by its multiplicative inverse is 1. The inverse of 3 is 3^{-1}. The inverse of -2 is $(-2)^{-1}$. The inverse of a^{-1} is a.

The inverse of a matrix

The inverse of a square matrix **A** is denoted by $\mathbf{A}^{-1}$ and
$$\mathbf{A.A}^{-1} = \mathbf{A}^{-1}.\mathbf{A} = \mathbf{I},$$
where **I** is the unit matrix of the same order as **A**.

■ Example

If $\mathbf{A} = \begin{pmatrix} 5 & 2 \\ 7 & 3 \end{pmatrix}$ and $\mathbf{B} = \begin{pmatrix} 3 & -2 \\ -7 & 5 \end{pmatrix}$

then $\begin{pmatrix} 5 & 2 \\ 7 & 3 \end{pmatrix}\begin{pmatrix} 3 & -2 \\ -7 & 5 \end{pmatrix} = \begin{pmatrix} 3 & -2 \\ -7 & 5 \end{pmatrix}\begin{pmatrix} 5 & 2 \\ 7 & 3 \end{pmatrix} = \begin{pmatrix} 1 & 0 \\ 0 & 1 \end{pmatrix}.$

∴ **B** is the inverse of **A** and **A** is the inverse of **B**.

Finding the inverse of a matrix
- Calculate the determinant of the matrix.
- If the determinant is 1, swap the elements of the leading diagonal and change the signs of the elements of the other diagonal.
- If the determinant is not 1 or 0, swap the elements of the leading diagonal, change the signs of the elements of the other diagonal and divide each element by the determinant.
- If the determinant is 0, no division is possible. In such a case, the matrix has no determinant and is called a singular matrix.

■ Examples

1. Find the inverse of $\begin{pmatrix} 2 & 1 \\ 4 & 2 \end{pmatrix}.$

   Determinant:
   $(2 \times 2) = (4 \times 1) = 4 - 4 = 0$
   Matrix is singular.
   No inverse can be found.

2. If $\mathbf{A} = \begin{pmatrix} 8 & 5 \\ 3 & 2 \end{pmatrix}$, find $\mathbf{A}^{-1}$.

   $|\mathbf{A}| = (8 \times 2) - (3 \times 5) = 16 - 15 = 1$

   $\mathbf{A}^{-1} = \begin{pmatrix} 2 & -5 \\ -3 & 8 \end{pmatrix}$

3. Given that $\mathbf{A} = \begin{pmatrix} 5 & 7 \\ 6 & 9 \end{pmatrix}$, find $\mathbf{A}^{-1}$.

   $|\mathbf{A}| = (5 \times 9) - (6 \times 7) = 45 - 42 = 3$

   $\mathbf{A}^{-1} = \frac{1}{3}\begin{pmatrix} 9 & -7 \\ -6 & 5 \end{pmatrix} = \begin{pmatrix} \frac{9}{3} & -\frac{7}{3} \\ -\frac{6}{3} & \frac{5}{3} \end{pmatrix} = \begin{pmatrix} 3 & -\frac{7}{3} \\ -2 & \frac{5}{3} \end{pmatrix}$

Exercise

1. State whether each of the following matrices has an inverse. If the inverse exists, find it.

   a) $\begin{pmatrix} 4 & 6 \\ 1 & 2 \end{pmatrix}$
   b) $\begin{pmatrix} 2 & 3 \\ 5 & 8 \end{pmatrix}$

   c) $\begin{pmatrix} 3 & 2 \\ 6 & 4 \end{pmatrix}$
   d) $\begin{pmatrix} 3 & -2 \\ 2 & -1 \end{pmatrix}$

   e) $\begin{pmatrix} 3 & 4 \\ 5 & 6 \end{pmatrix}$

2. Show that $\mathbf{I}$ is its own inverse, where $\mathbf{I} = \begin{pmatrix} 1 & 0 \\ 0 & 1 \end{pmatrix}$.

3. Show that each of the following matrices is its own inverse.

   a) $\begin{pmatrix} 1 & 0 \\ 0 & -1 \end{pmatrix}$
   b) $\begin{pmatrix} -1 & 0 \\ 0 & -1 \end{pmatrix}$

   c) $\begin{pmatrix} 0 & 1 \\ 1 & 0 \end{pmatrix}$
   d) $\begin{pmatrix} 0 & -1 \\ -1 & 0 \end{pmatrix}$

4. Given that $\mathbf{A} = \begin{pmatrix} 3 & 7 \\ 2 & 5 \end{pmatrix}$ find $\mathbf{A}^{-1}$.

Matrices and geometrical transformations

Matrices can be used to describe transformations in a plane.

Suppose P is the point (4, 3), that is P is the point with position vector $\begin{pmatrix} 4 \\ 3 \end{pmatrix}$.

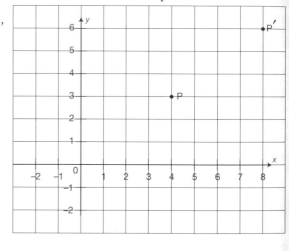

Multiply $\begin{pmatrix} 4 \\ 3 \end{pmatrix}$ by the matrix $\begin{pmatrix} 2 & 0 \\ 0 & 2 \end{pmatrix}$:

$$\begin{pmatrix} 2 & 0 \\ 0 & 2 \end{pmatrix} \begin{pmatrix} 4 \\ 3 \end{pmatrix} = \begin{pmatrix} 8 \\ 6 \end{pmatrix}.$$

$\begin{pmatrix} 8 \\ 6 \end{pmatrix}$ is another position vector, that is P′(8, 6).

The matrix $\begin{pmatrix} 2 & 0 \\ 0 & 2 \end{pmatrix}$ transforms point P(4, 3) into P′(8, 6).

This principle can be used to describe all the transformations you have worked with.

Reflection

Select the points (1, 0) and (0, 1). Find the images of these points under the given transformation. Write down the position vectors of these images. The position vector of the image of the first point becomes the first column and the position vector of the image of the second point is the second column of the required matrix.

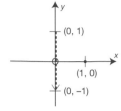

Reflection in the *x*-axis

(1, 0) → (1, 0)   The position vectors of these points are $\begin{pmatrix} 1 \\ 0 \end{pmatrix}$ and $\begin{pmatrix} 0 \\ -1 \end{pmatrix}$.

(0, 1) → (0, −1)

So the matrix of the transformation is $\begin{pmatrix} 1 & 0 \\ 0 & -1 \end{pmatrix}$.

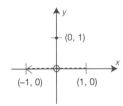

Reflection in the *y*-axis

(1, 0) → (−1, 0)   The position vectors of these points are $\begin{pmatrix} -1 \\ 0 \end{pmatrix}$ and $\begin{pmatrix} 0 \\ 1 \end{pmatrix}$.

(0, 1) → (0, 1)

So the matrix of the transformation is $\begin{pmatrix} -1 & 0 \\ 0 & 1 \end{pmatrix}$.

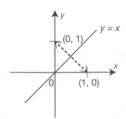

Reflection in the line *y* = *x*

(1, 0) → (0, 1)   The position vectors of these points are $\begin{pmatrix} 0 \\ 1 \end{pmatrix}$ and $\begin{pmatrix} 1 \\ 0 \end{pmatrix}$.

(0, 1) → (1, 0)

So the matrix of the transformation is $\begin{pmatrix} 0 & 1 \\ 1 & 0 \end{pmatrix}$.

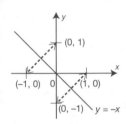

Reflection in the line *y* = −*x*

(1, 0) → (0, −1)   The position vectors of these points are $\begin{pmatrix} 0 \\ -1 \end{pmatrix}$ and $\begin{pmatrix} -1 \\ 0 \end{pmatrix}$.

(0, 1) → (−1, 0)

So the matrix of the transformation is $\begin{pmatrix} 0 & -1 \\ -1 & 0 \end{pmatrix}$.

Rotation

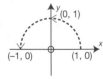

Rotation about the origin through 90° anti-clockwise (+90°)

(1, 0) → (0, 1)   The transformation matrix is $\begin{pmatrix} 0 & -1 \\ 1 & 0 \end{pmatrix}$.

(0, 1) → (−1, 0)

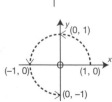

Rotation about the origin through 180° (half turn)

(1, 0) → (−1, 0)   The transformation matrix is $\begin{pmatrix} -1 & 0 \\ 0 & -1 \end{pmatrix}$.

(0, 1) → (0, −1)

Stretch

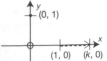

One-way stretch parallel to the x-axis, with scale factor k, with the y-axis invariant

$(1, 0) \rightarrow (k, 0)$ The transformation matrix is $\begin{pmatrix} k & 0 \\ 0 & 1 \end{pmatrix}$.
$(0, 1) \rightarrow (0, 1)$

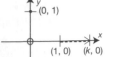

One-way stretch parallel to the y-axis, with scale factor k, with the x-axis invariant

$(1, 0) \rightarrow (1, 0)$ The transformation matrix is $\begin{pmatrix} 1 & 0 \\ 0 & k \end{pmatrix}$.
$(0, 1) \rightarrow (0, k)$

Enlargement

Enlargement with scale factor k, and centre of enlargement (0, 0)

$(1, 0) \rightarrow (k, 0)$ The transformation matrix is $\begin{pmatrix} k & 0 \\ 0 & k \end{pmatrix}$.
$(0, 1) \rightarrow (0, k)$

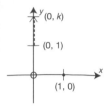

Shear

Shear in x-direction, x-axis invariant and scale factor k

$(1, 0) \rightarrow (1, 0)$ The transformation matrix is $\begin{pmatrix} 1 & k \\ 0 & 1 \end{pmatrix}$.
$(0, 1) \rightarrow (k, 1)$

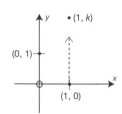

Shear in y-direction, y-axis invariant and scale factor k

$(1, 0) \rightarrow (1, k)$ The transformation matrix is $\begin{pmatrix} 1 & 0 \\ k & 1 \end{pmatrix}$.
$(0, 1) \rightarrow (0, 1)$

Translation

All the previous transformations are represented by a 2×2 matrix and, to find the image of a point, you have to pre-multiply the position vector $\begin{pmatrix} x \\ y \end{pmatrix}$ of the point by the appropriate matrix. It follows that the image of the origin $(0, 0)$ is the origin – in other words, the origin is invariant. This is not the case for a translation – every point moves, as there is no invariant point.

Translation is represented by a column vector or a column matrix of the order 2×1. To obtain the image of a point under the translation, you do not multiply by the matrix but you add the vector of the translation to the position vector $\begin{pmatrix} x \\ y \end{pmatrix}$ of the point.

■ Example

The figure on the left shows $\triangle PQR$.

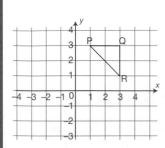

Show the new position of the triangle after the translation $\begin{pmatrix} -3 \\ +1 \end{pmatrix}$.

The position vector is P is $\begin{pmatrix} 1 \\ 3 \end{pmatrix}$.

The position vector of Q is $\begin{pmatrix} 3 \\ 3 \end{pmatrix}$.

The position vector of R is $\begin{pmatrix} 3 \\ 1 \end{pmatrix}$.

The image of P after the translation is $\begin{pmatrix} 1 \\ 3 \end{pmatrix} + \begin{pmatrix} -3 \\ +1 \end{pmatrix} = \begin{pmatrix} 1 + (-3) \\ 3 + 1 \end{pmatrix}$

$$= \begin{pmatrix} -2 \\ 4 \end{pmatrix}$$

The image of Q after the translation is $\begin{pmatrix} 3 \\ 3 \end{pmatrix} + \begin{pmatrix} -3 \\ +1 \end{pmatrix} = \begin{pmatrix} 3 + (-3) \\ 3 + 1 \end{pmatrix}$

$$= \begin{pmatrix} 0 \\ 4 \end{pmatrix}$$

The image of R after the translation is $\begin{pmatrix} 3 \\ 1 \end{pmatrix} + \begin{pmatrix} -3 \\ +1 \end{pmatrix} = \begin{pmatrix} 3 + (-3) \\ 3 + 1 \end{pmatrix}$

$$= \begin{pmatrix} 0 \\ 2 \end{pmatrix}$$

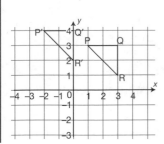

The new position of $\triangle PQR$ is $\triangle P'Q'R'$.

Exercise

1. Answer the whole of this question on a sheet of graph paper.
   a) (i) Draw x- and y-axes from -6 to $+6$, using 1 cm to represent 1 unit of x and y.
      (ii) Draw $\triangle ABC$ with vertices A(2, 1), B(5, 1) and C(5, 5).
   b) (i) Draw the image of $\triangle ABC$ under the transformation represented by the matrix $\begin{pmatrix} 0 & 1 \\ 1 & 0 \end{pmatrix}$ and label it $A_1B_1C_1$.
      (ii) Describe this single transformation.
   c) (i) Draw the image of $\triangle ABC$ under reflection in the line $y = -x$ and label it $A_2B_2C_2$.
      (ii) Find the matrix that represents this transformation.
   d) (i) Describe fully the single transformation that maps $\triangle ABC$ onto $\triangle A_2B_2C_2$.
      (ii) Find the matrix that represents this transformation.

2. Answer the whole of this question on a sheet of graph paper.
   a) Draw x- and y-axes from -6 to $+6$, using 1 cm to represent 1 unit on each axis.
      Draw the triangle whose vertices are A(2, 2), B(5, 2) and C(5, 3).
   b) **M** is the matrix $\begin{pmatrix} 0 & -1 \\ 0 & 0 \end{pmatrix}$ which represents the transformation T.
      Draw accurately the image of triangle ABC under the transformation T, labelling it PQR.
   c) **N** is the matrix $\begin{pmatrix} 1 & 0 \\ 0 & -1 \end{pmatrix}$ which represents the transformation U.
      Draw accurately the image of $\triangle ABC$ under the transformation U, labelling it XYZ.
   d) (i) Describe fully the single transformation that maps $\triangle PQR$ onto $\triangle XYZ$.
      (ii) Find the matrix that represents this transformation.
   e) (i) Calculate the matrix **NM**.
      (ii) This matrix represents the transformation V.
           Draw accurately the image of $\triangle ABC$ under transformation V, labelling it FGH.
      (iii) State whether the transformation V is equivalent to transformation T followed by transformation U or to transformation U followed by transformation T.

3. The matrix $\mathbf{M} = \begin{pmatrix} 3 & 2 \\ -2 & -1 \end{pmatrix}$ describes a transformation on $\triangle ABC$. The coordinates of the image are A'(2, 1), B'(0, 1) and C'(-3, 2). Find the following:
   a) the matrix $\mathbf{M}^{-1}$ that transforms $\triangle A'B'C'$ onto $\triangle ABC$
   b) the coordinates of A, B and C.

Check your progress

1. a) On a grid of squared paper, draw accurately the following transformations:
    (i)   the reflection of △A shown below left, in the *y*-axis, labelling it B
    (ii) the rotation of △A through 180° about the point (4, 3), labelling it C
    (iii) the enlargement of △A, scale factor 2, centre (4, 5), labelling it D.
    b) Describe fully the single transformation which maps △E onto △C.

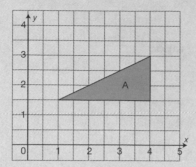 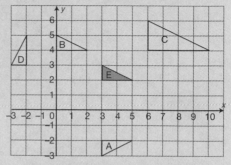

2. Describe fully the transformations of the shaded triangle onto triangles A, B, C and D in the diagram above right.

3. $m = \begin{pmatrix} 3 \\ -4 \end{pmatrix}$ and $n = \begin{pmatrix} -2 \\ 1 \end{pmatrix}$.

    a) Find:
    (i) $m + n$
    (ii) $3n$

    b) Draw the vector m on a grid, or squared paper.

4. △ABC is mapped onto △A′B′C′ by an enlargement.
    a) Find the centre of the enlargement.
    b) What is the scale factor of the enlargement?

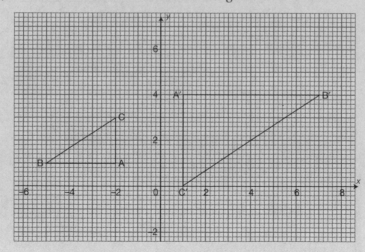

5. a) Write down the column vector of the translation that maps rectangle R onto rectangle S.

    b) Describe fully another single transformation (not a *translation*) that would also map rectangle R onto rectangle S.

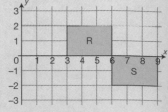

    c) (i)   Copy the diagram onto a grid.
              Enlarge R with centre of enlargement A(10, 2) and scale factor 2.

        (ii) Write down the ratio $\dfrac{\text{area of enlarged rectangle}}{\text{area of rectangle R}}$ in its simplest terms.

6. $\overrightarrow{OA} = \boldsymbol{a}$ and $\overrightarrow{OB} = \boldsymbol{b}$.

    a) $\overrightarrow{OC} = \boldsymbol{a} + 2\boldsymbol{b}$. Label the point C on a diagram.

    b) D = (0, –1). Write $\overrightarrow{OD}$ in terms of $\boldsymbol{a}$ and $\boldsymbol{b}$.

    c) Calculate $|\boldsymbol{a}|$, giving your answer to 2 decimal places.

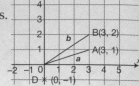

7. a) In each case, describe fully the single transformation that maps A onto:
        (i) B   (ii) C   (iii) D   (iv) E   (v) F.

    b) State which shapes have an area equal to that of A.

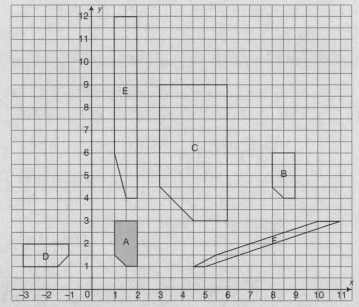

8. $\mathbf{A} = \begin{pmatrix} 1 & 2 \end{pmatrix}$, $\mathbf{B} = \begin{pmatrix} -3 \\ 4 \end{pmatrix}$ and $\mathbf{C} = \begin{pmatrix} -2 & 5 \\ -3 & 6 \end{pmatrix}$.

    a) Which one of the following matrix calculations is possible?
        (i) $\mathbf{A} + \mathbf{B}$ (ii) $\mathbf{AC}$ (iii) $\mathbf{BC}$

    b) Calculate $\mathbf{AB}$.

    c) Find $\mathbf{C}^{-1}$, the inverse of $\mathbf{C}$.

9. a) Solve for x and y $\begin{pmatrix} 3 & 2 \\ -1 & 6 \end{pmatrix} \begin{pmatrix} -3 \\ 2 \end{pmatrix} = \begin{pmatrix} x \\ y \end{pmatrix}$.

   b) Find the inverse of the matrix $\begin{pmatrix} 2 & -1 \\ 4 & 3 \end{pmatrix}$.

   c) Solve for t and u $\begin{pmatrix} 3t & u \\ -t & 3u \end{pmatrix} \begin{pmatrix} 1 \\ 2 \end{pmatrix} = \begin{pmatrix} 10 \\ -10 \end{pmatrix}$.

10. Answer the whole of this question on a sheet of graph paper.
    a) Draw axes from –6 to +6, using a scale of 1 cm to represent 1 unit on each axis.
       (i)   Plot the points A(5, 0), B(1, 3) and C(–1, 2) and draw △ABC.
       (ii)   Plot the points A′(3, 4), B′(3, –1) and C′1(1, –2) and draw △A′B′C′.
    b) (i)   Draw and label the line l in which △A′B′C′ is a reflection of △ABC.
       (ii)   Write down the equation of the line l.
       (iii) Find the values of p, q, r and s such that:

$$\begin{array}{ccc} \text{A} & \text{B} & \text{C} \end{array} \qquad \begin{array}{ccc} \text{A}' & \text{B}' & \text{C}' \end{array}$$
$$\begin{pmatrix} p & q \\ r & s \end{pmatrix} \begin{pmatrix} 5 & 1 & -1 \\ 0 & 3 & 2 \end{pmatrix} = \begin{pmatrix} 3 & 3 & 1 \\ 4 & -1 & -2 \end{pmatrix}.$$

       (iv)   What transformation does the matrix $\begin{pmatrix} p & q \\ r & s \end{pmatrix}$ represent?

    c) Reflect △A′B′C′ in the y-axis. Label the new triangle A″B″C″.
    d) If △ABC is rotated about the origin, it will map onto △A″B″C″. What is the angle of rotation?

Answers

PAGE 2:
1. 1; 3; 4; 5; 15
2. All
3. 23; 29; 31
4. 31; 83
5. No
6. No
7. Even
8. Odd
9. a) 36; 49; 64; 81; ...

   b) $\frac{1}{3}; \frac{1}{4}; \frac{1}{5}; \frac{1}{6}; ...$

   c) {−6; −5; −4; −3; −2; −1}

   d) 0,5; 0,75; 1,5

PAGE 3:
1. 25
2. 1.61 (to 2 dec.)
3. 0.55
4. 28 000 000
5. 2 187
6. −5.5

PAGE 4:
1. 3
2. 11
3. 14
4. 10
5. 0.5
6. 42
7. 28
8. 12
9. >/≠
10. </≠
11. >
12. =
13. $\sqrt{}$
14. >
15. </≈
16. <
17. >
18. =

PAGE 7:
1. a) 2×3^2
   b) 2^4
   c) 2^6
   d) 3^4
   e) $2^2 \times 5^2$
   f) $2^2 \times 3^2$
   g) 3×7
   h) 11
   i) $3^2 \times 5$
   j) $2^2 \times 3^3$
2. a) 36
   b) 36
   c) 120
   d) 72
   e) 15
3. a) 84
   b) 36
   c) 48
   d) 30

PAGE 8:
1. a) 10, 12
   b) 15, 18
   c) 17, 19
   d) −6, −9
   e) 18, 24
   f) 47, 76
2. 1 and 64
3. a) $1 + 3 + 5 + 7 + 9 + 11 = 36$
      $1 + 3 + 5 + 7 + 9 + 11 + 13 = 49$
   b) 100
   c) 44100
   d) sum = n^2
4. a) 15.21
   b) (i) $10 + 15 = 25$
           $15 + 21 = 36$
      (ii) Squares
      (iii) 121
5. a) 1, 7, 13, 19, 25, 31 (omit 16)
   b) 1, 4, 9, 16, 25 (omit 7)

PAGE 9:
1. b) {1; 3; 5; 7; 9}
   c) {w; x; y; z}
   d) {2; 4; 6}

PAGE 10:
2. a) The set of the first four prime numbers.
   b) The set of the last eight letters of the English alphabet.
   c) The set of the first five multiples of 5.
3. a) {4; 5; 6}
   b) {2; 4; 6}
   c) {7; 9; 11}
   d) {5; 9; 13; 17; 21}
   e) {(16; 3); (10; 5)}
4. a) {6; 7; 8; 9; ...}
   b) {1; 2; 3; 4; 5; 6; 7; 8; 9}
   c) {4; 5; 6; 7; 8; 9; 10}
   d) (2; 3; 5; 7; ...}

1. a) The set of people in your class.
   b) The set of letters of the alphabet.
   c) The set of even natural numbers.
   d) The set of animals.

PAGE 11:
1. a) {4; 5}
   b) { } or ∅
   c) {1; 3; 5; 7; ...}
   d) {o}
2. $A \cap B = \{3; 4\}$
   $B \cap C = \{5\}$
   $A \cap C = \{5\}$
   $\mathscr{E} \cap B = \{3; 4; 5\}$
3. a) {a; b; c; d; e; f}
   b) {x; y}
   c) {1; 2; 3; 4; 5; 6; 8; 10}
   d) (0; 1; 2; 3; ...}
4. a) True
   b) False
   c) False
   d) False

PAGE 12:
1. {*q; s; t*}
2. a) {2; 3; 5; 7; 11; 13; 17; 19}
   b) {1; 4; 6; 8; 9; 10; 12; 14; 15; 16; 18; 20}
3. The odd numbers and zero.
4. The set of positive numbers and zero.

PAGE 13:
1. a) A = {6; 12; 18; 24} B = {4; 8; 12; 16; 20; 24}   b) {12; 24}   c) {4; 6; 8; 12; 16; 18; 20; 24}
2. a) (i) {*a; b; c; d; e; f*} (ii) {*e; f; g; h*}   b) {*e; f*}   c) (i) {*i; j*} (ii) {*a; b; c; d*}
3. a)   b)

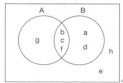   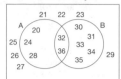

4. a) 6   b) 16   c) 16
5.

PAGE 14:
1. a) 36; 49; 64; 81   b) 343; 512; 729; 1000
2. a) 7; 11; 16; 2.83 (to 2 dec.)   b) 50; 12; 9
3. a) 169   b) 729   c) 16   d) –27   e) 10 000   f) 1 000 000

PAGE 17:
1. a) >   b) >   c) <   d) >   e) >   f) <
2. a) 8   b) 3   c) 4   d) –3   e) –5   f) 5
   g) –6   h) 1   i) –14
3. 66 °C
4. a) 18 °C   b) 11 °C   c) 24 °C   d) 13 °C
5. a) 330 m below sea level   b) 220 m below sea level
6. a) –15 °C   b) –26 °C

PAGE 18:
1. a) $\frac{2}{6}, \frac{3}{9}, \frac{10}{30}$ b) $\frac{4}{10}, \frac{20}{50}, \frac{200}{500}$ c) $\frac{10}{14}, \frac{20}{28}, \frac{25}{35}$
2. a) $\frac{1}{4}$ b) $\frac{2}{3}$ c) $\frac{1}{5}$
3. a) 9   b) 7   c) 4

PAGE 19:
1. $\frac{6}{7}$ 2. $\frac{4}{3}$ 3. $\frac{5}{8}$ 4. $\frac{51}{14}$ 5. $\frac{175}{36}$ 6. $\frac{293}{36}$ 7. $\frac{2}{15}$
8. $\frac{3}{8}$ 9. $\frac{2}{15}$ 10. $\frac{11}{8}$ 11. $\frac{15}{8}$ 12. $\frac{32}{7}$ 13. $\frac{30}{7}$ 14. $\frac{2}{15}$
15. $\frac{4}{7}$ 16. 7   17. 54   18. $\frac{1}{8}$ 19. 200   20. $\frac{55}{4}$ 21. $\frac{11}{12}$
22. $-\frac{5}{6}$ 23. $\frac{3}{200}$ 24. $\frac{61}{4}$

PAGE 20:
1. $\frac{2}{9}$ 2. $\frac{1}{4}$ 3. $\frac{2}{5}$ 4. 4   5. $\frac{5}{8}$ 6. $\frac{3}{2}$ 7. $\frac{6}{7}$ 8. 15   9. $\frac{49}{3}$

PAGE 21:
1. 750   2. 1 200   3. $500   4. 440 ℓ   5. $\frac{31}{8}$ 6. $\frac{7}{15}$ 7. 6
8. $\frac{2}{11}$ 9. a) $\frac{5}{12}$ b) $6 000   10. a) $\frac{11}{15}$ b) $\frac{4}{15}$ c) $133.33 (to the nearest cent)
11. 9   12. 19   13. $\frac{7}{15}$

PAGE 23:
1. a) 7   b) 10,02   c) 6.95   d) 9.082   e) 15.48   f) 0.97
    g) 0.0416   h) 0.072   i) 79.7
2. a) 0.75   b) 0.4   c) 0.27   d) 0.625

PAGE 25:
1. a) 29.712   b) 1.628   c) 202.916   d) 4.680   e) 0.004
    f) 1000.565   g) 0.625
2. a) 29.7   b) 1.63   c) 203   d) 4.68   e) 0.00353
    f) 1000   g) 0.625
3. a) 0.667   b) 0.714   c) 0.167   d) 0.727   e) 0.444
    f) 0.125

PAGE 26:
1. a) 3 : 1   b) 5 : 1
2. 52 hectares
3. $1600; $2800; $3600
4. 9 : 6 : 5
5. $125 and $100
6. 29 : 10 : 7 : 11

PAGE 27:
1. $440   2. $6,75   3. 1 hour   4. a)   320 g flour   64 g sultanas
                                                             80 g margarine   100 ml milk
                                                             32 g sugar   16 g salt
                                                   b)   4 : 1

PAGE 28:
1. 10 days   2. $2\frac{1}{2}$ days   3. Can't be solved by direct or indirect proportion.

PAGE 29:
1. 20%   2. 80%   3. 10%   4. 70%   5. 2%   6. 4%

7. 32%   8. 12.5%   9. $83\frac{1}{3}$%   10. $66\frac{2}{3}$%   11. $55\frac{5}{9}$%   12. 250%

13. $\frac{1}{25}$ 14. $\frac{1}{4}$ 15. $\frac{1}{2}$ 16. $\frac{3}{4}$ 17. $\frac{3}{5}$ 18. $\frac{5}{4}$

19. $\frac{5}{2}$ 20. $\frac{7}{30}$ 21. $\frac{1}{40}$ 22. $\frac{2}{3}$

PAGE 30:
1. 15   2. 1.25   3. $45   4. 8 kg   5. 30 m   6. 80   7. 14
8. 60   9. 13   10. $110

1. 50%   2. $33\frac{1}{3}$%   3. 5%   4. 9%   5. 90%   6. 80%   7. $\frac{7}{20}$
8. 40%   9. 94.34% (to 2 dec.)   10. 748

PAGE 31:
1. a) $5; 25%   b) $50; 10%   c) $0.30; 20%   d) $0.05; $16\frac{2}{3}$%
2. a) 25%   b) $13\frac{1}{3}$%   c) 5%   d) 10%
3. $66\frac{2}{3}$% profit.

PAGE 34:
1. a) 4 000 g   b) 5 000 m   c) 3.5 cm   d) 8.1 cm
    e) 7300 mg   f) 5.67 t   g) 210 cm   h) 2 000 kg
    i) 1.4 m   j) 2.024 kg   k) 0.121 g   l) 23 000 mm
    m) 35 mm   n) 8 036 m   o) 9.077 g
2. a) millilitre   b) kilogram
3. 32.4 cm; 3.22 m; $3\frac{2}{9}$ m
4. 125 ml; $\frac{1}{2}$ l; 0.65 l; 780 ml
5. 60
6. a) 14 230 mm; 0.01423 km   b) 19 060 mg; 0.00001906 t
    c) 2 750 ml; 275 cl   d) 4 000 000 mm²; 0.0004 ha
    e) 1 300 mm²; 0.0000000013 ha   f) 10 000 mm³; 0.00001 m³

PAGE 35:
1. $18.50   2. $4 163   3. £8 520   4. $331.53 (to nearest cent)   5. $2218.18 (to nearest cent)

PAGE 36:
1. 3 hrs 39 min.
2 a) 300 km   b) 120 km/h

3. 10 min. 3 sec.
4. 0230 hours, Monday, 10 February

PAGE 37:
1. a) 2 002 hours   b) 45 min.   c) 23 min.
2. a) 67 min.   b) 1110 – 1145 – 1159 – 1217   c) 1 425 hours

PAGE 38:
1. ≈ 4 2. ≈ 2 3. ≈ 10 4. ≈ 7 5. ≈ 0.4

PAGE 40:
1. a) 11.5 and 12.5   b) 7.5 and 8.5   c) 99.5 and 100.5
    d) 8.5 and 9.5   e) 71.5 and 72.5   f) 126.5 and 127.5
2. a) 2.65 and 2.75   b) 34.35 and 34.45   c) 4.95 and 5.05
    d) 1.05 and 1.15   e) –2.35 and –2.25   f) –7.25 and –7.15
3. $250 \leqslant m < 350$
4. $125 \leqslant L < 135$ $95 \leqslant W < 105$

PAGE 41:
1. a) 13 cm – 15 cm   b) 28 g – 30 g   c) 16.3 cm – 16.5 cm   d) 2.31 kg – 2.33 kg
2. 37.5 kg – 38.5 kg
3. a) $3.605 \leqslant L < 3.615$ $2.565 \leqslant B < 2.575$ b) $9.246825 \leqslant A < 9.308625$ c) $9.25 \leqslant A < 9.31$
4. a) 4.7 – 5.1   b) 27 – 30

PAGE 42:
1. $862.50   2. $3 375   3. $425   4. $211.20   5. $433.55

PAGE 43:
1. a) $100   b) $200   c) $340   d) $900
2. $5 000
3. $1 160
4. 4.5 ha
5. $75

PAGE 45:
1. a) $7.50   b) $160   c) $262.50   d) $480   e) $343.75
2. 4 years
3. 7%

PAGE 47:
1. a) $100   b) $60   c) $460
2. $2 850

Check your progress
1. a) 23   b) 12   c) 12
2. a) 64   b) 64
3. 57 °C
4. a) –30 cm   b) +10 cm   c) 100 cm
5. 1
6. a) $\sqrt{250}$ b) 25   c) none
7. $\frac{5}{16}$
8. a) 0.142857 142857 142857 14
   b) first 6 places recur
9. a) 5%   b) $\frac{6}{25}$ c) 722
10. a) >   b) =   c) >   d) <
11. $12.50
12. 8
13. 7.2 pesos
14. a) $12   b) $14.40
15. 29 975
16. 7.5%
17. 7.5%
18. $36
19. $635
20. a) $471   b) (i) 836   (ii) $447   c) Klaus   d) 206

PAGE 50:
1. a) 17   b) 3   c) 9   d) 3   e) 12   f) 15   g) 10   h) $\frac{5}{2}$
2. a) 3   b) 3   c) $\frac{8}{7}$ d) –11

3. a) 32   b) 16
4. a) 0   b) 6
5. a) $b > y$ b) $b = 2y$
6. $2n^2 - 15 = 377$
7. $x + 10 = \frac{3x + 10}{2}$
$x = 10$
8. A $= \frac{d\,\text{major} \times d\,\text{minor} \times \pi}{4}$

PAGE 51:
1. a) 3   b) 16.2
2. 88 m
3. a) 810   b) 6 750
4. 20.08 km (to 2 dec.)
5. 50 cm^3

PAGE 52:
1. a) –6   b) –16   c) –12   d) 45   e) 0   f) –16   g) –15
   h) –21   i) 6   j) 21   k) 0   l) 32
2. a) 1   b) –3   c) 10   d) 5   e) –9   f) –40   g) 6
   h) –5   i) –6   j) 9   k) 1   i) –1

PAGE 53:
1. a) 48   b) 432
2. a) $p + 2q$ b) $7n - 3$ c) $2x^3 + 2x^2 + 5x$ d) $3xy - 3 + y$
   e) $-c - 4d$ f) $x^2 - 3x + 6$
3. a) $16x - 12$ b) $14n + 35$ c) $48 - 16y$ d) $6x^2 + 9x - 15$
   e) $30pq - 6p - 12q$
4. a) $11n + 3$ b) $12x - 1$ c) $17y - 9$ d) $16d$
   e) $8x^2 + 3x - 14$

PAGE 55:
1. a) $x = 5$ b) $x = 6$ c) $x = \frac{7}{3}$ d) $x = \frac{15}{2}$ e) $y = \frac{1}{2}$ f) $n = 0$

2. a) $x = \frac{11}{4}$ b) $x = 2$ c) $y = \frac{4}{7}$ d) $m = -1$ e) $x = 1$ f) $n = \frac{6}{7}$

3. a) $x = -7$ b) $x = 5$ c) $m = \frac{9}{2}$ d) $y = -8$ e) $x = -\frac{9}{2}$ f) $p = \frac{1}{3}$

   g) $m = \frac{7}{2}$ h) $z = \frac{11}{3}$ i) $x = -\frac{6}{13}$ j) $x = 9$ k) $x = \frac{25}{3}$ l) $x = \frac{24}{5}$

   m) $x = \frac{5}{11}$ n) $x = \frac{8}{5}$ o) $x = 6$

4. $3n + 2 = 50 - n$, $n = 12$
5. a) 20 °C   b) 90 °F
6. $2x + 5 = x + 5 + 2$, Fatima is 4, sister is 2 years old.

PAGE 57:
1. a) $x = 13$ b) $x = \frac{1}{3}y - \frac{8}{3}$ c) $x = -\frac{1}{3}$

2. a) $x = 3$ b) $x = \frac{c - b}{a}$
3. $u = v - at$
4. $W = 2P - 12$

5. $y = -\frac{2}{3}x + 4$

6. $r = \frac{C}{2\pi}$

7. $H = \frac{2S^2}{3}$

8. $v = \pm\sqrt{\frac{2Fg}{W}}$

PAGE 58:
1. a) $p - s$ b) Mother is $p + 5$, daughter is $s + 5$ c) $p + s$ d) $2s$
2. a) $\frac{P}{12}$ b) $\frac{5P}{12}$ c) P

PAGE 60:
1. 23   2. 9   3. 95 g; 99 g; 106 g   4. 15 hours   5. 12 times

PAGE 61:
1. $2n + 3$ 2. $3n - 1$ 3. $-2n + 14$ 4. $\frac{1}{2}n + 1$ 5. n^3

PAGE 63:
1. A $= 400$ 2. $\frac{32}{5}$ 3. No, products not constant; $5 \times 30 \neq 8 \times 20$ 4. I $= 60$ 5. $t \propto s^3$

PAGE 64:
1. a) 1 000   b) 100   c) 1   d) 0.001
2. a) 10^4 b) 10^9 c) 10^{-6} d) 10^{-1} e) 10^{-2} f) 10^{-3}
   g) 10^7 h) 3.1×10^6

PAGE 65:
1. a) 7.89×10^4 b) 3.479×10^2 c) 5.8×10^{-5} d) 1.234×10^{-1}
2. a) 120   b) 0.0314   c) 0.0007605   d) 2 800 000 000
3. a) 4.8×10^{10} b) 1.44×10^4 c) 1.6×10^4 d) 5×10^{10}
   e) 7.28×10^5 f) 9.83×10^{-1} g) 2.27×10^4 h) 8.43×10^{-2}

PAGE 66:
1. a) 8^5 b) 4^8 c) 6^{12} d) 5^{14}
2. a) 3^6 b) 2^{15} c) 4   d) 7^6
3. a) 5^8 b) 5^8 c) 9^{24} d) 4^{30}
4. a) 2^5 b) 6   c) 3^8 d) 8^5

PAGE 67:
1. a) 405   b) 576   c) 118   d) 51
2. a) $12f^7$ b) $5y^8$ c) $6e^5$ d) $42p^3q^3$
3. a) $2p^4$ b) $7q$ c) $3y^2$ d) $2p^2q$
4. a) $4x^4$ b) $-15x^2y^2$ c) $2x^2$ d) $-2ay^3$
5. a) $64k^6$ b) $9p^8$ c) $3q^3 - 5q^2 - q$ d) $5y^6 + 10y^5 - 35y^4$

PAGE 68:
1. a) $\frac{1}{25}$ b) 1   c) $\frac{125}{8}$ d) 16   e) $\frac{1}{8}$ f) 1   g) $\frac{81}{16}$ h) 729

2. a) y^4 b) 1   c) p^{-3} d) q^6 e) p^6 f) $\frac{1}{q}$ g) $\frac{1}{e^3}$ h) 1

3. a) e^4 b) e^8 c) 2   d) k^4 e) y^2 f) y^8 g) $\frac{3}{4}$ h) k^7

4. a) $\frac{1}{x^7}$ b) $\frac{1}{4}$ c) x^a d) $\frac{1}{16}$ e) $\frac{1}{11}$ f) $\frac{b^4}{a^3}$ g) $\frac{1}{xy}$ h) $-\frac{1}{25}$

5. a) 1   b) 1   c) 4   d) 0   e) 1   f) 1

PAGE 69:
1. a) 4   b) 9   c) 125   d) 1 000
2. a) $\frac{125}{8}$ b) 2   c) 3   d) 8

3. a) $\frac{1}{9}$ b) $\frac{125}{27}$ c) e d) f^3

PAGE 70:
1. a) $-2y + 15$ b) $s - 18t$ c) $15z - 11$ d) $7p + 5q$
   e) $-5y + 31$ f) $11s - 11t$ g) $-18z - 13$ h) $p + 3q$
2. a) $11e - 6f$ b) $2t^2 - 3t - 2$ c) $3e + 2f$ d) $3t^2 - 3t + 11$

PAGE 71:
1. a) $t^2 - 9t + 20$ b) $6p^2 - p - 1$ c) $12s^2 - 7st - 12t^2$
2. a) $9p^2 + 12p + 4$ b) $16y^2 - 9$ c) $s^3 - 7s + 6$
3. a) $mx - my + nx - ny$ b) $6a^2 - 5ab + b^2$ c) $6a^2 + 14ab + 4b^2$
   d) $3a^2 - 5ab + 2b^2 - 3bc + 3ac$ e) $4x^2 - 16$ f) $9x^2 - 30x + 25$
   g) $x^2 + 4xy + 4y^2$ h) $-2x^2 + 4xy - 2y^2$ i) $x^2 - 3x - 14$
   j) $-x^2 - 7x + 10$ k) $-7x^2 + 28x$ l) $4x^2 + 16x + 41$

PAGE 72:
1. a) $5(2s + 3t + 4u)$ b) $3q(4p - 3r)$ c) $4y(4z - 1)$
2. a) $t(t + 5)$ b) $3y(3y - 2z)$ c) $2pq(3p + 7q - 5)$

PAGE 73:
1. a) $(b + 3)(y + 3z)$ b) $(2 - q)(2p + 3r)$ c) $(3x - 4)(x + 2)$ d) $(a - b)(c - d)$
2. a) $(x + 2)(x + 12)$ b) $(x + 3)(x - 2)$ c) $(x - 3)(x - 5)$
3. a) $(3x + 2)(x + 3)$ b) $(5x + 1)(x - 2)$ c) $(2x - 3)(2x - 1)$
4. a) 20 001   b) 200   c) 6
5. a) $(2x + 3y)(2x - 3y)$ b) $(6a + 5b)(6a - 5b)$ c) $(4c + 1)(4c - 1)$ d) $3(3x + 2y)(3x - 2y)$

PAGE 75:
1. a) $\frac{2p}{3q}$ b) $\frac{y}{y + 5}$ c) $\frac{x}{x - 5}$

2. a) $\frac{5a^2}{4}$ b) $\frac{x(x + 1)}{2(x - 2)}$ c) $\frac{3}{p - 3}$

3. a) $\frac{a}{20}$ b) $\frac{-5(x - 9)}{6}$ c) $\frac{5t + 7}{(t + 1)(t + 2)}$

4. $\frac{1}{x - 1}$

PAGE 77:

1. $x = \frac{r - pq}{p - 1}$

2. $y = \frac{b(a - x)}{a}$

3. $x = \frac{2y + 3}{y - 1}$

4. $f = \frac{2(s - ut)}{t^2}$

5. $q = \left(\frac{p + 4}{3}\right)^2$

6. $m = \frac{E}{c^2}$

7. $x = \frac{\sqrt{y - 2}}{3}$
   a) not defined, no solution
   b) $x \approx -0.26$

8. $r = \sqrt{\frac{3v}{\pi h}}$

9. $A = P(1 + r)^n$
   $P = \frac{A}{(1 + r)^n}$
   $r = 1 - \sqrt[n]{\frac{A}{P}}$

10. $v = \frac{uR}{2u - R}$

PAGE 79:

1. a) $x = 2; y = 2$ b) $p = 1; q = 3$ c) $u = 3; v = 1$

2. a) $s = \frac{5}{2}; t = 1$ b) $f = \frac{7}{3}; g = \frac{3}{2}$ c) $x = 1; y = -4$

3. $3n + 5p = 10; n + 10p = 10$ $n = \$2$ $p = \$\frac{4}{5} = 80c$

4. $m = -2$ $c = 16$

PAGE 80:

1. a) $x = 0$ or $x = -7$ b) $y = 4$ or $y = -4$ c) $t = \frac{5}{2}$ or $t = -3$

2. a) $x = 3$ or $x = 4$ b) $p = 6$ or $p = -1$ c) $n = -4$

3. a) $x = 4$ or $x = 9$ b) $y = -\frac{1}{2}$ or $y = 2$ c) $t = -6$ or $t = 2$

4. 3 years or 6 years

PAGE 82:

1. a) $x = -0.26$ or $x - 1.54$ (to 2 dec.)   b) $y = 4.24$ or $y = -0.24$ (to 2 dec.)
   c) $t = 1.30$ or $t = -2.30$ (to 2 dec.)

2. a) $n = 4.30$ or $n = 0.70$ (to 2 dec.)   b) $x = 8.48$ or $x = -2.48$ (to 2 dec.)
   c) $m = 0.43$ or $m = -0.77$ (to 2 dec.)

3. 48.54 cm (to 2 dec.)

PAGE 83:

1. a) $x > -2$ b) $y < 7$ c) $t = -3$

2. $-3 \leqslant x < 0$ 

3. $n = 1$ and $n = 2$

4. $x = 1, x = 2, x = 4$ and $x = 5$

PAGE 84: Check your progress

1. a) $c + w + m = 14$ b) $c > 3$ c) $c \neq m$ d) $w = m + 2$

2. 2.92 (to 2 dec.)

3. a) $S = 27$ b) $R = 7$ c) $R = \frac{S - 13}{5}$

4. $x = \frac{17}{2}$ b) $y = \frac{15}{4}$ c) $z = 5$

5. a)

no. of squares	1	2	3	4	5	6	7	8
no. of matches	4	7	10	13	16	19	22	25

   b) 19   c) no. of matches $= 3n + 1$

6. a) $3x + 2y = 20$ b) they stand for no. of journeys
   c) $x = 2$ and $y = 7$
       $x = 4$ and $y = 4$
       $x = 6$ and $y = 1$

7. $y = 24$

8. a) 1 600   b) 73   c) $\frac{6}{25}$ d) 300

9. a) $<$ b) 2.7×10^6 c) 3.38×10^5

10. a) $3p^6$ b) $25q^{24}$ c) $\frac{6}{y^3}$

11. 2.209032×10^9

12. 267 cm^2

13. a) (i) $\frac{1}{4e^5}$

       (ii) $\frac{6}{p^2}$

    b) (i) $x = 4$
       (ii) $x = -2$
       (iii) $x = 0$

14. $(+9) \div (-\frac{1}{3}); (-0.9)^3; (-2)^4; 2(-5)^2; (-3)^2 - 4(+2)(-6)$

15. a) 43   b) 15

16. a) $2y^3 - 6y^2 - 10y$ b) $-2(s + t)$

17. a) $3(5a - 3b + 4)$ b) $3xy(3x - 4y + 5)$

18. a) $3x^2 + 7x - 6$ b) $(2t + 1)(t - 2)$

19. a) $\frac{v + 2}{v(v + 1)}$ b) $h = \frac{3V - 2\pi r^3}{\pi r^2}$

20. $x = 3; y^2 - 2$

21. a) $a = \frac{1}{2}; b = 15$ b) $P = 65$

22. a) $10 \times 2 + 9$ b) (i) $10 \times q + p$

                       (ii) $10p + q - (10q + p)$
                             $= 10p + q - 10q - p$
                             $= 9p - 9q$

    c) (i) $9p - 9q = 18$
       (ii) $p = 8$ and $q = 6$
       (iii) 86

23. $y = 1$ or $y = -6$

24. a) $a = 6; b = -4$

    b) $\frac{9}{x} + bx = 16$

       $a + bx^2 = 16x$
       $bx^2 - 16x + a = 0$
       but $a = 6$ and $b = -4$
       $\therefore -4x^2 - 16x + 6 = 0$
       $2x^2 + 8x - 3 = 0$
    c) $x = 0.35$ or $x = -4.35$

25. $-5 \leqslant x < -2$

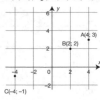

26. a) -1 b) 5   c) 50

PAGE 88:

1. a) $P(1; 3); Q(-2; 2) R(-1; -1); S(2; 0)$ b) Square

2. a)   b) x-axis at -2, y-axis at 1

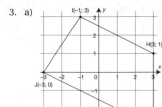

3. a)   b) parallel   c) trapezium

4. $S(-1; 2)$

PAGE 90:

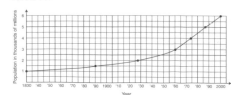

1. a) ≈ 1947
    c) ≈ 200 million
    b) ≈ 4 400 million
    d) increasing at a faster and faster rate

2. a) $\frac{5}{6}$th of a year
    c) \$4 800
    e) 1.9 years
    b) \$100
    d) 3.5 years
    f) 3 years

PAGE 92:

1. a) 145°F
    b) 62.5°F
    c) −18°C
    d) 36°C
2. a) 4 pounds
    b) 2 kg
    c) 34 kg
    d) 134 pounds
    e) (i)   30 kg = 70 pounds
       (ii)   18 pounds = 8 kg
       (iii) 60 pounds = 25 kg

PAGE 93:

1. a) 720 m
    b) 7 min.
    c) 0907 hours and 0921 hours
    d) to supermarket
2. a) 45 min.
    b) 1750 hours
    c) 1715 hours
3. a)
    b) 15 m
    c) 5 m

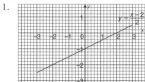

PAGE 95:

1. a) 6 min.
    b) 10 km/h
    c) 3 min.
    d) 3.$\dot{3}$ m/s

PAGE 97:

1. a) 1 500 m
    b) 2 m/s
    c) stopped
    d) 0.5 m/s
2. a) 2 m/s²
    b) 35 m
    c) 3.5 m/s
3. a) 1 m/s²
    b) 100 m
    c) 15 m/s

PAGE 98:

1.

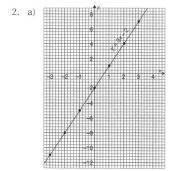

x	−2	0	2	4	6
$y = \frac{x-2}{2}$	−2	−1	0	1	2

2. a)

x	−3	−2	−1	0	1	2	3
$y = 3x - 2$	−11	−8	−5	−2	1	4	7

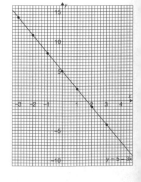

b)

x	−3	−2	−1	0	1	2	3
$y = 5 - 3x$	14	11	8	5	2	−1	−4

c)

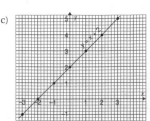

x	−3	−2	−1	0	1	2	3
y = x + 2	−1	0	1	2	3	4	5

d)

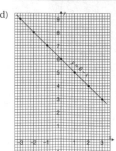

x	−3	−2	−1	0	1	2	3
y = 6 − x	9	8	7	6	5	4	3

e)

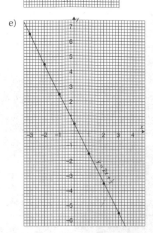

x	−3	−2	−1	0	1	2	3
$y = -2x + \frac{1}{2}$	6.5	4.5	2.5	0.5	−1.5	−3.5	−5.5

f)

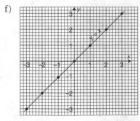

x	−3	−2	−1	0	1	2	3
y = x	−3	−2	−1	0	1	2	3

3.

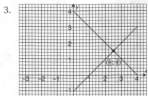

4. Parallel

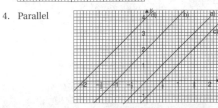

PAGE 101:

1.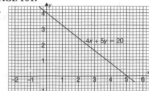

$m = -\frac{4}{5}$

2. a) $m = \frac{3}{4}$ b) $m = -\frac{4}{5}$

   c) $m = \frac{1}{2}$ d) $m = \frac{3}{4}$

4. $m = -\frac{5}{2}$

5. $m = \frac{3}{2}$

PAGE 102:

1. a) $m = -3; c = 4$ b) $m = \frac{1}{2}; c = 2$ c) $m = -1; c = 3$

2. a) $y = \frac{3}{5}x - 2$ b) $y = -\frac{1}{2}x + \frac{3}{4}$

3. A: $m = 1; c = 2$ B: $m = 2; c = -2$ C: $m = -1; c = 3$

    $y = x + 2$ $y = 2x - 2$ $y = -x + 3$

4. $y = 3x - 13$

PAGE 103:

1. EF = 17 units   2. AB = 15 units

3. CD = 26 units   4. GH = 10.82 units (to 2 dec.)

PAGE 104:

1.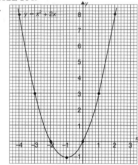

x	−4	−3	−2	−1	0	1	2
$y = x^2 + 2x$	8	3	0	−1	0	3	8

2.

x	−2	−1	0	1	2	3	4	5	6	7
$y = x^2 - 5x - 4$	10	2	−4	−8	−10	−10	−8	−4	2	10

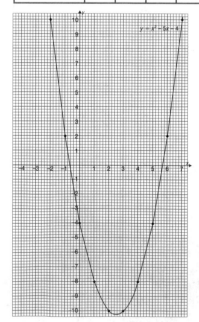

3.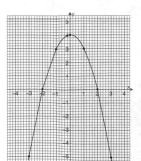

x	–3	–2	–1	0	1	2	3
$y = 4 - x^2$	–5	0	3	4	3	0	–5

PAGE 106:

1. a) $x = -1$; $y = 2$ b) $x = 4$; $y = 3$ c) $x = 3$; $y = -1$ d) $x = 3$; $y = 2$

   e) $x = \frac{5}{2}$; $y = \frac{3}{2}$ f) $x = -1$; $y = -2$

2. a) $x = 3$; $y = 2$ b) $x = -1$; $y = 2$

PAGE 108:

1. a) $x = -1$ or $x = 2$ b) $x = -2.4$ or $x = 3.4$ c) $x = -2$ or $x = 3$

2. a)

x	20	40	60	80	100	120
y	12	6	4	3	2.4	2

b) 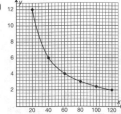   c) $y = \frac{240}{x}$

PAGE 110:

1. a)

x	–1	–0.5	0	0.5	1	1.5	2	2.5	3	3.5	4	4.5	5
$y = x^3 - 6x^2 + 8x$	–15	–5.6	0	2.6	3	1.9	0	–1.9	–3	–2.6	0	5.6	15

b)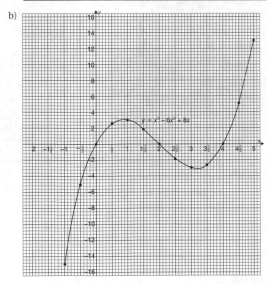

c) (i) $x = 0$ or $x = 2$ or $x = 4$ (ii) $x = 1$ or $x \approx 4.3$ or $x \approx 0.7$

2. a) 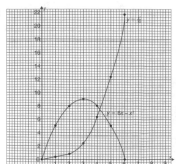　　　b) $x \approx 4.2$ or 0

PAGE 112:

1. a)

x	1	1.5	2	2.5	3	3.5	4	4.5	5
$y = 3x - \dfrac{12}{x}$	−9	−3.5	0	2.7	5	7.1	9	10.8	12.6

b)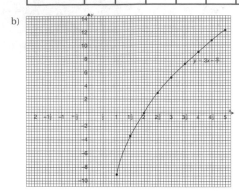

2. a) (i) 2　　(ii) 0.8

b)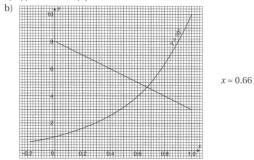

$x \approx 0.66$

3. a)

x	−3	−2	−1	0	1	2	3	4	5
$y = x^2 - 2x - 4$	−11	4	−1	−4	−5	−4	−1	4	11

b) (i)　$x \approx -1.25$ or $x \approx 3.2$
　　(ii)　$x \approx -1.8$ or $x \approx 3.8$
　　(iii)　$x = -1$ or $x = +3$

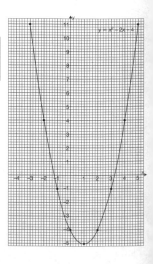

PAGE 115:

1. a) (i) 4　(ii) −2　　　b) (−1.5; 2.4)
2. Gradient ≈ -5
　　Population decreases by 5 per annum

PAGE 117:

1. a) $f(-1) = -5$ b) $f(0) = -1$ c) $f(0.5) = 1$ d) $f(-4) = -17$
2. a) $f(2) = 0$ b) $f(0) = -4$ c) $f(-3) = 5$ d) $f(0.25) = -\frac{63}{16}$
3. a) $f(2) = 0$ b) $f(-1) = -9$ c) $g(5) = -2$ d) $g(-2) = 5$
4. a) $h(2) = 16$ b) $h(-2) = 16$ c) $h(\frac{1}{2}) = 1$
5. a) $x = 3$ or $x = -2$ b) $x = 3, y = 6$
6. a) $f(1) = 2$ b) $f(3) = 6$ c) $f(a) = 2a$ d) $f(a + 2) = 2a + 4$
    e) $f(4a) = 8a$ f) $4f(a) = 8a$
7. a) $f(\frac{1}{2}) = 9$ b) $x = 2, y = 3$
8. a) $f(2) = 15$ b) $f(-2) = 3$ c) $f(0) = 1$

PAGE 118:

1. a) $gh(1) = 26$ b) $hg(1) = 7$ c) $gg(2) = 26$ d) $hh(5) = 29$

PAGE 120:

1. $f^{-1}(x) = \frac{x-3}{4}$
2. $g^{-1}(x) = 3x + 12$
3. a) $h^{-1}(10) = 8$ b) $hh^{-1}(20) = 20$ c) $h^{-1}h^{-1}(26) = 11$

PAGE 122:

1. 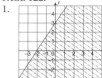   2.   3.

4. a) $y \geqslant 3x + 3$ b) $y < -x + 3$ c) $y \geqslant \frac{1}{3}x + 1$ d) $y \leqslant -\frac{3}{2}x$

PAGE 123:

1. 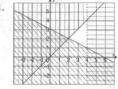   2. 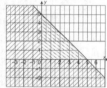   3.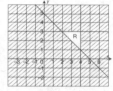

4. A: $x \leqslant 2$
    $x + y \leqslant 4$
    $y > 2x + 1$
5. (0, 2); (0, 3); (1, 0); (1, 1); (1, 2); (2, 0); (2, 1); (3, 0)
6.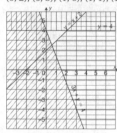

(0, 4); (1, 3); (1, 4); (2, 4)

PAGE 126:

1. min = 6, max = 30
2. a) 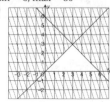   b) 12   3. min = 1, max = 4.5

PAGE 128: Check your progress

1. a) $70   b) 2.5 hrs   c) $p = 35; q = 20$
2. a) 4 °C   b) 100 °C   c) 14.6 min   d) 11.25 °C/min
   e) Gas, highest gradient
3. a) 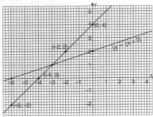   b) 23.68 (to 2 dec.)   c) 19 km

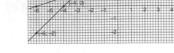

PAGE 129:

4. a) 90 km/h   b) 0.3 km/min²   c) 15 km   d) 2.5 min   e) 0.3 km/min
   f) 17.5 km
5. a) (i) 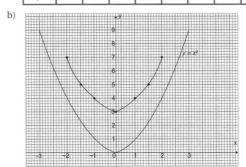   (ii) gradient = 1

   b) (i)

x	−6	−3	0	3
y	0	1	2	3

   c) (−3; 1)

6. a)

x	−2	−1.5	−1	−0.5	0	0.5	1	1.5	2
y	7	5.25	4	3.25	3	3.25	4	5.25	7

   b)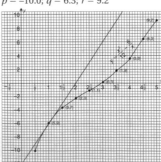

   c) No, $x^2 = x^2 + 3$ has no solution   d) (i) $x \approx \pm 2.45$ (ii) $x \approx 1.75$

7. a) gradient $= -\frac{1}{2}$ b)

x	0	2	4
y	−3	0	3

   c) $x = 3; y = \frac{3}{2}$

8. a) $p = -10.0; q = 6.3; r = 9.2$
   b)   c) $x \approx 2.9$ d) gradient $\approx \frac{12}{1.9} \approx 6.3$

9. a) (vi)   b) (ii)   c) (i)   d) (iv)

10. a) (i) $p = 160$; $q = 10$; $r = 2.5$

    (ii)   (iii) ≈ -40 g/min.   b) $t = 1$

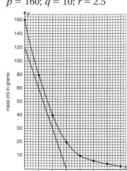

11. a) F   b) T   c) T   d) F

12. a) $f(-2) = 12$ b) $x = 1$ or $x = -\frac{1}{3}$ c) $x = 1.58$ or $x = -0.94$

    d) $x = 1$; $y = 1$ e) $y = \frac{4 - x}{3}$

13. a)   b) max = 11

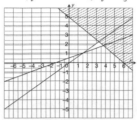

14. a) $60 - x$ Cluedo and $40 - y$ Fantasy   b) $x + y \leqslant 80$

      $x \geqslant 0$; $y \geqslant 0$ $100 - x - y \leqslant 55 \Rightarrow x + y \geqslant 45$

    c)   d) $x = 60$; $y = 0$

15. b) $5x + 6y \leqslant 60$ c) $x \geqslant 8$ d) (i)

                                           (ii)

    e) max profit = \$440

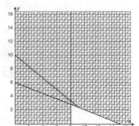

PAGE 138:

a) $a = 89°$
b) $b = 105°$
c) $c = 22°$
d) $x = 145°; y = 35°; z = 145°$
e) $p = 42°; q = 42°$
f) $x = 50°; y = 10°; z = 120°$
g) $i = 75°; j = 80°; k = 25°$
h) $a = 125°; b = 55°; c = 85°; d = 40°; e = 140°$
i) $x = 30°$

PAGE 139:

a) $a = 71°$
b) $b = 140°$
c) $c = 35°; d = 55°$

PAGE 142:

1. a) $x = 75°; y = 105°$
    b) $z = 75°$
    c) $a = 40°; b = 40°$
    d) $b = 260°$
    e) $a = 70°; b = 70°; c = 40°$
    f) $d = 73°; e = 36.5°$
    g) $f = 78°; g = 24°$
    h) $p = 67°; q = 50°; r = 63°$
    i) $x = 26°; y = 26°; z = 116°$
    j) $d = 35°; e = 70°; f = 40°$
    k) $c = 45°; d = 22.5°$
2. a) $e = 60°$
    b) $x = 57°; y = 53°$
    c) $c = 105°$
    d) $c = 55°$
3. $a = 132°$
4. 24 sides
5. a) $144°$
    b) $150°$
6. a) $1\,620°$
    b) $5\,400°$
7. a) 600 mm^2
    b) 340.90 mm^2 (to 2 dec.)
    c) 800 mm^2
8. a) $a = 60°; b = 120°$; Area $= 306$ mm^2
    b) $x = 70°; y = 110°; z = 40°$; Area $= 375.88$ mm^2 (to 2 dec.)
    c) Area $= 205.21$ mm^2 (to 2 dec.)

PAGE 143:

a: centre   b: radius   c: diameter   d: chord   e: circumference   f: tangent

PAGE 144:

1. $a = 130°$ (OS = OR)
2. AD̂C $= 90°$; BĈD $= 90°$ (AC and DB are diameters)

PAGE 145:

1. a) Equilateral triangle
    b) AÔB $= 60°$
    c) Six equilateral triangles are formed, each with angle 60°. The angles around the centre are $6 \times 60° = 360°$, thus not leaving a gap.
2. a) 3 cm
    b) $120° = $ AB̂C
    c) Figure is a regular hexagon
    d) All 120°
    e) Regular hexagon

PAGE 147:

5. BC ≈ 6.5 cm   CA ≈ 5.7 cm
6. PQ ≈ 6.3 cm   PR ≈ 3.2 cm
7. YZ ≈ 5.6 cm   Ŷ $\approx 69°$ Ẑ $\approx 51°$
8. D̂ $\approx 53°$ Ê $= 90°$ F̂ $\approx 37°$

PAGE 151:

1. a) 114 mm   b) 61 mm   c) 72 mm   d) 62 mm   e) 90 mm
2. a) 800 mm^2 b) 28 cm^2 c) 12 m^2 d) 100 m^2
3. a) 3 cm^2 b) 7.5 cm^2 c) 7.5 cm^2 d) 1.875 cm^2
4. a) 12 cm^2 b) 30 m^2 c) 36 cm^2 d) 30 m^2

PAGE 152:

1. a) 125.66 m (to 2 dec.)   b) 77.28 cm (to 2 dec.)   c) 179.07 mm (to 2 dec.)
2. 40 212.39 km (to 2 dec)
3. 65.03 m (to 2 dec.)
4. a) 42 units^2 b) 104 units^2 c) 108 units^2
    d) 37 units^2 e) 61.13 cm^2
5. 41.46 cm^2 (to 2 dec.)
6. a) Area $= 84.95$ cm^2 (to 2 dec.)   b) A $= 1.91$ m^2 (to 2 dec.)   c) A $= 1661.90$ mm^2 (to 2 dec)
    Perimeter $= 32.67$ cm (to 2 dec.)   P $= 4.90$ m (to 2 dec.)   P $= 144.51$ mm (to 2 dec.)
7. 176.71 m^2 (to 2 dec.)
8. 10

PAGE 153:

1. a) 10.47 cm (to 2 dec.)   b) 62.83 cm^2 (to 2 dec.)
2. a) 25.66 cm (to 2 dec.)   b) 89.80 cm^2 (to 2 dec.)
3. 4 895.65 cm^2 (to 2 dec.)
4. 80°
5. 191°
6. a) 76.50 m (to 2 dec.)   b) 380.50 m^2 (to 2 dec.)

PAGE 155:
1. a) 2.25   b) 1.5

PAGE 156:
1. a) 12   b) AD; QR   c) BQ; RC
2. a) Triangular prism   b) 5   c) 9
   d) PQ = ST; RQ = RS
3. a) 12   b) 6
4. a) △APB is isosceles   b) Equilateral   c) 60°
   d) 5 faces, 8 edges, 5 vertices
5. a) Cuboid   b) 6 faces, 12 edges, 8 vertices d) 12

PAGE 159:
1. a) 12 312 mm^2 b) 41 400 cm^2
2. 1 462 cm^2
3. a) 15.4 cm^3 b) 1 140.40 mm^3 (to 2 dec.)
4. a) 204 cm^2 b) 96 cm^3
5. a) (i) 0.8 m   b) 46.24 tonnes (to 2 dec.)
      (ii) 20.11 m^3 (to 2 dec.)
6. 7 238.23 cm^3 (to 2 dec.)

PAGE 160:
1. a) 65.4 cm^3 b) 15
2. a) Pyramid   b) 13 cm   c) 50 cm^3
3. a) 2.26 m^3 (to 2 dec.)   b) 7.24 m^2 (to 2 dec.)
4. a) 523.60 cm^3 (to 2 dec.)
   b) (i) 471.24 cm^2 (to 2 dec.)
      (ii) Volume = $\frac{2}{3}$ of capacity

PAGE 161:
a) AA corr. S   c) SAS   d) RHS

PAGE 162:
1. a) C   b) D
2. a) (ii) No, squares have only 90°-angles, therefore similar.   b) No
   c) △ABC ≡ △AFE; △ACD ≡ △AED

PAGE 163:
a) 2 lines of symmetry
c) Rotational symmetry of order 2
d) 1 line of symmetry
e) 4 lines of symmetry; rotational symmetry of order 2
f) 3 lines of symmetry; rotational symmetry of order 3
g) 5 lines of symmetry; rotational symmetry of order 5
h) Rotational symmetry of order 3
j) Rotational symmetry of order 2

PAGE 164:
2. a) Infinite   b) One

PAGE 165:
1. Drop a perpendicular from P to O, this is bisector of chord.
2. Construct ⊥ OXY to cut both circles. (O is centre)
   △OBX ≡ △OCX
   ∴ BO = OC
   △OAY ≡ △ODY
   ∴ AO = OD   AO – BO = OD – OC   ∴ AB = CD

PAGE 166:
b) (i) 70°   (ii) 20°   (iii) 70°

PAGE 168:
1. a) $p = 50°$; $q = r = 65°$ b) $b = 80°$ c) $c = 30°$; $d = 55°$; $e = f = 45°$
   d) $p = 85°$; $q = 105°$ e) $b = 60°$ f) $x = 94°$; $y = 62°$; $z = 24°$
   g) $p = 85°$; $q = 65°$
2. a) $2x$ b) $90° - x$ c) x
3. a) $a = 70°$ b) 125°   c) $c = d = 60°$; $e = 80°$; $f = 40°$
4. a) $90° - x$ b) $180° - 2x$ c) $2x - 90°$

PAGE 169:

1. length = 6.8 m; width = 5.2 m
2. a) 3 cm   b) 2.4 cm
3. a) 5.6 cm   b) 15°

PAGE 170:

1. b) $\hat{D}$ = 113.3° (to 1 dec.)   c) CD = 80.2 m (to 1 dec.)
   $\hat{C}$ = 91.7 (to 1 dec.)
2. a) 20°   b) 3.4 m (to 1 dec.)
3. a) approx. 14.8 m   b) approx. 21.6 m   c) approx. 34°

PAGE 171:

1. a) 270°   b) 135°   c) 045°
2. a) 262°   b) 135°
3. a) 110°   b) 050°   c) 230°
   d) 027°   e) 280°
4. a) approx. 107.5°   b) 287.5°   c) 147 (to nearest kilometre)
5. a) 9.6 km (to 1 dec.)   b) 270° (to nearest degree)

PAGE 174: Check your progress

1. a) (i) cuboid   (ii) triangular prism   (iii) pyramid
   b)
                                                              c) Faces + Vertices = Edges + 2

Solid	Faces	Vertices	Edges
A	6	8	12
B	5	6	9
C	5	5	8
D	4	4	6
E	7	10	15
F	9	10	17

2. a)   b)   c)

3. a) △AZE; △BYC; △DXC; △CZD; △DYE   b) ACD
4. a) 4   b) 5
5. a) 65°   b) 65°   c) 135°
6. $x = 32°$; $y = 121°$
7. $a = 90°$; $b = 53°$; $c = 90°$; $d = 53°$
8.   9.

PAGE 178:

1. a) $a = 13$ cm   b) $b = 24$ cm   c) $c = 3.4$ m   d) $d = 25$ mm
2. $x = 5.66$ cm (to 2 dec.)
3. BC = 4 km
4. 4.21 m
5. 110 km
6. a) tangent to centre   b) 8.06 cm (to 2 dec.)
7. a) 3.77 m (to 2 dec.)   b) 4.97 m (to 2 dec.)

PAGE 180:

1. a) 8.66 cm (to 2 dec.)   b) 43.30 cm² (to 2 dec.)
2. a) 28.28 cm (to 2 dec.)   b) 34.64 cm (to 2 dec.)
3. a) and d)
4. acute-angled

PAGE 181:

1. a) hypotenuse = 20; opp = 16; adj = 12   b) hypotenuse = 25; opp = 7; adj = 24
   c) hypotenuse = 13; opp = 12; adj = 5   d) hypotenuse = 85; opp = 84; adj = 13
2. hypotenuse = 200; opp (60°) = 173; adj (60°) = 100; opp (30°) = 100; adj (30°) = 173

PAGE 183:

1. c) $\frac{\text{opp (50°)}}{\text{adj (50°)}} \approx 1.2$
2. a) 1.1918   b) 0.7673   c) 1.3032   d) 5.1446
3. a) 0.0175   b) 0.1763   c) 1.0538   d) 10.0187
4. a) $\frac{8}{15}$ b) 0.5333

5. a) 2.14   b) 0.466
6. a) 1.0724   b) 32.17 m (to 2 dec.)
7. 32.32 m (to 2 dec.)
8. a) < 1 b) > 1
9. a) (i) 2 units   (ii) $\sqrt{3}$ units   b) 0.577 (to 3 dec.)   c) 0.5774

PAGE 185:
1. a) 40.4°   b) 51.0°   c) 74.3°   d) 84.3°
2. a) 21.8°   b) 37.9°   c) 38.0°   d) 70.0°
3. a) 35.0°   b) 77.5°   c) $c = 38.7°$; $d = 51.3°$
4. 71.8° (to 1 dec.)
5. 21.2° (to 1 dec.)
6. 053°
7. a) 13.3   b) 26.7

PAGE 187:
1. a) (i) $\frac{21}{29}$ (ii) $\frac{20}{29}$ b) (i) $\frac{8}{17}$ (ii) $\frac{15}{17}$ c) (i) $\frac{4}{5}$ (ii) $\frac{3}{5}$ d) (i) $\frac{13}{85}$ (ii) $\frac{84}{85}$
2. a) 0.0872   b) 0.5   c) 0.8660   d) 0.9962
3. a) 0.9962   b) 0.8660   c) 0.5   d) 0.0872
4. a) $\frac{q}{r}$ b) $\frac{e}{f}$ c) $\frac{IH}{IJ}$ d) $\frac{x}{r}$
5. a) 10.6 units   b) 4.23 units   c) 14.1 units   d) 53.1 units
6. a) 81.9°   b) 57.1°   c) 22.0°   d) 64.2°
7. a) 26°   b) 45°   c) 70°   d) 80°
8. 1.93 m (to 2 dec.)
9. a) 10.07 km (to 2 dec.)   b) 14.92 km (to 2 dec.)
10. a) 14.1 m (to 1 dec.)   b) 5.1 m (to 2 dec.)

PAGE 192:
1. a) 16.2°   b) 17.9 m
2. 13.86 cm (to 2 dec.)
3. a) 59.0° (to 1 dec.)   b) 1.75 m (to 2 dec.)   c) 4.05 m³
4. a) 020°   b) 281.91 m (to 2 dec.)   c) 98 667.73 m² (to 2 dec.)
5. a) 3.5 m (to 1 dec.)   b) 6.1 m (to 1 dec.)

PAGE 196:
1. a) 0.5736   b) −0.8192   c) 0.5   d) −0.8660
2. a) 17.23 cm² (to 2 dec.)   b) 22.75 cm² (to 2 dec.)   c) 24.25 cm² (to 2 dec.)
    d) 84.00 cm² (to 2 dec.)
3. 107.59 cm² (to 2 dec.)
4. a) (i) 56°   (ii) 124°   b) 118.7°
5. a) $25^2 < 52^2 + 63^2$ b 22.6°   c) 53.1°

PAGE 198:
1. $\hat{C} = 63°$; AC = 15.87 cm; BC = 21.35 cm (to 2 dec.)
2. $\hat{F} = 25°$; DE = 9.80 cm; EF = 14.90 cm (to 2 dec.)
3. $\hat{R} = 32.2°$ (to 1 dec.); QR = 7 cm $\hat{P} = 27.8°$ (to 1 dec.)
4. a) opposite side is less than 15 cm   b) $\hat{Y} = 30.9°$; $\hat{Z} = 109.1°$ c) 22.06 cm (to 2 dec.)

PAGE 201:
1. 8.62 cm (to 2 dec.)
2. 22.25 m (to 2 dec.)
3. 53.8°
4. a) 18.74 m (to 2 dec.)   b) 32.1° (to 1 dec.)   c) 52.9° (to 2 dec.)
5. a) 60°   b) 32.2°   c) 87.8°

PAGE 203:
1. a) 0.940 (to 3 dec.)   b) 0.574 (to 3 dec.)   c) −0.996 (to 3 dec.)
    d) −0.139 (to 3 dec.)   e) 0.643 (to 3 dec.)   f) −0.342 (to 3 dec.)
2. a) 65.0° and 115.0° (to 1 dec.)   b) 30° and 150° (to 1 dec.)   c) 185.0° and 355.0° (to 1 dec.)
3. a)   b)

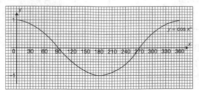

4. a)

x	0	15	30	45	60	75	90	105	120	135	150	165	180
y	−1	−0.71	−0.37	0	0.37	0.71	1	1.22	1.37	1.41	1.37	1.22	1

x	195	210	225	240	255	270	285	300	315	330	345	360
y	0.71	0.37	0	−0.37	−0.71	−1	−1.22	−1.37	−1.41	−1.37	−1.22	−1

5. a)

x	0	30	60	90	120	150	180	210	240	270	300	330	360
y	1	2	2.73	3	2.73	2	1	0	−0.73	−1	−0.73	0	1

b)

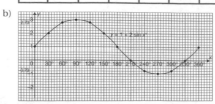

PAGE 207:

1. a) 25 cm b) 13.04 cm (to 2 dec.) c) 27.5° (to 1 dec.)
2. a) 111.68 m (to 2 dec.) b) 41.84 m (to 2 dec.)
3. a) 1 055.97 km (to 2 dec.) b) 49° c) 259°
    d) 1 112 hours
4. a) (i) 1.40 m (to 2 dec.) b) 2   4   7   10
        (ii) 6.60 m (to 2 dec.)   1.4   1.4   5.5   6.6
    c)

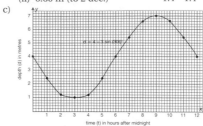

    d) Between 07h24 and 10h36
        (to nearest min.)

PAGE 208: Check your progress

1. a) 11.49066665 b) 11.5
2. AC = 9.83 m (to 2 dec.)
    BC = 6.88 m (to 2 dec.)
3. 284 mm
4. 9.9 m (to 1 dec.)
5. a) 10.07 m (to 2 dec.) b) 20.6° (to 1 dec.)
6. a) (ii) 78.3 m (to 3 s.f.) b) (i) 250.3 m (to 4 s.f.)
        (ii) 257.4 m (to 4 s.f.)
        (iii) 76.9 (to 1 dec.)
7. a) 5.16 m b) 3.11 m²
8. a) 7 cm b) 51.1° (to 1 dec.)
9. a) A (90°; 1) b) −1
    c) d) 2

10. a) (i) 107.31 km (to 2 dec.) b) (i) 5
        (ii) 66.6° (to 1 dec.)   (ii) 12 km/h
        (iii) 143.4° (to 1 dec.)

PAGE 214:

1. a)

People	Tally	Houses
1	‖	2
2	ЖЖ ЖЖ	10
3	ЖЖ ‖	7
4	ЖЖ ЖЖ ‖‖‖	13
5	ЖЖ	5
6	‖‖‖	3

b)

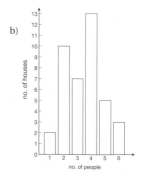

    c) 138
2. a) 6   b) 22

PAGE 216:

1.

2.
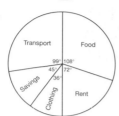

3. $300 000 wages; $240 000 raw materials; $80 000 fuel; $100 000 extras

4. a) $\frac{2}{9}$ b) 80°   c) 225 km²

5. a) $\frac{7}{24}$ b) 45°   c) 24

6. a) 75°   b) 26

7. a) 775   b) Week 4 ▢ ▢ ▢ ▢ ∴ 5.25 mm

8. a) Belgian   b) (i) $\frac{1}{6}$ (ii) 17%   c) 0.375   d) 36

PAGE 221:

1.

1	2	3	4	5	6	7	8	9
7	5	5	9	9	4	3	5	3

mean = 4.5

2. a) 5   b) 80   c) 3.5   d) 23.5   e) 42.5

3. a) 5   b) none   c) 2 and 5   d) 41 and 42

4. a) 0   b) 1   c) 0.85

PAGE 225:

1.
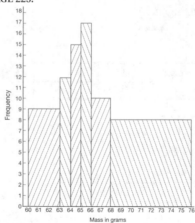

2. a)
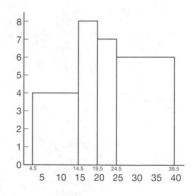

b)

Height in cm	Mid-pt (x)	Freq. (f)	fx
5–14	9.5	4	38
15–19	17	8	136
20–24	22	7	154
25–39	32	6	192
		25	520

mean height = 20.8 cm

PAGE 233:

1. a)

Height in cm	6–15	16–20	21–25	26–40
No. of plants	3	7	10	5
Cum. Freq.	3	10	20	25

b) 21–25   c) 23 cm

2. a) $36.25

b) $p = 12; q = 24; r = 35$

c)
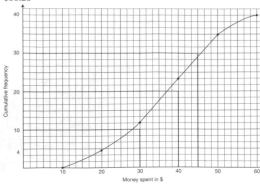

d) (i) $38
   (ii) Q ≈ $29; Q_3 ≈ $45,50
        Inter-quartile
        range ≈ $ 16.50

PAGE 234: Check your progress

1. a) $x = 126°$ b) 50   c) 40%
2. a) 2   b) 12   c) 25
3. a)

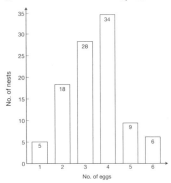

b) (i)   5
   (ii)   3
   (iii) 3.4

4. e.g. 3; 5; 10

5. a)

Mass (M) in grams	No. of fish	Classification
M < 300	3	Small
300 ≤ M < 400	9	Medium
M = 400	6	Large

b)
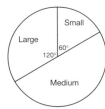

6. a) 1   b) 2   c) 2.12
7. a) (i) 5.28 min.
       (ii) 5 min 17 sec

b)

Time (t) in min.	Cumulative frequency
0	0
≤ 1	12
≤ 2	26
≤ 4	46
≤ 6	60
≤ 8	72
≤ 10	90
≤ 15	100

c)

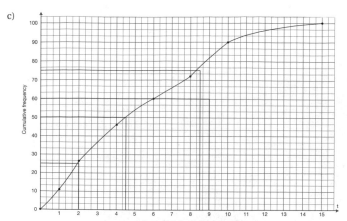

d) (i) 4.6 min.   (ii) $Q_3 \approx 8.4$ min.   (iii) ≈ 6.4 min.

PAGE 239:

1. $\frac{1}{13}$

2. $\frac{1}{2}$

3. a) $\frac{1}{5}$ b) $\frac{1}{2}$ c) $\frac{3}{10}$

4. a) $\frac{1}{4}$ b) 4

5. a)

No. of matches	39	40	41	42	43	44	45
Frequency	5	3	4	3	4	0	1

b) $\frac{3}{5}$

6. a) $\frac{2}{5}$ b) $\frac{1}{2}$

PAGE 240:

1. $\frac{2}{3}$ 2. $\frac{2}{13}$ 3. $\frac{2}{3}$ 4. a) $\frac{8}{15}$ b) $\frac{7}{15}$ 5. 0.875

PAGE 244:

1. $\frac{1}{6}$

2. a) $\frac{1}{45}$ b) $\frac{16}{45}$ c) $\frac{28}{45}$

3. a) $\frac{15}{33}$ b) $\frac{7}{22}$

4. a) (i)

  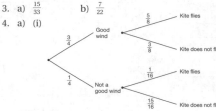

(ii) $\frac{15}{32}$ (iii) $\frac{33}{64}$ b) $\frac{31}{128}$

PAGE 245: Check your progress

1. a) $x = 150°$ b) 60   c) $\frac{1}{3}$

2. a) RBB, BRB, BBR   b) $\frac{2}{3}$

3. a)

Face	1	2	3	4
Prob.	$\frac{2}{9}$	$\frac{1}{3}$	$\frac{5}{18}$	$\frac{1}{6}$
	$\frac{4}{18}$	$\frac{6}{18}$	$\frac{5}{18}$	$\frac{3}{18}$

b) 2   c) 1   d) $\frac{13}{18}$

4. a) First die 6, second die 2.

b)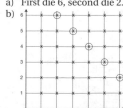

c) (i) $\frac{5}{36}$ (ii) $\frac{1}{6}$

$$\frac{2}{n-1} = \frac{1}{4}$$
$$n - 1 = 8$$
$$n = 9$$

5. a) $\frac{1}{4}$ b) $\frac{3}{5}$

6. a) (i) 50   (ii) 60   b) (i) $\frac{1}{2}$ (ii) $\frac{4}{9}$ (iii) $\frac{8}{15}$ (iv) $\frac{118}{795}$

7. a) (i) $\frac{1}{3}$ (ii) (1) $\frac{1}{5}$ (2) $\frac{4}{15}$

(iii)    (iv) $\frac{1}{3}$ b) $\frac{2}{5}$ c) 9

PAGE 248:

1. a) 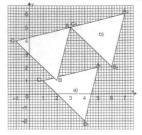   b)   2.

3.

PAGE 250:

1. a)   b)   c)

2. a)   b)   c)   d)

PAGE 252:

1. a)   b) 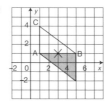   c)

2. a) 90° clockwise   b) 180°   c) 90° clockwise

PAGE 254:

1. 90° clockwise   2. 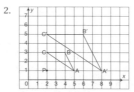   3.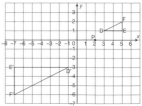

4. Scale factor = 2; centre (0, –1)
5. Scale factor = 1.5; centre (4, 2)

PAGE 256:

1. $a = \begin{pmatrix} 4 \\ 6 \end{pmatrix}$ $b = \begin{pmatrix} 4 \\ 2 \end{pmatrix}$ $c = \begin{pmatrix} -4 \\ 2 \end{pmatrix}$ $d = \begin{pmatrix} -4 \\ 2 \end{pmatrix}$ $e = \begin{pmatrix} 6 \\ -4 \end{pmatrix}$ $f = \begin{pmatrix} 0 \\ 4 \end{pmatrix}$

$g = \begin{pmatrix} 8 \\ 4 \end{pmatrix}$ $h = \begin{pmatrix} 4 \\ -2 \end{pmatrix}$

2.

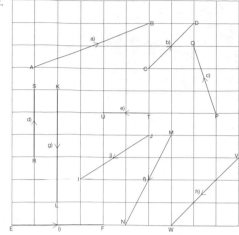

3. a) $\overrightarrow{AB} = \overrightarrow{DC} = \begin{pmatrix} 4 \\ 0 \end{pmatrix}$ b) $\overrightarrow{BC} = \overrightarrow{AD} = \begin{pmatrix} 1 \\ 3 \end{pmatrix}$

4. a) $\begin{pmatrix} 4 \\ 2 \end{pmatrix}$ b) $\begin{pmatrix} 5 \\ -1 \end{pmatrix}$ c) $\begin{pmatrix} 5 \\ -1 \end{pmatrix}$ d) $\begin{pmatrix} 0 \\ -3 \end{pmatrix}$

e) $\begin{pmatrix} 4 \\ -3 \end{pmatrix}$ f) $\begin{pmatrix} 5 \\ 2 \end{pmatrix}$

PAGE 260:

1. a) $\begin{pmatrix} 12 \\ -6 \end{pmatrix}$ b) $\begin{pmatrix} 3 \\ -5 \end{pmatrix}$

2. $\begin{pmatrix} 12 \\ -7 \end{pmatrix}$

3. a) $2a + 3b$ b) $a + \frac{3}{2}b$ c) b d) $a + \frac{1}{2}b$

PAGE 261:

1. $|OP| = 5$ units; $|OQ| = 13$ units; $|OR| = 17$ units

2. a) A(4, 2); B(−1, 3); C(6, −2)   b) $\overrightarrow{AB} = \begin{pmatrix} -5 \\ 1 \end{pmatrix}$; $\overrightarrow{CB} = \begin{pmatrix} -7 \\ 5 \end{pmatrix}$; $\overrightarrow{AC} = \begin{pmatrix} 2 \\ -4 \end{pmatrix}$

3. a) $-2p + 2q$ b) $2p + q$ c) $-p + q$

4. a) $\frac{1}{2}a$ b) $-\frac{1}{2}b$ c) $\frac{1}{2}a - \frac{1}{2}b$ d) $\frac{3}{4}(a + b)$

PAGE 264:

1.

2.

3. a) OA is invariant line of a shear with factor 2   b) OA is invariant line of a one–way stretch with factor 2

PAGE 266:

1. a) 2×2 b) 2×3 c) 3×2 d) 1×2 e) 1×3 f) 1×1

g) 2×1 h) 3×1 i) 3×3

2. **C** and **E** are equal. **G** and **H** are equal.

PAGE 267:

1. $(9 - 3)$ 2. $\begin{pmatrix} 15 \\ -6 \end{pmatrix}$ 3. $\begin{pmatrix} 10 & -6 \\ 0 & 2 \end{pmatrix}$ 4. $\begin{pmatrix} -4 & 8 \\ -12 & 0 \end{pmatrix}$

5. $\begin{pmatrix} 2 & 0 \\ 1 & \frac{1}{2} \end{pmatrix}$ 6. $\begin{pmatrix} 6a & -6b \\ 9a & 12b \end{pmatrix}$ 7. $\begin{pmatrix} 0 & 0 \\ 0 & 2 \end{pmatrix}$ 8. $\begin{pmatrix} 5 \\ 4 \end{pmatrix}$

PAGE 269:

1. a) (22)　　　b) (6)　　　c) (–17)　　　d) (–12)　　　e) $(a + 9)$

2. a) $\begin{pmatrix} 13 \\ 4 \end{pmatrix}$　　b) $\begin{pmatrix} 8 \\ 6 \end{pmatrix}$　　c) $\begin{pmatrix} 8 & 30 \\ 3 & 12 \end{pmatrix}$　　d) $\begin{pmatrix} 5 & -3 \\ 4 & 2 \end{pmatrix}$

3. a) $x = 4$　　b) $x = 1$　　c) $x = 7$

4. a) $\begin{pmatrix} -5 & 10 \\ 4 & 2 \end{pmatrix}$　b) $\begin{pmatrix} 6 & -1 \\ 4 & -9 \end{pmatrix}$　c) $\begin{pmatrix} 0 & 5 \\ 2 & -2 \end{pmatrix}$　d) $\begin{pmatrix} 0 & 5 \\ 2 & -2 \end{pmatrix}$　e) $\begin{pmatrix} -11 & -13 \\ -4 & -7 \end{pmatrix}$

5. (13)

6. a) (i) $\begin{pmatrix} -5 & 4 \\ 2 & 2 \end{pmatrix}$　　(ii) $\begin{pmatrix} -3 & 6 \\ 5 & 0 \end{pmatrix}$　　b) No

PAGE 270:

1. a) 17　　b) 0　　c) –5　　d) 21　　e) –1
2. $x = 9$
3. $x = -3$
4. $y = 4$
5. $p = 9$

PAGE 272:

1. a) $\begin{pmatrix} 1 & -3 \\ -\frac{1}{2} & 2 \end{pmatrix}$　　b) $\begin{pmatrix} 8 & -3 \\ -5 & 2 \end{pmatrix}$　　c) no inverse　　d) $\begin{pmatrix} -1 & 2 \\ -2 & 3 \end{pmatrix}$　　e) $\begin{pmatrix} -3 & 2 \\ \frac{5}{2} & -\frac{3}{2} \end{pmatrix}$

4. $\begin{pmatrix} 5 & -7 \\ -2 & 3 \end{pmatrix}$

PAGE 276:

1. a)　　　　　　　　　　　　　　　　　　　b) (ii) Reflection about line $y = x$

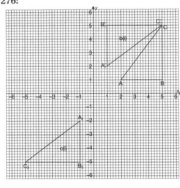

　　c) (ii) $\begin{pmatrix} 0 & -1 \\ -1 & 0 \end{pmatrix}$　　　　　　d) (i) Reflection in the line $y = -x$
　　　　　　　　　　　　　　　　　　　　　　　(ii) $\begin{pmatrix} 0 & -1 \\ -1 & 0 \end{pmatrix}$

2.

　　d) (i) Reflection in line $y = x$　　　e) (i) $\begin{pmatrix} 0 & -1 \\ -1 & 0 \end{pmatrix}$
　　　　(ii) $\begin{pmatrix} 0 & 1 \\ 1 & 0 \end{pmatrix}$　　　　　　　　　(iii) T followed by U

3. a) $\begin{pmatrix} -1 & -2 \\ 2 & 3 \end{pmatrix}$　　　　　　　　　b) A(–4, 7); B (–2, 3); C (–1, 0)

1.

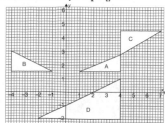

2. A : reflection in the x-axis

B : translation under $\begin{pmatrix} -3 \\ 2 \end{pmatrix}$

C : enlargement, scale factor 2, centre the origin

D : 90 anti-clockwise rotation around origin

3. a) (i) $\begin{pmatrix} 1 \\ -3 \end{pmatrix}$ (ii) $\begin{pmatrix} -6 \\ 3 \end{pmatrix}$ b)

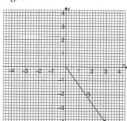

4. a) $(-1, 2)$

b) -2

5. a) $\begin{pmatrix} 3 \\ -2 \end{pmatrix}$

b) 180° clockwise rotation about $(6, 0)$

c) (i) (ii) 4

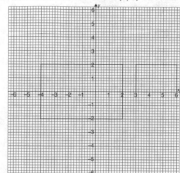

6. b) $a - b$

c) 3.16 units

7. a) (i) Translation under $\begin{pmatrix} 7 \\ 3 \end{pmatrix}$

(ii) Enlargement, scale factor 3, centre origin

(iii) 90° anti-clockwise rotation about origin

(iv) One-way stretch, invariant line the x-axis, Scale Factor 4

(v) Shear factor 3, invariant line the x-axis

b) B, D and F

8. a) (ii)

b) (5)

c) $\begin{pmatrix} 2 & -\frac{5}{3} \\ 1 & -\frac{2}{3} \end{pmatrix}$

9. a) $\begin{pmatrix} 3 & 2 \\ -1 & 6 \end{pmatrix} \begin{pmatrix} -3 \\ 2 \end{pmatrix} = \begin{pmatrix} x \\ y \end{pmatrix}$

$\dfrac{-9 + 4}{3 + 12} = \dfrac{x}{y}$

$\dfrac{-5}{15} = \dfrac{x}{y}$

$x = -5; y = 15$

b) Determinant of matrix $A = \begin{pmatrix} 2 & -1 \\ 4 & 3 \end{pmatrix}$

$$= (2 \times 3) - (4 \times -1)$$
$$= 6 + 4$$
$$= 10$$

$$A^{-1} = \frac{1}{10} \begin{pmatrix} 3 & 1 \\ -4 & 2 \end{pmatrix}$$

$$= \begin{pmatrix} \frac{3}{10} & \frac{1}{10} \\ -\frac{4}{10} & \frac{2}{10} \end{pmatrix}$$

$$= \begin{pmatrix} \frac{3}{10} & \frac{1}{10} \\ -\frac{2}{5} & \frac{1}{5} \end{pmatrix}$$

c) $\begin{pmatrix} 3t & u \\ -t & 2u \end{pmatrix} \begin{pmatrix} 1 \\ 2 \end{pmatrix} = -\dfrac{10}{10}$

$$\frac{3t + 2u}{-t + 6u} = \frac{10}{-1}$$

If $-t + 6u = -10$
$\therefore 6u + 10 = t$

$$3t + 2u = 10$$
$$\therefore 3(6u + 10) + 2u = 10$$
$$\therefore 20u + 30 = 10$$
$$\therefore 20u = -20$$
$$\therefore u = -1$$
$$\therefore t = 6(-1) + 10$$
$$= 4$$

10. a)

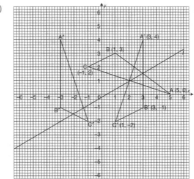

b) (ii) $y = \frac{2}{3}x - \frac{1}{3}$

(iii) $p = \frac{3}{5}$

$q = \frac{4}{5}$

$r = \frac{4}{5}$

$s = -\frac{3}{5}$

(iv) $\begin{pmatrix} p & q \\ r & s \end{pmatrix} = \begin{pmatrix} \frac{3}{5} & \frac{4}{5} \\ \frac{4}{5} & -\frac{3}{5} \end{pmatrix}$ – reflection in the x-axis

d) angle of rotation $= -90°$